The Values of Artificial Intell

THE DESIGN FLAW

Technical teams build capabilities but often lack business context. Business leaders demand solutions but rarely fully understand technical constraints. Data teams control the raw material but are often disconnected from both. And when things go sideways—as they often do—the finger-pointing begins.

This isn't a minor coordination issue; it's a fundamental **design flaw** in how organizations approach artificial intelligence (AI)—and a fundamental failure in how we create, capture, connect, and communicate its value. It is also a failure in our underlying understanding of how to get value from AI.

The Values of Artificial Intelligence shows how to fix this flaw—not through reorganization or the usual buzzwords, but through frameworks that make value creation truly broadly owned when it succeeds and accountable when it fails.

The tools in these pages are practical, tested, and deeply human—designed to bridge what technology delivers with what people ultimately value.

Author's Note

Everyone I talk to—from start-ups to governments, from classrooms to boardrooms – fears the same uneasy truth: "AI is moving faster than we can think, act, or adapt."

We're chasing progress but losing alignment. We're measuring speed – but losing direction.

This isn't just another technology revolution; it is a radical values reckoning.

For years I have helped organizations capture the business value of AI.

But every genuine breakthrough has revealed the same uncomfortable truth: **technology only scales when its values match its value.**

This realization – simple, inconvenient, and long overdue – is what gave birth to this book.

The Values of Artificial Intelligence isn't simply something to read.

It is a call to action for leaders who are rightly refusing to choose between commercial success and human significance.

It is about building systems that work and, more importantly, systems that ultimately matter.

"Having collaborated with Edosa in his professional capacity, and on his earlier books about unlocking value from data, I've witnessed firsthand his ability to turn complex ideas into practical guidance. This new work takes the journey even further, connecting AI's potential with the deep human values that guide meaningful transformation. It's a timely call for leaders to find the courage to make the right choices, and to lead with clarity, empathy, and conviction."
—**Amy Shi-Nash,** *Chief Analytics & Data Officer, Tabcorp and Data Board Member, MIT Sloan School of Management*

"Edosa Odaro offers a combination of clarity and depth on the critical role of values in AI. It's a blueprint for turning AI potential into measurable, human-centered impact. Through real-world cases, Odaro shows how to bridge the gap between technical capability and strategic value. Essential reading for any leader who wants AI to deliver on the values behind their initiatives."
—**Mohamed Zaki,** *Professor at the Institute for Manufacturing, Department of Engineering, University of Cambridge and Deputy Director of Cambridge Service Alliance*

"**The Values of AI** cuts through the noise to highlight what matters most, turning data and algorithms into meaningful impact. It is a practical guide and must-read for leaders who want AI to deliver results that are trusted and sustainable."
—**Julie Wall,** *Professor of AI & Advanced Computing, Centre for AI & Natural Language Technologies (CAINT), University of West London*

"AI technology is fast becoming a commodity but in order for companies to build a moat, leaders need to convey the strategic value to their stakeholders. Odaro's latest book plays a key part in explaining the concepts with business use cases leading to measurable results. This book can transform you into a well-versed business leader with a keen eye for cutting through the technical jargon and asking the tough questions required for your AI journey."
—**Peggy tsAI,** *Executive Director at JPMorganChase and Adjunct Faculty at Carnegie Mellon & University of Denver*

"It's easy to get swept up in the hype surrounding AI and its promise to transform organizations. But what does that transformation truly entail? What tensions arise when considering, selecting, and deploying AI solutions? How will value actually be created—and, more importantly, what must senior leaders and decision-makers put in place to ensure that value is sustained over time?

The Values of AI offers a fresh, practical perspective on the timeless challenge of aligning business strategy with IT strategy. It equips leaders with concrete, ready-to-use tools to guide critical decisions around AI adoption and integration. The result is a roadmap to AI value creation that is clear, adaptable, and comprehensive—bridging the gap between technological potential and human value."
—**Constantinos K. Coursaris,** *Professor at Department of Information Technologies, HEC Montréal and Chair, Advisory Board at SIG Human-Computer Interaction*

"I've had the privilege of benefiting immeasurably from Edosa's first two books, and I can confidently say this is his finest work yet. **The Values of AI** strikes at the heart of the AI revolution—posing some of the most pressing questions of our time.

Edosa first compels us to examine what we sacrifice when we adopt technology that doesn't just automate our work but actually 'automates values' as well. Then, like a skilled mentor, he leads us through this terrain of opportunities versus risks, consistently demonstrating that organisational value and proper ethical consideration need not be at odds—as we too often assume.

Finally, he delivers a comprehensive toolkit ensuring that both thinkers and practitioners across all fields continue to extract lasting benefit from this rich and defining work."
—**Ali Khan,** *Chief Data Officer at Finastra, Member CxO Council at Clear Ventures and Global Strategic Advisor*

"Edosa Odaro has once again, in this latest title, demonstrated his deep mastery of the technological pain points that organisations face—and how they can overcome them. By dissecting how companies can move beyond mere competitive parity in the current rush to adopt AI systems, and showing how they can create and sustain genuine, comprehensive value, he offers senior business leaders a practical lifeline. The understandings often effortlessly advanced in this text are particularly vital for those grappling with the challenge of justifying significant AI investments to an increasingly wary and sceptical set of stakeholders.

Among the many vexing questions the book addresses is how corporate leaders can transform substantial yet often poorly understood AI investments into sustained, business-wide growth, with demonstrable value replication. As Odaro points out, the real successes and opportunities in AI adoption emerge not merely from their technological advancements, but from the strategic, company-wide integration of models and frameworks that can enable the grander visions the technologies promise.

As every business leader who reads this work will attest, Odaro delivers not only a much-needed roadmap but also a highly replicable blueprint for unlocking value across diverse business scenarios and settings. This book is a must-read, both for the pressing issues it tackles and for its highly digestible frameworks and real-world examples."
—**Bob Enofe,** *Legal Director and Corporate Compliance, Trade and Public Policy Advisor*

"Finally, a complete desktop guide that cuts through the AI hype, challenges leaders' perceptions about AI. **The Values of AI** provides a practical roadmap to successfully implement those landmark technologies within their business and create real, sustainable value.

Odaro's deep-rooted, international and multi-sectoral experience shines through every blue print, framework and case study presented in the book.

It is clearly structured, easily readable and its comprehensive value creation toolkit is hugely valuable – pun intended – by itself!

It's one of those books you will be going back to and challenge your thinking and approach over time: a must read!"
—**Markus Krebsz,** *Founding Director at the United Nations (UN) ECE WP.6, Project Leader on AI at Stirling University and Honorary Professor at Woxen University*

"Cutting through the hype, **The Values of AI** reminds us to stop chasing the AI wave and focus instead on creating lasting value that matters. True to the style of his work, Edosa Odaro speaks with rare clarity on how to align technical capabilities with business goals, and he is never short on providing practical tools and frameworks that readers can put to work right away. A must-read for leaders who are serious about delivering on their AI promises."
—**Christina Stathopoulos,** *Data & AI Evangelist at Dare to Data, Global Keynote Speaker, Award-Winning Educator, and Adjunct Professor at IE Business School*

"As usual, Edosa dares to say what few will: AI's value isn't in the flashy tech. It's in cold, hard business transformations.

In **The Values of AI**, he dismantles the hype and equips leaders with the tools to tell a compelling AI value story.

For anyone tired of empty promises, this is THE blueprint for making AI work where it matters most for your business stakeholders."
—**Scott Taylor,** *The Data Whisperer and author of* Telling Your Data Story

"There is so much drivel and confused thinking around AI. But this book feels very real and well considered. Brutally honest about the challenges involved and humble in its recognition that nobody has all the answers."
—**Nick McEwen,** *Senior Manager Group Technology*

The Values of Artificial Intelligence
How Smart Leaders Capture and Connect AI Value to Human Values

Edosa Odaro

CRC Press
Taylor & Francis Group
Boca Raton London New York

CRC Press is an imprint of the
Taylor & Francis Group, an **informa** business
AN AUERBACH BOOK

Designed cover image: Web Large Image (Public)

First edition published 2026
2385 NW Executive Center Drive, Suite 320, Boca Raton FL 33431

and by CRC Press
4 Park Square, Milton Park, Abingdon, Oxon, OX14 4RN

CRC Press is an imprint of Taylor & Francis Group, LLC

© 2026 Edosa Odaro

Reasonable efforts have been made to publish reliable data and information, but the author and publisher cannot assume responsibility for the validity of all materials or the consequences of their use. The authors and publishers have attempted to trace the copyright holders of all material reproduced in this publication and apologize to copyright holders if permission to publish in this form has not been obtained. If any copyright material has not been acknowledged please write and let us know so we may rectify in any future reprint.

Except as permitted under U.S. Copyright Law, no part of this book may be reprinted, reproduced, transmitted, or utilized in any form by any electronic, mechanical, or other means, now known or hereafter invented, including photocopying, microfilming, and recording, or in any information storage or retrieval system, without written permission from the publishers.

For permission to photocopy or use material electronically from this work, access www.copyright.com or contact the Copyright Clearance Center, Inc. (CCC), 222 Rosewood Drive, Danvers, MA 01923, 978-750-8400. For works that are not available on CCC please contact mpkbookspermissions@tandf.co.uk

For Product Safety Concerns and Information please contact our EU representative GPSR@taylorandfrancis.com. Taylor & Francis Verlag GmbH, Kaufingerstraße 24, 80331 München, Germany.

Trademark notice: Product or corporate names may be trademarks or registered trademarks and are used only for identification and explanation without intent to infringe.

ISBN: 978-1-041-07386-4 (hbk)
ISBN: 978-1-041-07750-3 (pbk)
ISBN: 978-1-003-64205-3 (ebk)

DOI: 10.1201/9781003642053

Typeset in Times
by KnowledgeWorks Global Ltd.

Disclaimer

This book reflects the author's present recollections of experiences over time. Some names of persons and organizations have been changed to protect their privacy; some events and dialogues have been compressed. The frameworks, methodologies, and case studies presented are based on real-world implementations across multiple organizations and industries, though specific details may have been altered to maintain confidentiality.

The views expressed in this book are those of the author and do not necessarily reflect the official policies or positions of any organization with which the author has been affiliated. While every effort has been made to ensure accuracy, the rapidly evolving field of artificial intelligence means that some technological references may become outdated over time.

To my sons, Efe and Azuwa:

We live in a time when artificial intelligence (AI) can write poetry, diagnose diseases, and make decisions that shape human lives – yet it cannot love, hope, or dream the way you do every single day. As I watch you grow up in a world where machines can think but cannot feel, I am reminded that our greatest responsibility is not only to build smarter algorithms, but to ensure they serve the irreplaceable human qualities you carry so naturally.

When you ask me why some people don't have homes, or wonder aloud about the stars, or show kindness to a stranger, you remind me what no AI will ever replicate: the beautiful, messy, imperfect miracle of human consciousness.

This book is my promise to you and your generation. In a world racing towards an AI-powered future, I commit to ensuring that the technology we create amplifies rather than replaces the curiosity, empathy, and wonder that make us most human. May the AI of tomorrow serve the authentic intelligence you embody today.

And to my late mother, whose memories remind me that wisdom lies not in having all the answers, but in asking the right questions with love – and whose guidance continues to shape every decision I make about what truly matters.

Contents

Preface: The Invisible Value of AI xv
Acknowledgements xxiii
About the Author xxv

Chapter 1   The Multidimensional Values of AI 1

Chapter 2   The Data Foundation: Why Data Drives AI Value 26

Chapter 3   Spotting Signals of AI Value Failure 45

Chapter 4   The Value-First Approach to AI 58

Chapter 5   Creating a Cross-Organizational AI Strategy 75

Chapter 6   From Pilot to Scale: Building Sustainable AI Value 99

Chapter 7   Securing Executive Buy-In and Board Support 120

Chapter 8   Building Cross-Functional Support for AI Initiatives 149

Chapter 9   Fostering Employee Trust and Adoption 171

Chapter 10 Communicating AI's Societal and Ethical Values 200

Chapter 11 Measuring What Matters 238

Chapter 12 Creating Sustainable AI Value 260

Chapter 13 Future-Proofing AI Value 293

Chapter 14 The Data-to-Value Integration Blueprint 313

Chapter 15 Leading AI Value Creation 339

Chapter 16 From Chaos to Clarity: Your AI Value Creation Toolkit 380

Chapter 17 Conclusion: The Courage to Create Value That Matters 409

Index 413

Preface: The Invisible Value of AI

THE COURAGE TO CHOOSE

What would you sacrifice to make your AI projects work?

I asked this question to a room full of executives at a Fortune 500 company. They were investing millions in artificial intelligence (AI) initiatives, including chatbots, predictive analytics, and automation systems. The technology was impressive. The business cases were solid. The Return on Investment (ROI) projections looked promising.

But when pressed to define what they would sacrifice – what kind of trade-offs, side-effects, or other adverse consequences would be acceptable – the room went quiet.

Not because they didn't have answers, but because there were too many conflicting scenarios. Maximizing shareholder returns while protecting the livelihoods of people they'd worked with for decades. Delivering seamless customer experiences while preserving the craft and expertise that had been developed over countless years. Driving innovation that could make team members irrelevant while maintaining the stability their families depended on. Pursuing growth that might hollow out the communities they called home. Being seen as visionary leaders while being able to sleep at night knowing the human cost. Building something that would outlast them while hitting numbers that mattered only this quarter.

Later that same week, I was reminded why this question matters so much. I watched my five-year-old son ask his grandmother's voice assistant: "Alexa, why do some people not have a home?" The AI responded with housing statistics and economic factors. Technically accurate. But completely missing the point.

My son wasn't asking for data. He was grappling with something much deeper – why the world works the way it does. In this moment, I saw the same pattern I'd witnessed in that boardroom: AI doesn't just process our requests. It mirrors our values. When we build systems that optimize for efficiency over empathy, speed over understanding, metrics over meaning, we're not just creating technology – we are encoding what we believe matters.

Beyond struggling with an AI strategy question, the executives were confronting a similar challenge as the voice assistant: How do you respond meaningfully to what people actually need when you've been programmed to optimize for something else entirely?

AI has an uncomfortable way of forcing these questions into the open. When you build a customer service bot, what are you choosing to optimize – speed or understanding? When you implement predictive hiring systems, what do you prioritize – efficiency or equity? When you deploy automated decision-making, what do you entrust to machines and what requires human judgement?

What may look like straightforward technology decisions actually amount to moral choices. AI doesn't just automate processes, it automates values. The algorithms we build reflect the futures we're willing to accept.

The question isn't whether AI will change the world – it is whether we have the courage to ensure it changes the world in ways that reflect our deepest humanity with a view of future generations, not just our immediate priorities.

And here's what keeps me up at night: most organizations make these profound choices accidentally, reactively, by default rather than by design.

The silence in that boardroom wasn't due to lacking answers – it was due to a lack of courage to choose the answers that actually matter.

THE BOARDROOM REVELATION

I witnessed this dynamic play out again only a few weeks later in another boardroom, this time at a globally significant financial institution. The CTO was showcasing a groundbreaking AI fraud detection system, months in development, millions in investment, cutting-edge algorithms.

The numbers looked impressive: 89.7% accuracy. A less than 200-millisecond response time. An increase of over 30% in pattern detection compared to previous systems.

Then the CFO asked one simple question: "But by how much has it reduced our fraud losses?"

Chairs creaked. Nobody answered.

Despite all the technical brilliance, despite the project hitting every development milestone, nobody could clearly connect this AI system to what the organization ultimately cared about: reducing financial risk and protecting customers.

I've seen scenarios like this play out in organizations of every shape and size, from multinationals to ministries, private innovators to public institutions. Brilliant AI initiatives, packed with potential, ended up being crushed by vague strategies, poor data, misaligned teams, and worst of all, lost stories.

The boardroom incident mirrors a critical moment I experienced firsthand during a crucial presentation early in my career. I had every technical detail perfected. The architecture was flawless. The implementation plan was bulletproof. Then came the question from the CEO that stopped me cold:

What will this mean for our market position in three years?

In that moment, I realized something that would transform my entire approach: technical excellence alone captures only about 20% of potential value. The other 80% – the real opportunity – comes from something more subtle: the ability to see and seize strategic opportunities that others miss to be able to create a matrix of alignment that can enhance both organizational goals and AI capabilities.

THE VALUE PARADOX

Here's a big challenge nobody is talking about: Value really isn't what you think it is.

The unspoken secret at the heart of most AI disappointments is what I call the "value paradox." Everyone uses the word "value" constantly in AI discussions, yet nobody bothers to define it.

The truth is that value isn't a single destination – it often takes the form of multiple competing journeys happening simultaneously. The CEO may talk about market differentiation while the CFO fixates on cost reduction. The CTO may obsess over technical performance while legal worries about compliance risks. The chief operating officer demands efficiency gains while HR worries about workforce disruption. Customers simply expect magical experiences, and frontline employees just want tools that actually work.

These aren't just slightly different perspectives – they can be fundamentally conflicting worlds. What's valuable to shareholders often contradicts what's valuable to users. What's valuable to technical teams frequently gets lost in translation to business teams. What drives corporate profits may undermine community well-being. What advances company interests often clashes with what societies and regulators demand.

This conflict was painfully evident during that recent meeting I observed. Technical teams presented absolutely brilliant AI solutions. Strategy consultants discussed interesting market trends. Everyone was nodding. Yet I felt a familiar knot in my stomach, the same one I had experienced years ago when I was the technical leader in similar rooms. I knew what was about to happen: millions in potential value would remain unrealized – disappearing in the gap between technical insight and strategic action.

Preface: The Invisible Value of AI

Two Sources of Value Conflict

This mess can get even messier since value tends to split along two critical fault lines.

Firstly, there's the incentive-driven split, where value depends entirely on who benefits. A customer service AI agent might deliver tremendous value to customers while creating existential anxiety for employees whose jobs are changing. The same data might be "useless garbage" to one team and "pure gold" to another.

Then there's the ethics-driven split, where there are fundamental differences between what people value. Some prioritize speed over explainability. Others demand transparency over efficiency. These aren't just practical disagreements, they reflect deep differences in what people believe actually matters.

As Dr. Simona Soare, Professor of Strategy and Technology at Lancaster University, pointedly observes:

> Technology in itself is neither good nor bad, but it's also not neutral. The way that we understand the technology and the way that we understand the harms that the technology can pose very much depends on the context in which it is used, as well as the intentionality behind its use.

Both splits operate simultaneously across every AI initiative, creating a tangled web that most organizations never even try to unravel. Instead, we just pretend that everyone is aligned, launch the project, and then act surprised when it fails.

THE LEADERSHIP CROSSROADS

Now more than ever, failure is becoming harder to ignore.

We live in a world where AI doesn't just sit in data labs anymore. It headlines news cycles. It creates art that sells for millions and designs medicines that cure once-fatal diseases. It steers financial markets and whispers strategy into the ears of world leaders. But it also guides precision drone strikes, optimizes killing machines, and makes cold calculations that determine which companies thrive and which ones collapse.

From generative tools like ChatGPT to the ever-growing buzz around agentic AI, we're being told this isn't just another wave – it's a radical revolution.

But amidst all the headlines, all the pilots, all the prototypes, value still evades definition.

Because AI doesn't fail from lack of potential. It fails from lack of connection. It fails because its power doesn't translate into real, felt, and visible value.

"We definitely see that organisations leading the AI adoption, they are implementing it across the core business functions, they're rolling out AI into full production. And this has been enabled by very strong leadership, sponsorship, leadership engagement," explains Stela Solar, Microsoft's Global Head of Artificial Intelligence Solutions.

Yet there's a troubling disconnect between proclaimed and actual leadership engagement. As Greg Shove, CEO of Section, notes, despite C-suite executives trying to demonstrate commitment to AI, "many executives actually have their hands on AI and are not really using it." They "want to appear to be AI-savvy" without truly engaging with the technology.

In boardrooms across industries, a divide is forming. On one side stand the organizations where AI initiatives flourish, creating tangible value and competitive advantage. On the other side, companies where promising AI projects stall in pilot phases or deliver underwhelming results. The difference? It's not technology sophistication or investment size, it's the quality of executive leadership driving AI adoption.

WHY DRILLING DOWN ON VALUE MATTERS

If you're a business leader, data professional, policymaker, or advisor tasked with making AI "work" in your world, you already know this isn't about technology alone.

It's about pressure.

The pressure to show ROI on multi-million-dollar programs. The pressure to align technical teams with executive priorities. The pressure to explain complex systems in ways that boards, customers, regulators – and the public – can actually understand. And increasingly, the pressure to say something – anything – about how your organization is "using AI."

Everyone seems to have an AI strategy. But few have something worth showing. And even fewer can tell a story that resonates across tech, business, and other user groups.

The stakes have never been higher. AI has moved from a tech department curiosity to a boardroom imperative and front-page news. CEOs who couldn't spell "neural network" a couple of years ago now face relentless questions about their GenAI strategies.

Organizations face brutal economic pressure to do more with less while simultaneously transforming their entire business model. AI dangles the promise of meeting both demands, but only if it can deliver actual, measurable value.

You might find yourself in one of these high-stakes situations:

- You're a chief data, technology, analytics, or information officer who's built impressive AI capabilities but can't translate technical wins into business impact. Your budget is under scrutiny, your team's credibility is eroding, and your role feels increasingly precarious as executives question the millions spent with little visible return.
- You're a CEO facing relentless board questions about your AI strategy while competitors make bold claims about their AI transformation. You suspect much of it is smoke and mirrors, but you can't risk being left behind, yet you also can't afford another failed digital initiative.
- You're a business unit leader who sees AI's potential to save your division but watches in frustration as technical teams pursue their own agenda while your core problems remain unsolved – and your performance targets slip further out of reach.
- You're a data scientist whose brilliant models are gathering dust not for any technical deficiencies but because the organization can't implement them, leaving you questioning your career choice as your skills are wasted on proofs-of-concept that never see daylight.
- You're a policymaker or regulator trying to balance innovation with protection, knowing that today's policy recommendations will be out of date by the time they have a chance of being implemented.
- You're a consultant or advisor watching clients make the same expensive AI mistakes, burning through budgets on initiatives that look good in demos but fail to deliver actual value, risking your reputation and relationships when clients eventually question the return on your recommendations.

While these scenarios may sound wildly different, there is a common thread running through all of them: the struggle to make AI's value visible, tangible, and meaningful.

FROM CHAOS TO CALM: SIMPLIFYING SPEAKING SKILLS

We're living in a moment of history where AI is getting louder and louder.

From text-to-image generators and virtual copilots to autonomous agents that plan and act, we're told this is the dawn of intelligent machines that will change how everything works.

But most organizations aren't asking whether AI works. They're asking something much harder: "Where's the value – and how do we prove it?"

That's the gap this book looks to close.

Not by explaining the latest algorithms. But by going beyond the noise to look at the languages that enable us to move AI from hype to real impact.

We decode three languages, where fluency can enable AI to deliver real, lasting value as shown in Table 0.1.

Preface: The Invisible Value of AI

TABLE 0.1
The Three Languages of AI Value

Language	Core Question	Key Challenge	Value Outcome
Strategy	What are we trying to achieve with AI?	Aligning technical possibilities with organizational priorities	AI initiatives that advance genuine strategic goals, not just technical capabilities
Operations	How do we connect AI to daily realities?	Bridging the gap between data possibilities and organizational processes	AI systems that solve real problems for real people in real contexts
Influence	Why should people believe in and adopt our AI?	Building trust across diverse stakeholders with different concerns	AI initiatives that gain momentum, resources, and permission to scale

Each language requires different skills, different tools, and different mindsets. Yet most leaders are fluent in only one of them. The ability to become multilingual in AI value can empower us to translate between each of these critical domains.

THE OWNERSHIP QUESTION

Here's the most shocking part of AI implementation: Nobody owns the value creation process. When value fails to materialize, responsibility tends to mysteriously vanish.

Technical teams build capabilities but often have no clue about business context. Business leaders demand solutions but rarely understand technical constraints. Data teams control the raw material but sit separate from both. When things go sideways, the finger-pointing begins.

This isn't a minor coordination issue; it's a fundamental design flaw in how organizations approach AI.

I witnessed the human impact of such flaws, this firsthand, at a major insurance company where critical system failure occurred when a central motor insurance database wasn't properly updated. The result? Legitimate customers were being detained by police who found no record of valid insurance policies for the cars they were driving. Investigations revealed that "avoidable information gaps across support teams and systems were at the root of the mayhem – and had rendered team members powerless to avoid catastrophe."

These organizational silos eventually created a situation where "it took longer for webpages to respond to customer requests and less-than-optimal sluggishness was also creeping into mobile experience offerings" as the company approached a "cliff's edge."

This book shows how to create genuine shared ownership of AI value – not through some naive reorganization, but through frameworks that make value creation a truly collaborative effort with clear responsibilities that are embraced when things go well and can't be dodged when things get tough.

Paying attention to the three languages will enable you to:

- Capture AI's real strategic advantage
- Connect execution with intention across people and platforms
- Communicate value in ways that earn commitment, not just compliance

PRACTICAL TOOLS FOR REAL-WORLD IMPACT

Having gained many of these insights from practical experience, it made sense to present the material in a way that goes beyond theory. It's filled with practical frameworks and tools that have been battle-tested across industries and organizations. These aren't academic constructs, but working instruments you can apply, right now:

- **The AI Value Mapping Canvas:** A visual tool for aligning AI initiatives with strategic priorities across stakeholder groups
- **The Stakeholder Value Alignment Matrix:** A framework for identifying and resolving conflicting value expectations
- **The AI Narrative Framework:** A structured approach to communicating AI value to different audiences, from tech teams to boards to external stakeholders
- **The AI Value Dashboard:** A measurement system that captures multidimensional value beyond technical metrics
- **The Cross-Functional AI Roadmap:** A planning tool that bridges technical capabilities and business outcomes
- **The AI Adoption Flywheel:** A model for building momentum and engagement across your organization

Each of these tools addresses a specific aspect of the AI value challenge, and together they form a comprehensive toolkit for making AI's value visible, actionable, and sustainable.

THE POWER OF VALUE TRANSFORMATION

Since I've seen so many failed AI strategies, successful transformation journeys are all the more satisfying. When things come together, some kind of magic happens. Here are some examples:

Working on implementing a transformation plan with a global insurance company, we reduced claims reserve processing time from "over three weeks to deliver tangible insights" to "within three hours." That's the kind of transformation possible when you bridge the value gap.

As one financial services firm found, an initial $150K technical implementation project could be transformed into a strategic advisory engagement with a $9.5M value potential by shifting the conversation from implementation details to strategic value.

The firm's initial request was typical: help implement an AI strategy, provide technical expertise and best practices, with a budgeted $150K for a three-month phase. But during our first conversations, I heard something entirely different. The CTO was losing sleep over their competitors' AI advances. The CEO had promised significant progress to the board, but their teams were stuck, paralysed by competing priorities and technical choices. So, they were not just looking for implementation help – they needed quick wins to build momentum and confidence in a market where every delay meant losing ground to more agile competitors.

This context made three specific questions both relevant and urgent:

> **"What happens if your AI investments fail to deliver value?"** For this firm, failure wasn't just about wasted budget. In this highly regulated industry, failed AI initiatives could signal to regulators that the firm couldn't effectively manage emerging technologies, potentially triggering increased oversight. Meanwhile, competitors were already using AI to offer faster loan approvals and more personalized services, also making the cost of AI failure existential rather than merely financial.
>
> **"How much is six months of faster value capture actually worth?"** The firm was haemorrhaging market share to fintech startups that could approve loans in minutes while their process took days. Reaching the market six months faster with competitive AI capabilities could mean the difference between retaining the customer base and watching it evaporate. When we calculated the revenue impact of faster time-to-market in their specific competitive context, the numbers were staggering.
>
> **"What's the true cost of making the wrong technical choices?"** The firm had already witnessed other financial institutions make costly AI implementation mistakes that required expensive rework. In this regulatory environment, technical debt wasn't just inefficient, it

was also potentially compliance-breaking. The wrong choices could also lock them into architectures that couldn't scale or adapt to changing regulatory requirements, creating million-dollar problems further down the road.

These weren't generic value questions – they were precisely calibrated to the firm's specific strategic pressures. The resulting transformation ultimately delivering:

- Risk reduction in AI vendor selection ($2M saved)
- Acceleration of value capture (four months faster to market)
- Prevention of common scaling mistakes (avoided $800K rework)
- Confidence in board presentations (secured $5M additional funding)

The total value potential came to $9.5M, meaning that an increased consulting fee of $500K, which helped unlock this potential, represented less than 5% of the value created.

A ROADMAP TO AI VALUES

The AI wave is everywhere – and it will only gather momentum – but visible, lasting value still is elusive. That's why it is important to focus on turning potential into proof for leaders, including executives needing to prove AI impact without buzzwords, technical leaders trying to translate models into meaning, policymakers and advisors tasked with AI governance and trust, and high-stakes decision-makers who just do not have time for fluff.

Key steps for delivering results:

- Capture strategic AI opportunities worth pursuing.
- Connect across internal or external silos, functions, and stakeholders.
- Communicate value in a world that's oversaturated with noise.

AI isn't a magic bullet. But when you define and speak its value clearly, it becomes something else entirely: a value multiplier.

Acknowledgements

Writing *The Values of Artificial Intelligence* has been a journey through one of the most transformative periods in human history. As artificial intelligence (AI) capabilities have evolved from experimental tools to business-critical systems, I've had the privilege of working alongside remarkable individuals who have shaped both this book and my understanding of how AI can create genuine value.

The insights that form the foundation of this work emerge from collaborations with thousands of professionals across hundreds of organizations and their technology partners worldwide. I am deeply grateful to my colleagues, clients, and collaborators from international organizations spanning financial services, technology, insurance, telecommunications, government, and consulting – including teams from multinational corporations, government agencies, consulting firms, and technology vendors who have shared their expertise and opened their doors during critical transformation moments.

My experience serving on multiple boards has provided invaluable perspectives on AI's societal implications and potential. These experiences have reinforced my conviction that technology's greatest value lies in its service to human flourishing.

I owe special gratitude to the executives, practitioners, and teams who opened their organizations to me during critical moments of AI transformation. Your willingness to share both successes and failures has enabled the frameworks and insights that form this book's core. The courage to experiment, adapt, and persist in the face of uncertainty exemplifies the best of human leadership in an age of AI.

Susanne Yaro deserves recognition that extends far beyond her editorial contributions. Over multiple books and years of collaboration, she has evolved from editor to creative partner, helping shape not just the words on these pages but the thinking behind them. Her patience, insight, and unwavering commitment to excellence have been indispensable.

To my beta readers and reviewers: Your tough love and honest feedback have strengthened every chapter. The time you invested in reviewing draft manuscripts, challenging assumptions, and suggesting improvements has elevated this work beyond what I could have achieved alone.

My acquiring editor deserves special thanks for believing in this project from its earliest conception and providing steady support throughout the publication process. Your understanding of both the technical and human dimensions of AI transformation has been invaluable.

To my colleagues in the AI and data community: Our ongoing conversations, debates, and collaborations continue to shape my thinking. The field of AI advances through collective intelligence, and I am honoured to contribute to that shared endeavour.

My friends have provided essential balance, reminding me that the most important relationships remain fundamentally human. Your encouragement, perspective, and willingness to pull me away from work when needed have sustained me through the challenges of writing while managing demanding professional responsibilities.

To my father and siblings: Your support, critique, and belief in the importance of this work have been constant sources of strength. Our family conversations about technology's role in society have shaped my thinking about AI's responsibilities and possibilities.

To my sons, Efe and Azuwa: You inspire me daily with your curiosity, creativity, and natural ability to see possibilities where others see problems. Your questions about how things work and why they matter remind me that the future we're building must serve the wonder and potential you represent.

Finally, to my wife Anja: This book exists because of your support, sacrifice, and unwavering belief in its importance. You have shouldered far more than your share to create the space and time this work required. From late-night writing sessions to weekend research, you have been my partner in every sense. Your keen insights, honest feedback, and emotional support have made this book possible. Thank you for understanding that creating something meaningful for others sometimes requires everything we have to give.

Ultimately, my hope is that the insights and frameworks presented here will help others create AI implementations that truly serve human flourishing while delivering sustainable business value. In an age of AI, our greatest opportunity lies not in what machines can do, but in how thoughtfully we choose to apply their capabilities to the challenges that matter most.

About the Author

Edosa Odaro stands at the forefront of AI and data transformation, helping organizations worldwide turn AI potential into a measurable business impact. As Executive Advisor for Data & AI and with a Chief Data Analytics & Privacy Officer background, he has guided more than 35 international organizations—including Barclays Group, AXA, AIG, and the European Commission—through complex transformations that delivered substantial, sustainable results.

With an academic foundation spanning economics, computer science, business, technology, and innovation, Edosa brings a rare combination of technical depth and strategic vision to the AI field. His work has generated nine-figure efficiency gains, created entirely new business models powered by intelligent interventions, and improved the quality of countless lives.

Recognized as a *Financial Times* Top 100 Most Influential Leader and a Global 100 Data Activator, Edosa has earned acclaim not just for his technical expertise but also for his ability to solve the human challenges that determine whether AI works for us—or works against us. His philosophy is simple but radical: **sustainable AI value emerges where technological capability meets human understanding.**

1 The Multidimensional Values of AI

Chapter Roadmap

In this chapter, we'll explore why AI value isn't singular but multidimensional. We'll discover six key value dimensions of AI implementations and how organizational silos create the "AI value fog." From there, we'll turn to game theory and gamification approaches to value alignment – as well as practical frameworks for translating value across stakeholder groups to discover actionable steps to create cross-functional value clarity.

THE MANY DIMENSIONS OF VALUE

"That's not what we asked for."

The Chief Technology Officer's (CTO) face was flushed as he stared at the presentation on the screen at the quarterly executive review of a $3.7 million AI initiative of his organization, a Fortune 500 financial services firm with over 14 million customers, hundreds of millions of transactions, and a level of regulatory scrutiny that made every technology decision a potential headline. The meeting was fast descending into open conflict.

"Actually, it's exactly what you asked for," the data science lead shot back. "And we built a system with 94% prediction accuracy – currently the best across our industry."

The Chief Finance Officer (CFO) cut in, "But where are the cost savings? We're six months in and we haven't seen a penny of the $12 million in operational efficiencies you promised."

"That's because the business units aren't implementing the recommendations," the CTO countered. "We're delivering insights and they're just ignoring them."

The Chief Operating Officer (COO) shook her head. "Your insights are just not actionable. And they don't fit our operational workflows."

"And customers are complaining about the new process," added the Chief Marketing Officer (CMO). "Our satisfaction scores dropped eight points last month."

I watched this executive team – smart, experienced leaders who genuinely wanted the best for their organization – talk past each other for 45 painful minutes. Despite spending millions and deploying some genuinely impressive technology, they couldn't agree on whether their AI investment was succeeding or failing.

This wasn't a minor communication issue; it was the perfect example of AI's value problem.

PATTERN RECOGNITION: Notice how each executive in this scenario is judging success through an entirely different lens. The CTO sees technical excellence, the CFO sees missed financial targets, the COO sees operational friction, and the CMO sees customer dissatisfaction. All are right, yet they can't align.

Here's the truth most AI conversations miss: value isn't just one thing. It's many things happening simultaneously, often pulling in opposite directions. When we talk about "AI value" as if it's singular and universal, we're setting ourselves up for confusion, disappointment, and ultimately failure.

In my work with organizations across sectors, I've discovered that AI value splits along multiple dimensions.

Economic Value: Beyond Simple ROI

Most conversations about AI value start and end with dollars and cents. That's understandable but dangerously incomplete.

Economic value from AI typically manifests in three ways:

- **Cost Reduction:** Automating tasks, reducing errors, optimizing resources
- **Revenue Growth:** Creating new offerings, improving targeting and market reach, enabling price optimization
- **Margin Expansion:** Improving operational efficiency, reducing waste, enhancing productivity

But even within this financial lens, value perceptions diverge wildly. The CFO might focus on immediate cost savings while the CEO considers long-term market positioning. The head of sales sees revenue potential while operations can be fixated on efficiency gains.

Take the experience of an insurance client who implemented an AI-powered fraud detection system. The risk department celebrated reduced fraud losses (approximately $167M annually). The operations team valued the 73% reduction in false positives that was lowering their workload. Customer service appreciated the 29% drop in legitimate transaction declines. Same system, three entirely different value perceptions – but crucially, all three departments experienced genuine improvements.

This represents the ideal scenario for multidimensional AI value, where the system creates positive outcomes across different stakeholder groups simultaneously. The challenge becomes far more complex when AI creates winners (on the one hand) and losers (on the other hand) – when one department's efficiency gains come at the expense of another's workflow disruption, or when cost savings in operations translate to service degradation for customers. In those situations, the "different value perceptions" become competing interests that can derail even technically successful implementations.

This isn't just about different opinions – it's about fundamentally different mental models of what matters.

Executive Insight

> We kept measuring our AI's technical performance and wondering why executives weren't impressed. Then we realized we were showing them the equivalent of a car's horsepower when they cared about journey time and fuel economy. Once we translated technical metrics into business outcomes, everything changed.
>
> — CTO,
> *Global Financial Services Firm*

KEY INSIGHT: Microsoft's research demonstrates this value perception gap clearly. In an "EY Microsoft Tech Directions" podcast, Stela Solar, Microsoft's Global Head of Artificial Intelligence Solutions, found that organizations often struggle until they align leadership around a unified understanding of AI value: "These AI projects, because leadership is involved, they become much more tightly aligned to business goals, business priorities. And so that then tends to deliver the kind of results that start to create this spiraling effect where the AI projects are aligned to business outcomes and goals."

REFLECTION QUESTIONS: How do different stakeholders in your organization measure AI success? Have you created clear translations between technical metrics and business outcomes?

Strategic Value: Competitive Position and Market Differentiation

Beyond immediate financial returns, AI creates strategic value that's harder to quantify but often more important:

- **Competitive Advantage:** Creating brand recognition or capabilities competitors can't easily replicate

- **Market Differentiation:** Offering unique experiences or solutions that stand out
- **Innovation Acceleration:** Compressing development cycles and enabling experimentation

A healthcare system I worked with invested in AI-powered decision support not primarily for cost-savings but because it saw this as strategically essential to its future. It believed healthcare was moving towards personalized precision-based services, and without AI capabilities, it would eventually become non-competitive regardless of its current market position.

Its CFO couldn't calculate a clean ROI, but its board recognized that sometimes strategic necessity trumps immediate returns.

This strategic perspective is well illustrated by real-world implementations. Cristina Busmales, Former Chief Revenue Officer at Benevolent, explains how AI transforms strategic capabilities: AI is "identifying novel targets, validating them, and optimizing clinical trials through protocol design, patient recruitment, and site selection." This represents strategic value through fundamentally reimagining core business processes rather than just improving existing ones. Bryce Hall, McKinsey's Associate Partner, presents a landscape focused view and observes: "We saw nearly a 25 percent year-over-year increase in the use of AI in standard business processes. We've moved beyond the phase of 'Is AI a shiny object?' toward broader mainstream adoption and actual value creation." This signals that strategic necessity is driving adoption across industries.

Executive Insight

> At BMW, we recognized early that artificial intelligence is the key technology in the process of digital transformation. But for us, the focus remains on people. AI supports our employees and improves the customer experience.
>
> — **Michael Würtenberger,**
> *Head of BMW Group Project AI*

OPERATIONAL VALUE: THE ENGINE ROOM OF TRANSFORMATION

AI creates tremendous operational value that directly impacts how work gets done:

- **Efficiency:** Completing tasks faster with fewer resources
- **Quality:** Reducing errors and inconsistencies
- **Decision Improvement:** Making better choices with better information

A utilities client implemented an AI visual inspection system that detected defects that human inspectors missed over 30% of the time. The operational value wasn't just in catching more defects – it was in learning why those defects occurred in the first place, leading to process improvements upstream.

Operational value often creates compound effects. Better quality reduces rework, which improves throughput, leading to enhanced capacity utilization. This creates a value chain that's greater than the sum of its parts.

QUICK WIN: Map the trickle-down benefits of your AI implementations. What additional values emerge when the initial improvements (like error reduction) trigger subsequent improvements (like capacity utilization)? This "value chain" thinking often reveals benefits that aren't visible in initial ROI calculations.

CUSTOMER VALUE: EXPERIENCE, PERSONALIZATION, AND TRUST

AI has the potential to create distinct value dimensions for customers:

- **Experience Enhancement:** Making interactions smoother, faster, and more satisfying

- **Personalization:** Tailoring offerings to individual needs and preferences
- **Trust Building:** Creating consistency and reliability that strengthens relationships

A telecom company implemented AI-powered network optimization that reduced outages by 37%. Customers didn't know or even care that this was powered by AI – they simply experienced better service. The value wasn't the technology itself but the improved reliability it created.

When customers benefit from AI, they rarely appreciate the technology – they just value the outcome. This distinction is crucial when measuring and communicating AI's impact.

Employee Value: Augmentation, Development, and Satisfaction

AI can create specific value dimensions for the workforce:

- **Augmentation:** Enhancing human capabilities and productivity
- **Skill Development:** Creating opportunities for growth and learning
- **Satisfaction:** Removing tedious tasks and enabling more meaningful work

A legal services firm implemented AI document review that handled routine contracts. Its lawyers initially feared job losses but came to value the system when they realized that it freed them to work on more complex, interesting cases while still meeting billing targets.

WARNING SIGN: Research from Writer's 2025 Survey reveals just how critical this dimension is. According to industry analysis by eMarketer, when employees fear displacement rather than augmentation, 31% admit to deliberately undermining AI initiatives, with one in ten tampering with performance metrics to make AI systems appear ineffective. This employee resistance doesn't just slow adoption – it actively reverses the potential value.

The employee value dimension is often overlooked in AI planning but proves especially critical to adoption. People support what benefits them and resist what threatens them – regardless of the technology's brilliance. It is also clear that when employees are unhappy, this tends to lead to unhappy customers.

Societal Value: Ethics, Community, and Sustainability

The final value dimension extends beyond organizational boundaries and relates to AI's potential to create better outcomes for society.

- **Ethical Considerations:** Fairness, transparency, and responsible use
- **Community Impact:** Benefits to broader society, beyond direct customers
- **Sustainability:** Environmental protection and social responsibility improvements

A city government deployed AI traffic optimization that reduced congestion by 17%, but the real societal value was in the reduced carbon emissions and improved air quality for residents.

This dimension becomes increasingly important as AI regulations are becoming more prevalent – and consumers demand responsible technology practices. What was once "nice to have" is rapidly becoming expected and standard practice.

PATTERN RECOGNITION: Notice how the most successful AI implementations address multiple value dimensions simultaneously? The traffic optimization system delivered operational efficiency, economic value, customer experience improvements, and societal benefits. This multi-dimensional impact is what separates transformative AI from merely incremental improvements.

NAVIGATING THE AI VALUE FOG

With all these competing value dimensions, it's no wonder many organizations struggle to define and communicate AI's impact. The result is what I call the "AI value fog" – a persistent state of confusion that can undermine even the most technically sound initiatives.

The fog occurs for four specific reasons:

STRATEGIC VAGUENESS

Many AI initiatives begin with impressive-sounding but fundamentally vague objectives:

- "Becoming AI-driven"
- "Leveraging data for competitive advantage"
- "Modernizing our decision-making"

These statements sound good but lack the specificity needed to guide implementation, sustain progress, or measure success. They create a fog where anything could be considered valuable – or nothing at all.

A global consumer goods company started an AI initiative with the stated goal to "harness data for growth." A year and several million dollars later, they had impressive models but couldn't explain how these connected to actual business growth. Without specific value targets, technical success can become strategic failure.

Jeremy Foster, Vice President of Cloud Infrastructure and Software Group at Cisco, explains: "Communicating in silos is a trap that you can sometimes fall into. How do you keep people visible? Good visibility across the entire project as you work on it."

ACTION STEP: Before approving any AI initiative, test the strategic clarity by asking: "If this succeeds, what specific metrics will improve, by how much, and for whom?" If the answer lacks numerical or other measurable specificity, push for greater clarity before proceeding.

OPERATIONAL DRIFT

As AI initiatives progress, technical complexities often take centre stage. Teams can become focused on model accuracy, technical performance, and data challenges – gradually drifting away from the original business objectives.

One of my financial services clients began an AI project to improve customer retention but became so consumed with data quality issues that they lost sight of the customer problem they were trying to solve. They built a technically impressive system that ultimately addressed the wrong problem.

NARRATIVE GAPS

Most organizations lack a consistent language for discussing AI value across different functions. Technical teams talk performance metrics, business teams discuss financial outcomes, and leadership considers strategic implications – all without a shared vocabulary to connect these perspectives.

In a pharmaceutical company, the data science team celebrated their algorithm's superior predictive power while executives wondered why drug development timelines weren't improving. Both were right within their respective frames of reference, but without a bridging narrative, they talked past each other.

Matthew Leiggi, Senior Revenue Enablement Manager at Rectangle Health, explains:

> Cross-collaboration is always paramount and that includes hosting alignment meetings, workshops, and what have you with sales, marketing, the product teams to make sure that everybody is on the same page, we're all aligned, and understand exactly what we're supposed to be marching towards.

KEY INSIGHT: According to Microsoft's research with enterprise AI implementations, organizations that establish a "common vocabulary" across technical and business teams are 2.7 times more likely to achieve their AI value targets. This vocabulary isn't about technical jargon – it's about creating shared-understanding of what success actually looks like.

Perception Mismatch

Different stakeholders experience AI's impact differently. What feels like a modest improvement to developers might represent a frustrating change for end-users or a strategic breakthrough for executives.

A large wealth management firm implemented an AI dealmaker system that produced statistically better results, but field sales executives perceived it as undermining their expertise. The value was real but invisible to key stakeholders whose buy-in was essential.

How AI Success Gets Lost in the Value Fog

The four fog factors don't just create confusion – they set in motion specific mechanisms that actively erode AI's potential value.

Strategic Vagueness Creates Misdirected Effort

When objectives lack specificity, organizations invest resources without clear direction. RAND research reveals that senior leaders often underestimate the time required to train effective AI models, expecting results in days rather than weeks or even years. This mismatch creates tension with data science teams struggling to deliver results against unrealistic timelines.

As one organizational leader told RAND researchers:

> The biggest challenge is usually getting buy-in from all relevant parties. The problem is usually threefold. First, there's some distrust in AI, especially when it comes to more complex use cases that involve automated decision-making. Secondly, there are almost always some upfront costs that many businesses are not prepared for or are just not willing to pay. And, thirdly, there's still a lack of education around AI, especially among employees.

Operational Drift Diverts Resources from Value Creation

But even when organizations overcome these initial alignment challenges, another pattern emerges. As initiatives progress, teams naturally gravitate towards solving the technical problems they understand rather than the business problems that matter. The same RAND study found that leaders often instruct data science teams to solve the wrong problem with AI, resulting in models that have minimal business impact despite technical excellence.

This creates a double bind: leaders set unrealistic expectations while simultaneously providing unclear direction about what problems actually need solving. The study also identified that leaders frequently switch priorities and sometimes instruct teams to move forward prematurely, impacting progress and diverting projects before they demonstrate real results.

Narrative Gaps Prevent Effective Collaboration

When different functions lack a shared language for discussing AI, critical collaboration breaks down. According to RTS Labs, the lack of alignment on what constitutes value – whether it's user adoption, business objectives, or innovation opportunities – poses a significant barrier to successful AI deployment.

When one team's perception of value is too narrow, focusing solely on technical features, while another prioritizes broader business outcomes, the resulting solutions may not address the full spectrum of organizational needs.

Perception Mismatches Destroy Adoption and Trust
When stakeholders perceive AI's impact differently, adoption suffers regardless of technical quality. Cross-functional misalignments often lead to solutions that are technically sound but practically irrelevant.

PATTERN RECOGNITION: Notice how these four factors create a reinforcing cycle? Strategic vagueness leads to operational drift, which creates narrative gaps. This, in turn, causes a perception mismatch, which then perpetuates strategic vagueness. Breaking this cycle requires addressing all four factors simultaneously, not just fixing one at a time.

WARNING SIGN: When different stakeholders can't agree on whether an AI initiative is succeeding or failing, you're experiencing the AI value fog. This isn't just a communication problem – it is a fundamental misalignment in how value is perceived and measured.

Breaking Through the Fog

Organizations that successfully navigate the value fog employ specific strategies to create clarity:

1. **Explicit Value Definition:** According to RAND's research on AI implementation, successful organizations clearly articulate what success looks like across multiple value dimensions before implementation begins. The study found that when leaders instruct data science teams without specifying business objectives, AI projects often solve the wrong problems despite technical excellence.
2. **Cross-Functional Alignment:** Christopher Mael, Director of Sales and Operations Planning at Peloton, describes how his organization overcame departmental conflicts: "It's creating the space for conversations. Fundamentally, I think it's asking tough questions... It's less about tools, it's more about getting people together within a closed door and saying alright, we're gonna talk about tough things. Let's do it!"
3. **Continuous Narrative Connection:** Sarah Fury, a marketing leader at Pantheon, explains how her organization created a common language across departments:

    We all had people from other departments who showed up for a couple of days of in-person training. What we walked away with was a common alignment on how we can manage projects better but also a vocabulary that we were all using that was the same.

4. **Value Translation Tools:** The most effective organizations build systematic frameworks that show how improvements in technical metrics (like model accuracy) translate to operational improvements (like process efficiency) and ultimately to business outcomes (like cost savings or customer satisfaction). These translation mechanisms prevent the common pattern where technical teams celebrate algorithmic achievements while business stakeholders question the investment's value.

These approaches aren't just theoretical – they're practical solutions implemented by organizations that have successfully broken through the fog. As one chief analytics officer from a global telecommunications company told me: "The technology wasn't our biggest challenge – it was maintaining clarity about what value we were trying to create and for whom. Once we solved that, the technical implementation became much more straightforward."

NAVIGATING THE "VALUE VS. VALUES" TENSION

Beyond the difficulty of defining "value" in AI initiatives lies a deeper tension that creates significant organizational friction: the distinction between "value," often equated with economic benefits, and "values," which go beyond economic impact to include ethical and societal principles.

During a recent healthcare AI implementation workshop I facilitated, this tension erupted in real time. The finance team presented ROI projections for a new diagnostic AI, calculating value in reduced staffing costs. The patient support team immediately pushed back, arguing that patient care values must remain paramount. Meanwhile, the compliance officer worried about privacy values, and the innovation team spoke of competitive value. The room crackled with cross-purposes. This happened not because anyone was wrong, but because everyone was right – from their perspective.

This tension between value and values isn't just conceptual – it's a visible manifestation of the invisible fault line beneath many AI implementation failures.

KEY INSIGHT: In the "Value versus Values" discourse, analysed by Andy Polaine, this tension is framed as a false dichotomy. Polaine argues against the notion that businesses should prioritize shareholder value above all else, stating this approach is "both ethically and economically flawed" since "economies operate within societies that uphold various values."

Harvard Business School Professor Ranjay Gulati explains this predicament:

> I think there has been a growing sentiment among business leaders for at least a decade, if not longer, that business has a larger role to play than simply shareholder value. But that story of what that role is has been unclear.

Research from PWC's Asia Pacific Sustainability team reveals how organizations increasingly frame their values-based initiatives in value-creation language. Director Dean Alborough notes: "From a financial value creation perspective, clients are looking to better understand the sustainability levers within their own business and how these levers influence the financial bottom line."

Some leaders attempt to resolve this tension by treating value and values as complementary rather than competing elements. Professor Gulati suggests: "I also came to realize that businesses with purpose exist not only to serve society but also to serve shareholders and make money. Businesses have to connect the commercial with their social logic. It's not an either-or."

REFLECTION QUESTIONS: In your organization, do discussions about AI typically separate economic value from ethical values – or are they integrated? How might reframing this as a "both/and" rather than an "either/or" question change your approach?

The clearest indicator of this tension appears in cross-functional teams trying to deploy AI. As healthcare expert Denise Wiseman, Chief Community Manager and Founder of The PX Community, observes: "Our primary focus too often is on competition, whether it is among individuals, teams and departments, one organization to another.... We're focused on the red-green scorecards rather than how we're looking holistically."

REGIONAL AND CULTURAL DIMENSIONS

The tension between value and values takes on additional complexity in global organizations or cross-cultural contexts. Different regions approach this balance through distinct frameworks.

In the United States, the emphasis often falls on market-driven innovation and competitive advantage, with ethical considerations addressed through corporate social responsibility initiatives. According to strategist Brian Eastwood, AI adoption in the United States is primarily market-driven, with executives focusing on innovation and competitiveness above all else.

European leaders see a stronger regulatory emphasis on AI as important. The EU AI Act exemplifies this approach, urging organizations to treat compliance as a core strategic pillar of AI initiatives. As noted by Paola Cecchi-Dimeglio, Senior Fellow at Harvard University, this requires embedding ethical and regulatory considerations into the foundation of design and deployment – not as afterthoughts, but as conditions for long-term legitimacy.

The Multidimensional Values of AI

Meanwhile, regions like the Middle East demonstrate particularly high AI confidence, with 88% of GCC CEOs having adopted generative AI in the past year. As one regional leader noted in a 2025 National interview, "We're lucky to be in a region where AI is a top priority – not just for CEOs, but also for country leaders."

African leadership approaches often emphasize collaborative efforts and capacity-building alongside innovation, with a growing focus on developing local AI capabilities that enable inclusive economic growth and social transformation.

This regional variation creates additional complexity in multinational organizations, where different cultural approaches to the value-values balance must be reconciled. As we'll explore in Chapter 10, successful global AI implementations require frameworks that can accommodate these diverse regional perspectives while maintaining a coherent ethical foundation.

Overall, the value-values conundrum can have a dual impact on organizational silos. While these silos can heighten the tension, they can also be reinforced, when each department feels compelled to tightly hold on to their version of value or values. For example, engineering teams optimize for technical excellence, compliance teams for risk mitigation, commercial teams for market impact, and executive teams for strategic positioning – each convinced their lens is most critical.

In companies that successfully navigate this tension, leaders recognize that AI implementations must honour multiple value dimensions simultaneously, not sequentially. Their secret isn't prioritizing one dimension over another, but creating frameworks that connect them, building bridges rather than ranking priorities.

ACTION STEP: In your next AI planning meeting, explicitly map both economic value targets and ethical values considerations. Create a "Value-Values Matrix" that shows how your implementation addresses multiple dimensions, with specific metrics for each.

This tension between value and values will continue to emerge throughout your AI journey, particularly as you build cross-organizational strategies (referenced in Chapter 5) and address external stakeholder concerns (see Chapter 10). The organizations that succeed don't merely manage this tension – they leverage it to create more robust and sustainable AI implementations.

UNDERSTANDING VALUE THROUGH DIFFERENT STAKEHOLDER LENSES

To navigate the value fog, we must understand how different stakeholders view AI's impact. Rather than being slightly different perspectives, these can represent fundamentally different evaluative frameworks.

BOARD AND CEO PERSPECTIVES

At this level, AI value primarily centres on:

- Strategic positioning and market differentiation
- Long-term competitive advantage
- Risk management and organizational resilience
- Innovation capabilities and future readiness

Boards ask questions like: "How does this change our competitive position?" and "Does this create sustainable advantages?" rather than focusing on technical details or even short-term returns.

A retail board I worked with was less interested in the specifics of their new AI merchandizing system than in how it positioned them against competitors like Amazon and Walmart. They evaluated the initiative through a strategic lens that technical teams rarely consider.

BUSINESS UNIT LEADER PERSPECTIVES

These leaders often view AI value through:

- Performance improvement against business metrics
- Resource efficiency and allocation
- Team capabilities and productivity
- Customer and market impact

They ask: "How does this help me hit my targets?" and "What problems does this solve for my team right now?"

The sales leader at a software company implemented an AI lead scoring system. While the data team focused on prediction accuracy, she cared only about whether her reps spent time on the right prospects and closed more deals. The model's complexity was irrelevant to her value assessment.

TECHNICAL TEAM PERSPECTIVES

Technical stakeholders typically evaluate AI through:

- Model performance and accuracy
- Data quality and availability
- System integration and scalability
- Implementation elegance and innovation

They ask: "Does this perform to specification?" and "Is this technically sound, scalable and sustainable?"

A data science team at a manufacturing company built a predictive maintenance system with 94% accuracy – a technical triumph. But they struggled to understand why operational managers weren't enthusiastic about their achievement, not recognizing that maintenance supervisors valued different things: crew scheduling flexibility, parts availability, and production impact minimization.

EMPLOYEE PERSPECTIVES

Frontline workers and end-users assess AI based on:

- Job impact and security
- Workflow changes and convenience
- Learning requirements and support
- Trust and control factors

They ask practical questions: "Does this make my job easier or harder?" and "Do I have to learn another system?" and the crucial "What happens to my role if this works as planned?"

A hospital implemented AI-assisted diagnosis, but adoption lagged because physicians saw it as questioning their expertise rather than enhancing it. The value proposition didn't address this perspective until it was reframed as "decision support" rather than "diagnostic assistance."

KEY INSIGHT: Extensive research on AI's Societal and Ethical Impacts reveal a stunning 30 percentage point gap between how executives and employees perceive AI value. While 70% of executives felt their company's approach to AI had been strategic and successful, only 40% of employees shared this positive view. This perception gap doesn't just affect morale – it directly impacts adoption and ultimately determines whether AI investments deliver their promised value.

CUSTOMER PERSPECTIVES

Customers rarely care about AI itself – they instead evaluate outcomes:

- Service quality and reliability
- Problem resolution and ease
- Personalization and relevance
- Trust and transparency

They ask: "Does this make my experience better?" and "Can I rely on the information provided?" not "Is this powered by sophisticated algorithms?"

A telecommunications company proudly marketed their "AI-powered customer service," but customers were unimpressed with the technology claim. When they reframed it as "Get answers in seconds without waiting on hold," satisfaction scores jumped – same technology, different value framing.

EXTERNAL STAKEHOLDER PERSPECTIVES

Regulators, communities, and other external parties focus on:

- Fairness and non-discrimination
- Transparency and accountability
- Social and environmental impact
- Privacy and data protection

These stakeholders ask: "Is this technology being used responsibly?" and "What are the broader implications beyond the organization?"

A lending institution implemented an AI credit scoring system that improved approval rates and reduced defaults – clear internal value. But they faced regulatory scrutiny over potential bias, a value dimension they hadn't adequately considered in their implementation.

WARNING SIGN: Research shows serious environmental concerns emerging that many organizations overlook. Studies from the Markkula Center for Applied Ethics reveal that "generating a single AI image consumes as much energy as charging a smartphone," while research by George and colleagues documents "substantial water requirements for cooling data centres that run large language models." These environmental impacts often remain invisible in traditional value assessments but increasingly matter to regulators, communities, and environmentally conscious consumers.

Executive Insight

> The most valuable AI implementations we've done aren't the ones with the most sophisticated algorithms. They're the ones that connected across functional boundaries and created value that no single department could achieve alone.
>
> — **Chief Digital Officer,**
> *Global Healthcare System*

THE EXECUTIVE LENS: C-SUITE VALUE ALIGNMENT

While we've explored board and CEO perspectives, the broader C-suite provides a fascinating microcosm of AI's value challenge. Research shows that executives with aligned AI value perspectives achieve 3.5 times greater returns on AI investments compared to those with fragmented views. Yet in my experience, C-suite leaders often have dramatically different value orientations:

- **The Revenue Champion (CEO, CRO)** prioritizes AI that drives top-line growth, such as new offerings, customer acquisition, and market expansion. They evaluate AI primarily through growth metrics and competitive positioning.
- **The Efficiency Advocate (CFO, COO)** focuses on cost reduction, productivity improvement, and resource optimization. Their value lens emphasizes ROI, cost savings, and operational streamlining.
- **The Risk Guardian (CISO, CLO)** views AI through risk management, compliance, and governance. They measure value in terms of reduced errors, improved compliance, and mitigated threats.
- **The Technology Catalyst (CTO, CDO)** prizes capabilities development and future positioning. Their value framework emphasizes building long-term competitive advantage through cutting-edge applications.

A global financial services firm I worked with found that their AI initiatives consistently underperformed until they created an "executive value alignment" process that required explicit discussion of these different perspectives before approving major AI investments. The CTO explained, "We realized we weren't speaking the same language. The CFO was asking about cost reduction while I was talking about innovation capacity – both valuable, but on completely different timescales and measurement frameworks."

This pattern has become so prevalent that many organizations now establish dedicated AI leadership roles. Pedro Uria Recio, chief data analytics and AI officer at True Digital, explains:

> The leader of AI in an organization has to be at the C-level. It is also relevant to highlight a McKinsey & Company study about the top factor that differentiates winners from laggards in analytics. The top factor cited in the research was the presence of a tech-savvy C-level champion.

Research by Foundry indicates that 11% of midsize to large organizations have already designated chief AI officers (CAIOs) and 21% are actively seeking one. This reflects the growing recognition that executive alignment is crucial for AI success.

In some organizations, AI leadership is formalized through the CAIO role. Colin Reeves explains that the CAIO role involves "modernizing processes with AI, ensuring that AI is used with ethics and governance in mind, and in building an AI-first culture."

ACTION STEP: Evaluate your organization's executive alignment by conducting an "AI value perception audit." Ask each C-suite member to independently define what successful AI implementation looks like. Where you find significant divergence, create specific translation mechanisms that connect their perspectives.

The most effective executive teams recognize that AI isn't just another technology investment – it's a strategic capability that requires unprecedented alignment across traditionally separate domains. As we'll explore further in Chapter 7, this alignment starts with establishing shared value definitions that acknowledge and connect these diverse executive perspectives.

BREAKING DOWN SILO BARRIERS

"The organization gets larger and larger, we kind of just end up becoming and working in our echo chamber and we get disconnected from what the bigger goal is."

These words from Jay Patel, CEO of Amtech, capture perhaps the greatest threat to AI's multidimensional value: organizational silos – the invisible walls separating departments, functions, and teams. These silos don't just hinder implementation; they fundamentally limit what AI can achieve.

Silos emerge naturally as organizations scale, with departments developing specialized languages, metrics, and priorities. When AI initiatives cross these boundaries – especially those highly valuable ones that could have desirable impacts on other parts of the organization – these differences can become barriers.

Christopher Mael, Director of Sales and Operations Planning at Peloton, describes the executive challenge: "Our executive meetings were a little bit like Thunder Dome. Everybody was going after their own KPIs or metrics, driving their struggle to get anything in forward motion because those metrics were either adversarial or contrasting."

The impact of such silos on AI initiatives can be profound:

- **Value Blindness:** Teams can't see benefits outside their functional area
- **Implementation Resistance:** People resist changes that don't optimize for their metrics
- **Data Fragmentation:** Critical information remains locked in departmental systems
- **Competing Priorities:** Teams compete for resources and contradictory objectives get in the way of progress

PATTERN RECOGNITION: In organizations with persistent silos, AI initiatives tend to create local optimizations that sub-optimize the whole. For example, an AI-powered marketing system might maximize lead generation while overwhelming sales capacity, or an operations AI might maximize efficiency while degrading customer experience. Breaking the barriers associated with silos means optimizing for integrated outcomes, not departmental metrics.

Organizations that successfully break down these barriers employ specific strategies:

CREATE CROSS-FUNCTIONAL VALUE TEAMS

Rather than leaving value integration to chance, establish dedicated teams responsible for connecting perspectives. A global insurance company formed an "AI value team" with rotating membership from operations, finance, engineering, and HR. Every AI initiative required sign-off from this cross-functional team, ensuring that decisions were informed by multiple value perspectives.

DEVELOP VALUE TRANSLATION TOOLS

Build bridges across different value languages. A healthcare system created a "value translation matrix" showing that each 1% improvement in their diagnostic AI's accuracy meant 17 additional correct early diagnoses monthly, $213,000 in reduced treatment costs, and six fewer readmissions quarterly. This translation made technical improvements meaningful to both clinical and administrative stakeholders.

ALIGN INCENTIVES AND METRICS

Shared KPIs create shared destiny. Marketing, sales, and customer experience teams will work together more efficiently when they are unified around a single objective, like increasing customer retention. In this setup, each department bears equal responsibility for meeting the goals, lowering conflict and promoting shared ownership of achievement.

FOSTER CROSS-POLLINATION

Enable team members to experience different value perspectives firsthand. At Cisco, the HR strategy welcomes input from lower-level managers in top-level decision-making. This approach makes it possible for teams and divisions to plan together, facilitating the sharing of new technology and product advancements across the organization.

Executive Insight

> Cisco has created a bridge between siloed architecture experts and customer-facing staff in diverse groups across the company.
>
> — **Sean Worthington,**
> *Vice President of IT, Operational Excellence and Service Enablement, Cisco*

BUILD SHARED LANGUAGE

Create common vocabulary that bridges specialized jargon. The most successful AI implementations establish a unified language that allows technical teams to communicate with business stakeholders and vice versa. This isn't about dumbing down technical concepts or oversimplifying business requirements – it is about creating translation mechanisms that preserve meaning that is accessible across functions.

Without shared language, organizations fall into a pattern where data scientists celebrate "95% model accuracy" while business leaders ask "but did sales increase?" While neither perspective is wrong, teams often operate in different linguistic universes that prevent effective collaboration.

QUICK WIN: Create a one-page "AI value glossary" for your organization that defines key terms across technical, operational, and business domains. Include translations that show how technical concepts (like "model accuracy") relate to business outcomes (like "decision quality"). Invite input from disparate teams to ensure diversity in representation and distribute the resulting glossary before cross-functional AI meetings.

When organizations resolve barriers related to silos, AI's value multiplies. Examples show that cross-functional integration of AI, such as across research and development, supply chain, production, after-sales service, and administration can lead not just to improved performance of individual functions but also create connections between them that drive the kind of competitive advantages that no single department could achieve alone.

As we'll explore further in Chapter 2, data silos often mirror organizational divisions, creating technical barriers on top of cultural ones. Later, in Chapter 8, we'll provide comprehensive frameworks for building cross-functional support that systematically addresses these barriers and creates lasting alignment.

EXECUTIVE LEADERSHIP APPROACHES TO BREAKING SILOS

C-suite executives play a decisive role in breaking down the organizational silos that limit AI value. Effective leaders address silos through these approaches:

Strategic Budget Allocation

Innovative C-suite leaders design funding models that incentivize cross-functional collaboration. A global insurance company implemented a "cross-boundary budget" where 20% of each department's AI allocation could only be accessed through joint initiatives with other functions. The organization's CFO explained, "When we made collaboration a financial necessity rather than just a cultural aspiration, behavioural changes followed quickly."

According to Christian Barnard, executives must ensure they have buy-in from key stakeholders across departments such as IT and operations before finalizing AI budgets. Without their

The Multidimensional Values of AI

endorsement and collaboration, AI initiatives risk encountering resistance or misalignment with business processes.

Strategic Hiring and Partnerships

Forward-thinking executives build teams and partnerships specifically designed to bridge departmental boundaries. Research by Marc Schmitt reveals that CAIOs are essential in ensuring AI deployment aligns with organizational objectives across departments, fostering a unified approach to AI implementation.

The case of Parminder Bhatia, CAIO at GE HealthCare, demonstrates this approach. Bhatia leads AI deployment by focusing on AI strategy across multiple functions, using AI to streamline radiology workflows, reduce scan times, improve diagnosis, and automate measurement. As Bhatia explains,

> We're driven by a mission to revolutionize healthcare interfaces by integrating voice, text, and the latest in AI visualizations. This approach isn't just about technological novelty. It's also about creating user-centric tools that aim to redefine how medical professionals interact with and leverage technology, improve their efficiencies, and ultimately improve patient experience and outcome.

C-Suite Modelling

Perhaps most importantly, successful executives model cross-functional thinking in their own behaviours and decision processes. As Jay Patel, CEO of Amtech, observed, "As leaders, we need to look beyond individual success to holistic success, creating a truly shared vision."

REFLECTION QUESTIONS: How do your executive team meetings model the cross-functional collaboration you want to see throughout your organization? Are AI initiatives discussed in functional silos, or as integrated strategies with multiple value dimensions?

In Chapter 15, we'll provide a comprehensive framework for developing these executive capabilities, enabling leaders to drive value integration across their organizations.

THE VALUE MULTIPLIER EFFECT

When properly aligned, AI doesn't just create incremental improvements – it creates multiplicative effects that transform organizational capabilities.

This happens through four specific mechanisms:

TRANSCENDING TRADITIONAL TRADE-OFFS

Conventionally, organizations face hard choices between competing priorities, for example:

- Quality vs. speed
- Personalization vs. scale
- Consistency vs. flexibility

AI can sometimes resolve these trade-offs, creating value by eliminating previously necessary compromises.

A manufacturing giant had always balanced thorough quality inspection against production speed. With an AI visual inspection system, the company achieved both simultaneously: 100% inspection at full production speed. This wasn't just an improvement; it was a transformation of what was possible.

Cross-Functional Value Amplification

When AI bridges organizational silos, it creates value beyond what any single function could achieve alone.

A retailer implemented AI demand forecasting that connected merchandizing, supply chain, and store operations. Each function saw specific improvements, but the combined effect – reducing out-of-stock merchandize by 21% while decreasing inventory by 17% – was only possible because the system crossed traditional boundaries.

The Compounding Effect of Data-Driven Learning

Unlike traditional technologies that depreciate over time, AI systems often improve with use. As they process more data and incorporate feedback, their value compounds.

A medical diagnostics company found that its AI imaging system improved its accuracy by 4–7% annually through continued learning without additional development investment – creating a widening advantage over static systems.

KEY INSIGHT: My research, as presented in the book *Value-Driven Data*, revealed that this compounding effect is even more powerful when organizations focus on measuring differential value – the gap between baseline performance and AI-enabled performance – rather than absolute metrics. This differential measurement makes the exponential improvement curve visible, helping justify continued investment in AI capabilities.

Enabling Entirely New Capabilities

The most powerful AI value comes from enabling what was previously impossible, not just improving what already exists.

An insurance client used AI to create personalized policy bundles in real time during customer conversations, something that previously required days of actuarial work. This wasn't just faster; it was a fundamentally new capability that changed their customer acquisition model entirely.

ACTION STEP: For each AI initiative in your organization, identify at least one example from each of these four multiplier categories. If you can't find examples where you can "transcend trade-offs" or "enable new capabilities," you may be underestimating your AI's transformative potential.

CASE STUDIES: MULTIDIMENSIONAL VALUE IN PRACTICE

Let's examine how these value dimensions play out in real organizational settings:

Retail Transformation: From Value Conflict to Value Integration

A national specialty retailer implementing an AI demand forecasting system initially faced near-failure due to cross-functional value conflicts. As the organization's chief digital officer confided, "We nearly killed a tremendously valuable system because we couldn't agree on what 'good' looked like. Supply chain was celebrating inventory reductions while stores complained about stockouts for promotional items."

This wasn't merely miscommunication – it was a fundamental clash of value perspectives:

Supply Chain Team Values

- Inventory efficiency (22% inventory reduction, saving $322M)
- Operational streamlining (31% reduction in manual interventions)
- System optimization metrics

The Multidimensional Values of AI

Store Operations Team Values

- Product availability for customers
- Staff time savings (5–8 hours weekly per manager)
- Customer satisfaction scores

Marketing Team Values

- Promotion execution integrity
- Brand perception
- Campaign effectiveness metrics

The breakthrough came when the retailer created a cross-functional "value council" that established shared metrics acknowledging each department's priorities while creating balance:

Primary Cross-Functional Metrics

- Perfect order rate (combining availability, timeliness, and accuracy)
- Days of inventory with minimum availability standards
- Promotional execution compliance

The retailer's EVP of Supply Chain explained the transformation: "The magic wasn't in the algorithm – it was in getting everyone to see how their piece connected to everyone else's. We stopped optimizing for our individual metrics and started optimizing for the customer experience and financial outcome together."

This cross-functional integration created multidimensional value that would have been impossible through siloed implementation:

Financial Impact

- $322M inventory reduction
- $28M reduction in markdown costs
- $41M increase in full-price sales

Operational Impact across Functions

- Supply chain: 31% reduction in exceptions requiring manual intervention
- Merchandizing: Planning cycle reduced from 6 weeks to 10 days
- Store operations: 5–8 hours weekly of manager time reallocated to customer service
- Marketing: Promotional waste reduced by 23%

Customer Impact

- 17% increase in product availability
- 28% reduction in "substitute items" in online orders
- Net promotor score (NPS) improvement of 12 points on product availability

This retailer's journey illustrates the cross-functional integration that we'll examine more thoroughly in Chapters 5 and 8. The success hinged on creating shared value definitions and metrics – principles we'll formalize into implementation frameworks in Chapter 14.

QUICK WIN: Create a simple visual map showing how each department's KPIs affect other departments. For example, how marketing promotions impact inventory requirements, which affect supply chain metrics. This "interdependency map" helps everyone see how their decisions affect others, building the foundation for shared value understanding.

Healthcare Example: Clinical Outcomes, Patient Experience, and Operational Efficiency

A regional healthcare system deployed AI in its emergency department:

Clinical Value: 11% improvement in triage accuracy, ensuring critical patients were prioritized appropriately
Patient Value: Average wait time reduction of 24 minutes (22%)
Operational Value: Resource utilization improved by 18%, allowing more patients to be seen
Financial Value: $4.2M annual savings through reduced boarding time and optimized staffing
Strategic Value: Improved emergency care metrics raised their national ranking, attracting specialist physicians

KEY INSIGHT: Different stakeholders cared about different metrics, but the system had to deliver across all dimensions to succeed.

Manufacturing: Supply Chain Optimization, Quality Improvement, and Innovation

A discrete manufacturing company implemented AI across its operations:

- **Production Value:** Defect reduction of 32% through early detection and intervention
- **Supply Chain Value:** 41% reduction in material shortages through predictive planning
- **Employee Value:** Workplace injury reduction of 27% through anomaly detection
- **Financial Value:** $18M annual savings through combined efficiency improvements
- **Innovation Value:** New product development cycle shortened by 40% through simulation

This approach explicitly mapped different value expectations across stakeholders, ensuring the implementation addressed multiple dimensions simultaneously.

Executive Insight

> We wasted a year arguing about whether our AI initiative was successful because finance, operations, and customer service all used different measuring sticks. Creating a shared value framework didn't eliminate those perspectives – it connected them.
>
> — **SVP Operations,**
> *Retail Organization*

GAME THEORY: A FRESH APPROACH TO AI VALUE ALIGNMENT

When departments clash over AI priorities, we're witnessing more than personality conflicts – we're seeing classic game theory scenarios play out in real time. Game theory, the study of strategic interactions among rational decision-makers, offers powerful insights into aligning different stakeholders' interests during AI implementation.

Dr. Thuc Vu, entrepreneur and co-founder of OhmniLabs and Kambria, explains the connection: "If you think of collaboration in business as a game, people are just playing a game. It's like a one-shot game – a bit of a dilemma – and at Nash equilibrium, everyone is going to cheat in a way, non-collaborate."

This perfectly describes what happens in many AI implementations. When departments optimize for their own metrics (their "winning strategy"), this likely leads to a collectively suboptimal outcome. Finance pushes for cost efficiency that might undermine customer experience. IT prioritizes system integration that might slow market responsiveness. Marketing advocates for features that might increase technical complexity.

The breakthrough comes when organizations reframe AI implementation from a one-time game (where non-cooperation is rational) to a repeated game with aligned incentives.

"But if we remodel it and repeat the game so that people will play the game over again," Dr. Vu continues, "there are some very interesting game theoretical approaches which can enforce the behaviour out of this game, so that we can not only incentivize but force collaboration between the players."

This explains why the most successful AI implementations establish ongoing cross-functional governance rather than one-time approval processes. When departments know they'll continue working together on AI evolution – and when their success metrics include collaborative outcomes – cooperative behaviour becomes the rational choice.

PATTERN RECOGNITION: Notice how traditional project management approaches often set up AI initiatives as "one-shot games" where departments can optimize for their local benefit? By restructuring governance as continuous and establishing shared success metrics, you move from the so-called prisoner's dilemma (where defection is rational) to an iterated game (where cooperation becomes the optimal strategy).

A healthcare company I worked with applied game theory principles by creating a shared risk/reward pool for their AI diagnostic system. Rather than having IT judged solely on implementation timeliness, clinicians on diagnostic accuracy, and finance on cost metrics, they established a bonus structure where 30% of each department's performance evaluation came from the overall system's success. This created a "repeated game" where cooperation became the optimal strategy for all players.

The Nash equilibrium shifted from departmental optimization to system-level success, and their AI implementation delivered value across all dimensions.

This game theory lens offers a powerful way to reconceptualize the organizational dynamics around AI implementation. We'll revisit and expand on these principles in Chapter 8, where we'll explore specific mechanisms for using game theory and gamification to align departmental interests and create sustainable cross-functional collaboration.

GAMIFICATION: TURNING CROSS-FUNCTIONAL COLLABORATION INTO AN ENGAGING EXPERIENCE

Beyond formal structures and frameworks, pioneering organizations are applying gamification techniques to transform cross-functional AI collaboration from an organizational challenge into an engaging experience.

Gamification – applying game-like elements to non-game contexts – creates powerful incentives for breaking down silos and fostering the collaboration essential for multidimensional AI value. According to Assembly CEO Jonathan Fields, gamification improves teamwork by "cultivating a sense of accomplishment among employees and promoting collaborative efforts towards shared goals."

Here's how leading organizations apply these principles to AI value alignment:

VISUALIZATION DASHBOARDS

A global manufacturer created an "AI value nexus" – a real-time dashboard showing how technical performance metrics connected to operational improvements, financial outcomes, and customer experience measures. Teams competed to create the most balanced impact across all four dimensions, with quarterly recognition for the most "value-balanced" initiative.

CROSS-FUNCTIONAL CHALLENGES

A financial services firm established monthly "value-a-thons" where cross-functional teams competed to identify new AI use cases with impact across multiple value dimensions. The winning ideas received fast-tracked funding and implementation resources, while team members earned recognition and development opportunities.

Value Translation Competitions

An insurance company gamified the "translation challenge," rewarding team members who could most effectively explain technical concepts to business stakeholders and business requirements to technical teams. This created a culture where communication across the value dimensions became a valued skill, not an administrative burden.

Collaborative Recognition Systems

Several organizations have implemented point systems rewarding cross-functional collaboration, with leaderboards and recognition for teams and individuals who most effectively bridge traditional silos. As one retail banking executive explained, "We made breaking down boundaries as rewarding as exceeding individual targets."

Egge Haak, a supply chain expert, describes the transformative effect: "Imagine a space where the finance head understands the implications of logistics, where marketing sees the ripples of their campaigns on inventory. This is the world that gamification opens up."

When properly designed, these approaches do more than make collaboration fun – they fundamentally reshape how teams interact around AI value.

QUICK WIN: Create a simple "AI value translation challenge" where team members earn points for connecting technical achievements to business outcomes. For example, how does a 5% improvement in model accuracy translate to customer experience metrics? This gamified approach makes value translation a team sport rather than an administrative burden.

VALUE TRANSLATION FRAMEWORK

To help organizations connect different value perspectives, I've developed this value translation framework that shows how metrics in one domain affect outcomes in others (see Table 1.1). Table 1.2 presents examples of these translations.

TABLE 1.1
AI Value Translation Framework

Technical Metrics	Operational Metrics	Business Outcomes
Model accuracy	Process error rate	Cost savings
False-positive rate	Resolution time	Customer satisfaction
Response speed	Resource utilization	Productivity gain
Data quality	Decision quality	Risk reduction

TABLE 1.2
Translation Examples

Technical Improvement	Operational Impact	Business Value
10% improvement in model accuracy	22% reduction in process errors	$1.7M annual savings
Reduction in false positives	15% faster customer resolution	8-point NPS increase
30% faster model response	12-minute workflow reduction	$820K labour reallocation
25% enhanced data completeness	18% better decision quality	$3.2M risk reduction

# The Multidimensional Values of AI

EXECUTIVE-SPECIFIC VALUE TRANSLATIONS

For CFO

- 10% algorithm accuracy improvement → 4.2% reduction in operational costs
- Reduced false positives → 17% decrease in manual intervention requirements

For COO

- Enhanced prediction capability → 22% improvement in resource utilization
- Model responsiveness → 9-minute reduction in average process completion time

For CMO

- Personalization enhancement → 14% increase in customer engagement metrics
- Recommendation accuracy → 18% improvement in conversion rates

This translation framework provides a starting point for connecting different value dimensions. In Chapter 11, we'll build on this approach with comprehensive measurement frameworks tailored to different industries and AI applications, creating a complete system for quantifying and communicating AI's multidimensional impact.

ACTION STEP: Create a customized version of this framework for your next AI initiative. Work with stakeholders from different functions to identify the specific chains of impact from technical metrics to business outcomes in your context.

Executive Value Translation

For C-suite leaders, specialized translation frameworks are particularly valuable. These frameworks convert technical metrics to executive priorities:

For CFO (Cost and Efficiency Focus)

- Algorithm accuracy improvements → Operational cost reductions
- Reduced false positives → Decreased manual intervention requirements

For COO (Resource and Process Optimization)

- Enhanced prediction capability → Improved resource utilization
- Model responsiveness → Reduced time to process completion

For CMO (Customer Engagement and Growth)

- Personalization enhancement → Increased customer engagement
- Recommendation accuracy → Improved conversion rates

A manufacturing executive I worked with created an "AI value card" for each initiative that explicitly showed the connections between technical improvements and their impact on each C-suite member's priority metrics. "Before these cards, we'd spend half our AI steering committee meetings just trying to understand each other," she explained. "Now we can immediately focus on decisions rather than translations."

When executive teams share a common translation framework, they can move beyond definitional debates to strategic decision-making. This alignment becomes increasingly important as AI assumes more strategic significance within the organization.

RECOGNIZING YOUR VALUE ALIGNMENT CHALLENGE

Before you can address AI's multidimensional value challenge, you need to assess where you stand. Ask yourself these diagnostic questions:

1. Can everyone in your AI initiative clearly articulate what success looks like in specific, measurable terms?
2. Do technical and business stakeholders use the same language when discussing the project's goals?
3. Have you explicitly mapped how different stakeholders will experience and evaluate the initiative?
4. Are you measuring success across multiple value dimensions, not just technical or financial metrics?
5. Do you have mechanisms to reconcile conflicting value priorities when they emerge?

If you answered "no" to any of these questions, you likely have a value alignment gap that puts your AI investment at risk.

WARNING SIGN: In my recent "AI Advisor Accelerator" research, I found that in 70% of struggling AI initiatives, stakeholders couldn't consistently describe what success would look like. This "divergent vision problem" predicted implementation failure more accurately than technical challenges or funding limitations.

EXECUTIVE-DRIVEN AI SUCCESS FACTORS

Research across diverse organizations shows that executive leadership is the primary differentiator between successful and unsuccessful AI initiatives. The following factors are particularly critical:

1. **Clear Strategic Vision and Alignment with Business Goals:** Organizations with executives who articulate specific connections between AI initiatives and strategic objectives achieve 3.2 times higher implementation success rates than those with vague goals like "becoming AI-driven."
2. **Strategic Investment in Talent and Infrastructure:** Leaders who allocate resources strategically – balancing immediate ROI with longer-term capability-building – create sustainable AI advantages. As one retail CEO noted, "We learned to view our AI investments as creating organizational muscles, not just solving immediate problems."
3. **Cross-Functional Leadership Engagement:** When executives from multiple functions actively engage in AI initiatives (not just IT or innovation leaders), implementation success rates increase by 61%. Leading organizations have established cross-functional AI governance teams at the executive level, with representation from each major business unit.
4. **Ethical Governance Integration:** Executives who integrate ethical considerations into AI governance from the start avoid costly retrofits and reputation damage. A Deloitte study found that C-level leaders increasingly prioritize ethical decision-making in AI development, recognizing that "ethically designed governance structures are important to hold both leaders and employees accountable."
5. **Cultivation of AI Literacy among Leadership:** Organizations where executives invest in their own AI understanding – not just delegating to technical teams – are 2.4 times more likely to achieve successful implementations. As one healthcare CEO explained, "I don't need to understand the algorithms, but I do need to understand the possibilities and limitations."

The Multidimensional Values of AI

REFLECTION QUESTIONS: How does your organization currently evaluate the success of AI initiatives? Is it primarily through technical metrics, financial returns, or a balanced approach that includes multiple value dimensions?

These executive-driven success factors form a crucial foundation for addressing the value alignment challenges we've identified. In Chapter 7, we explore specific strategies for securing and maintaining executive buy-in through these approaches.

FIRST STEPS TOWARDS CROSS-FUNCTIONAL VALUE CLARITY

Addressing the multidimensional value challenge starts with these practical actions:

CONDUCT A VALUE EXPECTATION AUDIT

Before implementation begins, document how different stakeholders and functions will interact with and benefit from the AI system. Create a simple matrix as shown in Table 1.3.

This value map becomes your alignment baseline and prevents the siloed thinking that undermines many AI initiatives.

CREATE VALUE TRANSLATION MECHANISMS

Develop tools that help stakeholders understand how impacts in one area affect others. A translation guide might show:

- How a 10% improvement in prediction accuracy translates to operational metrics
- How operational improvements connect to financial outcomes
- How technical performance links to customer experience

A retail banking leader told me: "The single most important thing we did was creating a value translation office that helped technical teams understand business priorities and helped business teams understand technical constraints. They became the bridge between worlds that otherwise wouldn't connect."

ESTABLISH CROSS-FUNCTIONAL GOVERNANCE

Create formal structures where different value perspectives must be reconciled:

- **Value council** with representatives from each stakeholder group
- **Balanced scorecard** showing multiple value dimensions simultaneously
- **Trade-off framework** for making decisions when value dimensions conflict

TABLE 1.3
Simple Stakeholder Identification Matrix

Stakeholder Group	Primary Value Expected	Key Metrics	Potential Concerns
Finance	Cost reduction	ROI, OpEx reduction	Implementation costs
Operations	Efficiency	Throughput, error rates	Process disruption
Customers	Experience	Satisfaction, NPS	Privacy, personalization
Employees	Augmentation	Productivity, satisfaction	Job security, training

A manufacturing company formed an "AI value team" with rotating membership from operations, finance, engineering, and HR. Every AI initiative required sign-off from this cross-functional team, ensuring multiple value perspectives were considered.

BUILD MULTI-DIMENSIONAL SUCCESS REVIEWS

Replace single-dimension progress reports with integrated reviews that show:

- Technical performance metrics
- Operational impact measures
- Financial outcomes
- Employee experience indicators
- Customer impact evidence

A financial services firm replaced its traditional project updates with "value realization reviews" that required teams to report impacts across all five dimensions, preventing the common pattern of technical teams reporting only on model performance.

FOSTER A VALUE-INCLUSIVE CULTURE

Create an environment where cross-functional collaboration is expected and rewarded:

- Celebrate wins that cross traditional boundaries
- Recognize individuals who bridge different value perspectives
- Share stories that highlight multidimensional impact
- Create opportunities for immersion in different departmental perspectives

The organizations that succeed with AI aren't necessarily those with the most advanced technology. They're the ones that maintain crystal clarity about the multiple values they're pursuing and how those values manifest for different stakeholders.

ENGAGE EXECUTIVES AS VALUE TRANSLATORS

Beyond establishing structures, successful organizations actively engage their executive team as value translators across the organization:

Executive Value Storytelling

Train executives to tell compelling value stories that connect technical achievements to business outcomes. When leaders consistently translate AI's impact across different value dimensions in their communications, the value narrative permeates throughout the organization.

A pharmaceutical CEO I worked with made it a practice to highlight one AI initiative at each company town hall, explicitly showing how it created value across multiple dimensions. "We don't just say 'our new algorithm is 20% more accurate,'" she explained. "We say 'our new algorithm helps researchers identify promising compounds 20% faster, which reduces our development timeline by approximately six months, potentially helping patients access new treatments sooner while improving our competitive position.'"

Executive Learning Journeys

Create structured opportunities for executives to experience AI value from different stakeholder perspectives. A financial services firm instituted quarterly "perspective shifts" where C-suite members spent half a day experiencing their AI implementations from different vantage points – as customers, employees, and partners.

"Until I sat with frontline staff trying to incorporate AI recommendations into their workflow, I completely misunderstood the adoption challenges," their COO admitted. "That experience fundamentally changed how we approach implementation."

QUICK WIN: Schedule "value walk-throughs" where executives experience your AI initiatives from multiple perspectives. Have them interact with the system as users, view the monitoring dashboards used by technical teams, and see how the outputs impact operational workflows. These experiential learning journeys create empathy and understanding that abstract reports never can.

THE VALUE INTEGRATION IMPERATIVE

As we've explored, AI's true power lies not in any single value dimension, but in its ability to create impact across multiple dimensions simultaneously. Yet most organizations continue approaching AI through fragmented, siloed lenses. While this is understandable, given traditional organizational structures, it is increasingly untenable as AI permeates every aspect of business.

The multidimensional value challenge isn't merely an implementation issue, it's a strategic imperative that will separate AI leaders from laggards in the coming years.

Organizations that master value integration will develop compounding advantages. Their AI initiatives will move faster, deliver broader impact, and create more sustainable competitive differentiation. Most importantly, they'll build institutional capabilities that make each subsequent AI implementation more effective than the last.

Determining values also helps provide a compass to ensure organizations are aligned internally and move in the right direction.

"The challenge in scaling is to never lose sight of your core values," warns Davin Salvagno, bestselling author and founder. "As you grow, it's essential to constantly reinforce your values and make sure everyone is aligned – because without that, culture becomes toxic."

The same applies to AI value. Without alignment across value dimensions, implementations become toxic – technically impressive but organizationally divisive, financially promising but culturally destructive, strategically ambitious but operationally impractical.

KEY INSIGHT: In *Value-Driven Data*, I cited a financial technology company case study that revealed how their rapid growth (30% annually across EMEA regions) created an increasingly complex environment where different perspectives on value became increasingly disconnected. Despite handling billions of transactions, the company struggled to connect online and offline experiences effectively, illustrating how fragmented value perspectives become more problematic as organizations grow and AI systems proliferate.

The path forward requires more than technical excellence or even executive commitment. It demands a fundamentally new approach to how we conceive, implement, and evaluate AI's impact across organizations. The frameworks and approaches outlined in this chapter provide a starting point, but the real work happens when you apply them to your specific organizational context.

These approaches to value integration set the foundation for success, but they require ongoing reinforcement through appropriate leadership, governance, and communication strategies.

2 The Data Foundation
Why Data Drives AI Value

Chapter Roadmap

In this chapter, we highlight how data limits or enables AI potential. Organizations that build appropriate data foundations before rushing to implement advanced algorithms avoid many common failure patterns. The data readiness assessment framework provides a structured approach to identifying and addressing critical gaps.

THE CRITICAL DATA-AI VALUE CONNECTION

There is a common assumption that sufficient investments in cutting-edge Artificial Intelligence (AI) tools and talent are enough. In one global financial institution, which had just invested over $15 million, I witnessed the Chief Technology Officer (CTO) beaming with pride in a leadership meeting as he showcased the new generative AI platform that would transform customer interactions.

The self-congratulatory moment was cut short when a senior risk manager asked a simple question: "But how will this work with our current customer data that's split across six different systems, with different formats and quality issues?"

The immediate response was far from convincing. In its rush to build advanced AI capabilities, the team had overlooked the most fundamental truth: AI is only as good as the data it consumes.

This wasn't an isolated incident. I've seen this pattern repeat across industries and organizations of all sizes. Everyone wants AI's magic, not platforms, not technology, but few want to face the unglamorous reality of what makes it possible: solid, accessible, high-quality data.

PATTERN RECOGNITION: Throughout this chapter, watch for the recurring theme of organizations investing heavily in AI technology while underestimating the foundational data work required. This disconnect leads to predictable failure patterns. We explore these patterns and illustrate how to avoid them.

THE EXECUTIVE-PRACTITIONER DISCONNECT

Perhaps the most alarming aspect of the scenario where projects get derailed due to an unsound data foundation is how common it is. Despite growing executive enthusiasm for AI, the people closest to the data often tell a different story. Even when leaders express confidence in their data readiness, IT and data teams report the opposite – citing the persistent quality, access, and integration challenges that dominate their daily work. The resulting gap isn't theoretical; it shows up in project delays, spiralling costs, and erosion of credibility.

As John Armstrong, CTO of Worldly, puts it: "[Data readiness is] a big, big issue, because if not done right, your organization could spend literally millions of dollars on the wrong solution set to achieve the wrong outcome."

The disconnect between executive perception and technical reality creates a fundamental tension that undermines AI success. While business executives envision transformative AI applications, technical teams are stuck in the trenches of data preparation, unable to build on foundations that don't yet exist.

WARNING SIGN: If your technical teams consistently need more time than expected to deliver AI results, it may not be a technology problem, it's likely a data foundation issue. Ask them directly: "What percentage of your time is spent on data preparation versus model development?"

THE HARD TRUTH ABOUT AI VALUE

Here's the hard truth that many executives don't want to hear: your data foundation determines your AI value ceiling. No matter how sophisticated your algorithms or how talented your data scientists are, poor data foundations will limit what's possible.

Think of it this way: would you build a skyscraper on a foundation designed for a small house? Of course not. Yet organizations routinely attempt to build advanced AI capabilities on data foundations never designed to support them.

Throughout my work with organizations implementing AI, I've discovered a direct correlation between data maturity and AI success. Organizations with strong data foundations achieve results faster, scale more effectively, and create more sustainable value than those rushing to implement AI on shaky data ground.

As was bluntly stated by one respondent cited in a 2024 RAND Corporation study: "80% of AI is the dirty work of data engineering." This unglamorous reality contradicts the exciting narratives about AI that dominate executive discussions, but ignoring it guarantees disappointment.

REFLECTION QUESTIONS: In your organization, who owns data quality? If the answer is "everyone" or "no one specifically," you've identified your first critical gap in AI readiness.

THE HIDDEN COSTS OF POOR DATA FOUNDATIONS

The costs of building AI on weak data foundations go far beyond the obvious technical limitations:

- **Opportunity Cost:** A healthcare provider spent nine months having data scientists clean and integrate patient data before they could even begin building their clinical risk prediction model. That's nine months of delayed value and competitive advantage lost.
- **Resource Drain:** A retail client had data scientists spending 70% of their time on data preparation tasks rather than actual model development. Their most expensive technical talent was doing work that could have been addressed through better data architecture.
- **Trust Erosion:** A financial services firm rushed an AI-powered investment recommendation system to market despite data quality issues. When customers received clearly inappropriate recommendations, trust in all their digital tools plummeted, damaging their broader digital transformation efforts.
- **Scale Limitation:** A manufacturing company successfully piloted an AI quality inspection system in one plant but couldn't scale it to others because each location stored production data differently. Their pilot success became an expensive dead end.

These costs rarely show up in project budgets or post-mortems, but they represent the true price of neglecting data foundations.

According to Gartner research, poor data quality costs organizations approximately $12.9 million annually. That's before we even factor in the specific impacts on AI initiatives.

KEY INSIGHT: The true cost of poor data quality goes beyond the failures it causes – to the opportunities it prevents. As Melody Chien, Senior Director Analyst at Gartner, explains: "Good quality data provides better leads, better understanding of customers and better customer relationships. Data quality is a competitive advantage that D&A leaders need to improve upon continuously."

Building on Solid Ground: Lessons from Value-Driven Data

In my previous book, "Value-Driven Data," I outlined how organizations must align their data capabilities with business objectives. This principle becomes even more critical when AI enters the picture.

AI doesn't just use data – it magnifies both its value and its flaws. When data is inaccurate, AI makes inaccurate predictions. When data is biased, AI perpetuates and potentially amplifies that bias. When data lacks context, AI draws misleading conclusions.

The relationship between data maturity and AI success follows a clear pattern I've observed across dozens of organizations (see Table 2.1).

Most organizations attempt to implement AI at level 1 or 2, then wonder why they struggle to move beyond pilots. They're building advanced technology on foundations never designed to support it.

Accenture's "AI: Built to Scale" research confirms this pattern, showing that 72% of strategic AI scalers have a core data foundation compared to only 63% of their non-scaling counterparts. The difference in return on AI investments between companies in the proof of concept (PoC) stage and strategic AI scalers averages $110 million – a compelling business case for getting the foundation right.

QUICK WIN: Assess your organization's data maturity level using the Table 2.1. Simply knowing where you stand can prevent misaligned expectations and help you plan appropriate foundation-building steps before major AI investments.

Let me share a real example that illustrates this principle in action:

A retail banking client spent two years trying to develop an AI-powered customer recommendation engine with limited success. When they paused to assess their data maturity, they discovered they were at level 1 – with customer data fragmented across deposit, lending, and investment systems. They spent the next six months focusing on data integration before returning to AI development. Their subsequent implementation took just four months and delivered over three times the original value targets.

The lesson? Sometimes slowing down to strengthen your foundation allows you to later move towards your ultimate goals much faster.

ACTION STEP: Before launching your next AI initiative, allocate time and resources for a data readiness assessment. For every month you plan to spend on AI development, budget at least one week for data foundation evaluation.

TABLE 2.1
Data Maturity vs. AI Success

Data Maturity Level	Typical AI Outcomes
Level 1: Fragmented Data exists in silos with inconsistent definitions and limited accessibility	Limited to isolated use cases with extensive manual preparation; most initiatives stall at proof of concept
Level 2: Organized Basic data governance and some integration across systems	Successful departmental applications but struggle to scale across functions
Level 3: Connected Enterprise data architecture with consistent definitions and accessibility	Cross-functional AI initiatives become feasible with reasonable implementation timelines
Level 4: Strategic Data managed as a strategic asset with clear ownership and quality controls	AI can be rapidly deployed to address emerging opportunities; significant competitive advantage

ESSENTIAL DATA FOUNDATIONS FOR AI VALUE

Now that we understand why data foundations matter, let's explore the five critical dimensions that determine your AI readiness. Each dimension represents a specific capability that enables AI value, and weaknesses in any area can undermine your entire initiative.

Data Accessibility and Democratization

AI thrives on data access. When data is locked in departmental silos or legacy systems, AI's potential is severely constrained.

A manufacturing client failed to extract value from their predictive maintenance initiative until they addressed a fundamental access problem: maintenance history data was locked in an Enterprise Resource Planning (ERP) system, sensor data lived in operational technology systems, and parts information existed in supply chain databases. Once they created a unified access layer, their time-to-value dropped from months to weeks.

Effective data democratization for AI requires:

- **Technical Accessibility:** Systems and Application Programming Interfaces (APIs) that enable appropriate data flow
- **Governance Accessibility:** Clear permissions and access controls that enable responsible use
- **Knowledge Accessibility:** Documentation and metadata that makes data discoverable
- **Skills Accessibility:** Tools and interfaces that match users' technical capabilities

True accessibility isn't just about technical connectivity – it's about making data usable by the people and systems that need it.

According to the Vorecol Editorial Team's 2024 research, organizations with unified data access frameworks saw a 47% increase in cross-departmental collaboration compared to organizations with siloed data structures. This demonstrates how data accessibility directly influences collaboration, which in turn enables more comprehensive AI solutions.

KEY INSIGHT: Successful data democratization balances accessibility with appropriate governance. As Leo Oliemans of TechFabric emphasizes: "It is crucial to implement centralised data management procedures that guarantee quality and accessibility across departments."

Quality and Usability Considerations

Data quality directly determines AI quality. Yet many organizations treat quality as an afterthought rather than a foundational requirement.

A financial services firm tried implementing an AI risk assessment system but achieved only 62% accuracy – far below expectations. Investigation revealed the root cause wasn't their algorithm but underlying data quality issues: missing fields, inconsistent definitions and outdated information. When they addressed these issues, accuracy jumped to 91% without changing the model itself.

For AI readiness, quality must be assessed across multiple dimensions:

- **Accuracy:** Does the data reflect reality?
- **Completeness:** Are critical fields populated consistently?
- **Timeliness:** Is the data updated with the frequency AI requires?
- **Consistency:** Are definitions and formats standardized?
- **Relevance:** Does the data capture what matters for the specific use case?

Different AI applications have different quality requirements. A product recommendation engine might tolerate some data gaps, while a clinical decision support system demands near-perfect accuracy. Understanding these requirements before implementation is crucial to success.

QUICK WIN: Identify your organization's "gold standard" datasets – those that are most trusted and reliable. Understanding why these datasets work well can provide a template for improving others.

INTEGRATION ACROSS DISPARATE SOURCES

AI's greatest value often comes from connecting insights across traditionally separate data domains. This requires thoughtful integration that preserves context and meaning.

A telecommunications company struggled with customer churn until they integrated data across previously separate systems: billing, service usage, support interactions, and network performance. This 360-degree view enabled AI to identify subtle patterns – like the relationship between payment timing, data usage patterns, and network congestion – that predicted churn with 84% accuracy.

Effective integration for AI requires:

- **Identity Resolution:** Connecting entities (customers, products, and locations) across systems
- **Temporal Alignment:** Ensuring time-based data can be properly sequenced
- **Semantic Consistency:** Maintaining consistent meaning across different sources
- **Relationship Preservation:** Capturing how different data elements relate to each other

True integration goes beyond simply combining datasets. It requires preserving the context and relationships that give data its meaning.

At AstraZeneca, this integration approach has transformed the company's ability to find novel drugs. As Cindy Hoots, Chief Digital Officer and Chief Information Officer (CIO), explains:

> We have things that we call knowledge graphs, which are an ability to kind of link unrelated or seemingly unrelated information, and taking vast quantities of data and making sense of it in a way that as individuals doing our day-to-day work, use it to really find more novel drugs.

KEY INSIGHT: Knowledge graphs and other advanced integration techniques don't just combine data – they create new value by revealing relationships that were previously invisible. These relationships often contain the most valuable insights for AI.

GOVERNANCE AND ETHICAL USE

As AI's impact grows, so do requirements for responsible governance. Strong data governance isn't a constraint on AI value – it's an enabler of sustainable success.

A healthcare provider implemented AI-based treatment recommendations but faced physician resistance due to concerns about how patient data was being used. By implementing transparent governance – clear data usage policies, explainable algorithms, and regular ethics reviews – they transformed resistance into enthusiastic adoption.

For AI readiness, governance must address:

- **Compliance:** Meeting regulatory requirements across jurisdictions
- **Ethics:** Ensuring responsible use aligned with organizational values
- **Transparency:** Making data lineage and usage visible to stakeholders
- **Accountability:** Assigning clear ownership for data quality and use

The Data Foundation

As AI regulation increases globally, strong governance becomes not just an ethical choice but a business necessity. Organizations that build governance into their data foundation avoid costly remediation and create trust that accelerates adoption.

PATTERN RECOGNITION: Organizations often view governance as a constraint that slows innovation. In reality, clear governance creates the trust and clarity that enable faster adoption. Without it, stakeholders rightfully hesitate to embrace AI systems they don't understand or trust.

SCALE AND PERFORMANCE FOUNDATIONS

Many AI initiatives succeed as pilots but fail at scale because their data foundations can't handle production demands.

A retail client piloted an inventory optimization AI with impressive results in five stores. When they attempted to scale to 500 stores, their data infrastructure collapsed under the volume. What worked for gigabytes failed utterly for terabytes. They had to rebuild their entire data pipeline before realizing enterprise value.

According to Cisco's 2024 AI Readiness Index, most businesses have infrastructure with "inadequate scalability to accommodate the increasing needs for AI-powered technologies." This infrastructure limitation becomes a critical bottleneck as organizations attempt to move beyond pilots.

Planning for scale requires addressing:

- **Volume Capacity:** Can systems handle the full production data volume?
- **Velocity Requirements:** Can data be processed at the speed AI needs?
- **Variety Management:** Can the system incorporate all relevant data types?
- **Operational Resilience:** Can data systems maintain performance under stress?

Scale issues rarely appear during pilots but can completely derail production implementations. Building with scale in mind from the beginning prevents costly rework later.

WARNING SIGN: If your pilot AI implementation was configured to use a subset of data (like three months of history instead of three years, or five stores instead of 500), you likely haven't tested true scaling requirements. Plan for five times data growth before moving to production.

ACTION STEP: For any AI pilot, create a "scale readiness checklist" that requires explicit consideration of how data volumes and processing needs will change in production. Have both technical and business stakeholders review and sign off on the scaling plan.

COMMON DATA CHALLENGES THAT UNDERMINE AI VALUE

Beyond the foundational elements, several specific challenges consistently undermine AI value creation. Understanding these challenges can help you anticipate and address them before they derail your initiatives.

SILOED DATA ENVIRONMENTS AND ORGANIZATIONAL BOUNDARIES

Data silos aren't just a technical problem – they reflect and reinforce organizational boundaries that limit AI's potential.

A global bank implemented AI risk models separately in their retail, commercial, and investment divisions. Each built similar capabilities using different approaches, wasting resources and creating inconsistent customer experiences. More importantly, they missed the cross-divisional insights that would have provided the greatest value.

Breaking down silos requires:

- **Executive Sponsorship:** Leadership that prioritizes enterprise value over departmental control
- **Incentive Alignment:** Rewards for data sharing and collaborative outcomes
- **Business Process Integration:** Workflows that cross traditional boundaries
- **Technical Connectivity:** Architecture that enables appropriate data flow

Those implementing the most successful AI implementations don't just connect data technically – they create organizational structures and incentives that encourage collaboration around shared goals.

Maureen Holland, Manager of Assurance Services & US Privacy at AstraZeneca, captures this challenge perfectly:

> We gotta start breaking down those silos. Because when you actually talk to stakeholders and you build these relationships, and even though they're doing different things with the data, 95% of the time, a lot of the data sources are the same. And so instead of everybody working independently, we need to start bringing these different teams together.

This observation reveals a striking paradox: even though teams are often working with the same data sources, organizational structures and cultural norms keep them operating in isolation. The waste is enormous, both in terms of redundant effort and in terms of missed opportunities for synergy.

KEY INSIGHT: Data silos are organizational problems that manifest as technical challenges. Trying to solve them with technology alone rarely succeeds. Effective solutions address both the technical connections and the organizational dynamics that created the silos.

The Data Black Hole Phenomenon

Perhaps the most alarming manifestation of data silos is what I have described as the "data black hole phenomenon." I first encountered this at a major financial institution in the early 2000s, where a process buried deep within the maze of data transformations had embedded exclusion logic. This filtered disagreeable data into a specified data location for further processing, which was undocumented and unmonitored, effectively becoming a black hole, home to countless millions of dollars worth of missing data.

This example illustrates how siloed environments don't just prevent data sharing – they can actively hide critical data problems, creating blind spots that undermine AI effectiveness. The black hole metaphor is apt, as these hidden processes absorb valuable information without any visibility to the broader organization.

WARNING SIGN: If your organization uses terms like "exception handling," "error queue," or "variance reporting" without clear ownership and resolution processes, you may have data black holes forming. These are often the first places to look when AI systems produce unexpected results.

Data Quality and Consistency Issues

Data quality problems create a garbage-in, garbage-out dynamic that tends to undermine AI value regardless of algorithmic sophistication.

A retail client implemented an AI-powered customer segmentation system but results varied wildly across regions. Investigation revealed that each region interpreted customer attribute fields differently. The Northeast region coded high-value customers as "A," while the Southwest used "Tier 1." The AI couldn't recognize these as equivalent, creating inconsistent recommendations.

The Data Foundation

In *Value-Driven Data*, I documented a particularly sobering example of quality issues: in one financial services case, team members spent five to seven days each month just validating data, with some team members spending ten days per month on sales analysis. Cost reconciliation issues required the equivalent of ten full-time staff daily, demonstrating the massive inefficiency caused by data foundation weaknesses.

Addressing quality for AI requires:

- **Proactive Assessment:** Understanding quality issues before they impact AI performance
- **Root Cause Remediation:** Fixing processes that create quality problems, not just their symptoms
- **Automated Monitoring:** Continuous quality tracking to catch issues early
- **Clear Standards:** Defined quality expectations aligned with AI requirements

Quality isn't a one-time fix but an ongoing commitment. Organizations that build quality into their data processes avoid the expensive cycles of AI deployment and remediation that plague many implementations.

QUICK WIN: Create a simple data quality scorecard for each dataset critical to your AI initiatives. Include metrics like completeness percentage, consistency score, and update frequency. Update and review monthly to track improvements and identify emerging issues.

Governance and Compliance Complexities

As data use expands and regulations grow, governance becomes increasingly critical to AI success.

I witnessed a financial services firm build an advanced customer targeting AI but being unable to deploy it because they couldn't verify that all data usage complied with privacy regulations across their operating regions. Their technical success became a practical failure.

This tension between data security and accessibility creates what Leo Oliemans describes as competing priorities where IT security teams focus on protection, while data science teams prioritize accessibility and flexibility. The resolution requires what Oliemans calls "cross-functional alignment," which balances these legitimate but opposing concerns through shared governance models.

Navigating governance complexities requires:

- **Regulatory Tracking:** Staying current on evolving requirements
- **Privacy by Design:** Building compliance into data architecture
- **Metadata Management:** Maintaining clear lineage and usage records
- **Risk-Based Governance:** Applying appropriate controls based on sensitivity

As AI becomes more powerful, governance becomes not just a compliance requirement but a trust-building necessity. Organizations that build governance into their foundation avoid deployment delays and instil stakeholder confidence.

REFLECTION QUESTIONS: Does your organization view data governance as primarily a compliance requirement or a value enabler? How might shifting this perspective change your approach to AI implementation?

Legacy Systems and Technical Debt

Aging systems often contain critical data but weren't designed for the access patterns AI requires.

One example comes from a manufacturing company, where maintenance history resided in a 15-year-old system that couldn't handle the query volume its subsequent predictive maintenance

AI required. Every analysis caused system-wide slowdowns, making the AI practically unusable despite its technical accuracy.

Another example, documented in *Value-Driven Data*, comes from a multinational services company, where lack of data accessibility severely limited business capabilities. A senior leader in the retail products team explained that they were forced to work with Excel sheets containing 35 million entries that were delivered months after the events they recorded. This outdated approach made timely analysis impossible, especially for new market entries, where understanding regional performance differences was critical.

Managing legacy constraints requires:

- **Realistic Assessment:** Understanding limitations before AI development begins
- **Strategic Modernization:** Prioritizing updates based on value potential
- **Data Abstraction:** Creating access layers that shield AI from underlying complexity
- **Hybrid Architectures:** Combining legacy and modern systems effectively

Legacy systems don't necessarily prevent AI success, but ignoring their limitations guarantees problems. Organizations that realistically assess and address technical debt create a more sustainable path to value.

KEY INSIGHT: The most successful approach to legacy systems isn't wholesale replacement (which often fails due to complexity and cost) but strategic abstraction – creating modern data access layers that shield AI applications from underlying complexity while gradually modernizing core systems.

Cultural Resistance to Data Sharing

Perhaps the most persistent challenge isn't technical but cultural: resistance to sharing data across organizational boundaries.

A healthcare system's departments treated patient data as departmental property rather than an organizational asset. Their AI diagnostic assistant failed because it couldn't access the complete patient history needed for accurate recommendations.

In the "Breaking Silos" podcast, by Denise Wiseman and Raj Nagra, the suggestion is that healthcare organizations are particularly affected by data siloing, creating barriers to effective patient care. Wiseman refers to the "siloed nature of our healthcare, and how those silos are really preventing us from making the change that so many people are speaking up about."

In my work at a global omnichannel insurance company that we will refer to as TAG, the siloing effect was glaringly obvious. Teams were progressively calcified in their approaches, with some refusing to meet with others even within their own teams – seemingly because they had originally started in different subsidiaries of the firm or were based in different locations. This extreme example demonstrates how entrenched organizational boundaries can become, creating nearly impenetrable barriers to effective data sharing.

Overcoming cultural resistance requires:

- **Value Demonstration:** Showing tangible benefits of data sharing
- **Risk Mitigation:** Addressing legitimate concerns about misuse
- **Recognition Systems:** Rewarding collaborative behaviours
- **Leadership Modelling:** Executives demonstrating data-sharing mindsets

Technical solutions to sharing rarely succeed without addressing the underlying cultural issues. Organizations that build data-sharing cultures create a foundation for AI that transcends specific technologies.

The Data Foundation

ACTION STEP: Implement a "data asset of the month" program that highlights a specific dataset valuable across multiple departments. Recognize both the team that created/maintained it and teams that found innovative ways to use it. This simple recognition creates visibility and encourages sharing.

DATA REGULATION AND PRIVACY RESTRICTIONS

As privacy regulations increasingly come into effect, this can create complex constraints on how data can be used for AI.

A global retailer built a customer behaviour AI only to discover they couldn't legally use it in European markets due to General Data Protection Regulation (GDPR) restrictions on data usage. What worked in North America was essentially worthless in Europe.

Navigating regulatory complexity requires:

- **Jurisdictional Awareness:** Understanding requirements across operating regions
- **Consent Management:** Building systems that respect user choices
- **Data Minimization:** Using only what's necessary for specific purposes
- **Privacy-Enhancing Technologies:** Implementing anonymization and other protective measures

Regulations aren't going away – they're increasing. Organizations that build privacy considerations into their data foundation avoid costly rework and market limitations.

WARNING SIGN: If your AI teams can't clearly explain how they're addressing privacy regulations like GDPR, California Consumer Privacy Act (CCPA), or industry-specific requirements, you have a significant governance gap that could prevent deployment.

BRIDGING DATA AND BUSINESS STRATEGIES

Creating a value-driven data foundation requires explicit connection between business objectives and data investments. This connection transforms data from a technical concern to a strategic asset.

ALIGNING DATA INVESTMENTS WITH AI VALUE TARGETS

Too many organizations invest in data capabilities without clear links to business outcomes. This creates sophisticated data environments that don't address what matters most.

A telecommunications company spent millions on a data lake but still couldn't predict customer churn effectively. They had built technical capacity without focusing on the specific data needs of their highest-value AI use case.

Effective alignment requires:

- **Value-Driven Prioritization:** Focusing on data that enables specific business outcomes
- **Use Case Mapping:** Explicitly connecting data capabilities to AI applications
- **Investment Staging:** Building capabilities in sequence with value delivery
- **Outcome Measurement:** Tracking how data improvements drive business results

Organizations that align data investments with specific value targets avoid the "data for data's sake" trap that consumes resources without delivering returns.

At a global universal banking group, I introduced an approach to value assessment that reduced the time needed for initial client engagements by 60% while increasing identified value by three times. This dramatic improvement came not from better technology but from better alignment between data capabilities and business outcomes.

KEY INSIGHT: The most effective data investment strategies start with clear business outcomes rather than technical capabilities. Instead of asking "What data capabilities should we build?," ask "What business outcomes do we need to achieve, and what data capabilities would enable them?"

Setting Data Priorities Based on Business Impact

Not all data deserves equal attention. Organizations have to set priorities based on value potential.

A retailer spent months cleaning and integrating store demographic data for an AI location planning system. Meanwhile, they neglected product availability data that would have driven much greater near-term revenue through inventory optimization. They prioritized data that was interesting rather than impactful.

In my previous publications, I have emphasized the importance of setting priorities based on business impact. The vision, obstacles, value (VOV) model provides a framework for identifying and measuring data value, ensuring resources are directed towards high-impact opportunities.

Effective prioritization requires:

- **Impact Assessment:** Estimating the value of different data improvements
- **Feasibility Analysis:** Understanding the effort required for each improvement
- **Dependency Mapping:** Identifying what must come first to enable later capabilities
- **Portfolio Balancing:** Combining quick wins with longer-term foundations

Organizations that are strict about setting priorities based on business impact create faster pathways to value while building sustainable capabilities.

QUICK WIN: Create a simple 2 × 2 matrix with "business value" on one axis and "implementation effort" on the other. Plot your potential data improvement initiatives on this matrix to identify "quick wins" (high value and low effort) and "strategic investments" (high value and high effort) to prioritize.

Measuring the Return on Data Investments

Data investments often lack the clear Return on Investment (ROI) metrics applied to other business initiatives. This makes them vulnerable when there are budget constraints.

A financial services firm struggled to justify ongoing investment in data quality until they began measuring its direct impact: 23% faster AI development cycles, 17% higher model accuracy and $3.2 million in avoided remediation costs. These concrete metrics transformed data from a cost centre to a value driver in executive discussions.

Effective measurement requires:

- **Direct Value Metrics:** Quantifying immediate business impacts
- **Enablement Metrics:** Measuring how data improvements accelerate other initiatives
- **Risk Reduction Metrics:** Calculating avoided costs and compliance benefits
- **Capability Metrics:** Assessing improved organizational abilities

Organizations that measure data's contribution to value gain sustainable support for ongoing investment.

ACTION STEP: For your next data improvement initiative, establish at least one quantifiable business metric that will improve as a result. Track this metric before and after implementation to demonstrate tangible value.

The Data Foundation

CREATING A VALUE-DRIVEN DATA ROADMAP

Moving from current state to AI-ready data requires a clear roadmap that balances short-term wins with long-term foundations.

A healthcare provider created a three-year data transformation plan with no visible benefits until year three. When budget pressures hit in year two, the initiative was cancelled, effectively wasting all previous investment.

In contrast, a global insurer transformed its approach to data by implementing principles like "immutability by design" and "timestamp separation," enabling the firm to move from systems that took over three weeks to deliver tangible insights to producing results within three hours. This dramatic improvement created visible value that sustained support for ongoing investment.

Effective roadmaps require:

- **Value Staging:** Sequencing work to deliver incremental benefits
- **Dependency Management:** Building foundations before capabilities that require them
- **Risk Balancing:** Combining safe improvements with higher-risk, higher-reward initiatives
- **Flexibility:** Adapting to changing business priorities and technical realities
- **Clear Milestones:** Creating visible progress points that maintain momentum

Organizations that create value-driven roadmaps maintain support through visible wins while building towards larger transformations.

KEY INSIGHT: The most sustainable data roadmaps follow a "value-first" sequence: start with high-visibility quick wins, use those successes to fund more foundational work, and then leverage the improved foundation to enable transformative capabilities. This creates a virtuous cycle of increasing investment based on demonstrated value.

DATA MATURITY ASSESSMENT FOR AI READINESS

Before launching AI initiatives, smart organizations assess their current data capabilities against their ambitions. This assessment prevents painful surprises and enables realistic planning.

EVALUATING YOUR CURRENT DATA CAPABILITIES

Many organizations overestimate their data readiness, leading to painful discoveries mid-implementation.

A retail banking client confidently began developing an AI-powered fraud detection system, only to discover three months in that their transaction data lacked the granularity needed for effective pattern recognition. What they thought would be a six-month project stretched to 18 as they rebuilt core data systems.

The disconnect between perception and reality can be striking. According to recent research on AI scaling and infrastructure, while 90% of business leaders believe their organizations are ready for AI at scale, 84% of IT practitioners disagree – reporting that they still spend the majority of their time resolving data issues. This perception gap leads to unrealistic expectations, missed milestones, and rising frustration across project teams.

Effective assessment requires honest evaluation across multiple dimensions:

- **Data Availability:** Do you have the data AI needs to deliver the intended value?
- **Data Quality:** Is the data accurate, complete, and consistent enough for reliable AI?
- **Data Accessibility:** Can the right people and systems access data when and how they need it?

TABLE 2.2
AI Data Readiness Diagnostic Tool

Dimension	Score 1	Score 3	Score 5
Availability	Critical data missing or uncollected	Most data available but with gaps	All necessary data available
Quality	Significant accuracy or consistency issues	Moderate quality suitable for some AI use cases	High quality suitable for critical applications
Accessibility	Data locked in silos with limited access	Basic integration with some access limitations	Comprehensive access with appropriate controls
Governance	Limited or ad hoc governance	Governance defined but inconsistently applied	Comprehensive governance aligned with AI needs
Infrastructure	Cannot support AI data volumes or velocity	Can support limited AI applications	Fully scalable for enterprise AI needs

- **Data Governance:** Are appropriate controls in place for responsible AI use?
- **Technical Infrastructure:** Can your systems handle AI's data processing requirements?

I've developed a simple diagnostic tool (see Table 2.2) that helps organizations quickly identify their AI data readiness. Score each dimension from 1 (major gaps) to 5 (fully capable).

Your lowest score typically determines your overall readiness. A bank scored 4s across most dimensions but just 1 on governance. As a result, the organization's AI credit scoring initiative stalled completely due to compliance concerns.

ACTION STEP: Use the AI data readiness assessment Table 2.2 to score your organization's readiness for your highest-priority AI use case. Have both business and technical stakeholders complete the assessment separately, then compare results to identify perception gaps.

IDENTIFYING CRITICAL GAPS FOR AI SUCCESS

Once you've assessed your current state, identify the specific gaps most critical to your AI ambitions.

When a manufacturing client conducted a thorough data assessment before launching a predictive maintenance AI, the most critical gap that was discovered wasn't data availability but time synchronization – sensors and maintenance records operated on different time systems, making it impossible to correlate events. Fixing this specific issue unblocked the entire initiative.

Effective gap analysis requires:

- **Use Case Specificity:** Identifying data needs for particular AI applications
- **Impact Ranking:** Prioritizing gaps based on their effect on intended outcomes
- **Feasibility Assessment:** Understanding what can realistically be addressed
- **Root Cause Analysis:** Looking beyond symptoms to underlying issues

The most important gaps aren't always the most obvious. A retailer focused on cleaning customer demographic data when the real barrier was inconsistency in product hierarchy, which prevented meaningful recommendations.

KEY INSIGHT: The most critical data gaps often aren't pure technology issues but integration challenges where context is lost between systems. For example, time synchronization problems between maintenance records and sensor data, or customer identity mismatches between online and

The Data Foundation

in-store systems. Identifying these "connection points" is often more valuable than improving any single data source.

Prioritizing Investments Based on Value Potential

With limited resources, organizations must prioritize data investments that create the greatest AI value.

A financial services firm identified 12 data quality issues affecting their AI initiatives. Rather than tackling all simultaneously, they prioritized the three that would unlock 80% of the value in their customer retention models. This focused approach delivered benefits months earlier than a comprehensive solution.

Effective prioritization requires:

- **Value Quantification:** Estimating the business impact of addressing each gap
- **Dependency Mapping:** Understanding which capabilities enable others
- **Effort Estimation:** Assessing the resources required for each improvement
- **Quick Win Identification:** Finding high-value, low-effort opportunities

Organizations that are disciplined about setting priorities based on value potential can score quicker wins related to AI success while, in parallel, building long-term capabilities.

REFLECTION QUESTIONS: What's your organization's current approach to prioritizing data improvements? Is it driven primarily by technical considerations, business value, or simply whoever shouts the loudest? How might a more systematic value-based approach change your priorities?

Setting Realistic Expectations and Timelines

Perhaps the most important outcome of data assessment is setting realistic expectations about what's possible and when.

A healthcare provider initially planned to implement an AI diagnostic assistant in a six-month time window. After assessment revealed significant data integration challenges, they reset expectations to 18 months. Rather than being a failure, this step resulted in clarity, prevented disappointment and enabled proper resource allocation.

Businesses with infrastructure that limits the scalability needed to accommodate the increasing needs for AI-powered technologies have to build timelines that include foundational work before AI value can be realized.

Effective expectation setting requires:

- **Executive Education:** Helping leaders understand data prerequisites for AI
- **Timeline Honesty:** Communicating realistic implementation horizons
- **Value Staging:** Identifying what benefits can be delivered when
- **Resource Clarity:** Defining what investments will be necessary

Organizations that set realistic expectations based on data readiness avoid the disappointment and lost trust that come from overpromising and underdelivering.

QUICK WIN: Create a simple "data foundation timeline" that shows how long it typically takes to address different types of data challenges (e.g., quality improvements, integration projects, and governance implementation). Share this with stakeholders when planning AI initiatives to set realistic expectations.

ACTION STEP: For your next AI project, build a timeline that explicitly includes data foundation work as a distinct phase with its own milestones rather than assuming data will be ready when needed.

CASE STUDIES: DATA AS THE VALUE FOUNDATION

Let's examine how organizations across sectors have addressed data foundations to enable AI success. These cases illustrate both the challenges and the transformative potential of getting data right.

Financial Services: How Data Transformation Unlocked $50M in AI Value

A global bank struggled for two years with an AI-powered risk management initiative. Despite hiring top data science talent and investing in cutting-edge algorithms, they achieved only modest results, which led executives to question the feasibility of continuing to fund the endeavour.

Analysis revealed the core issue: risk data was fragmented across 14 systems with inconsistent definitions, poor quality controls, and limited accessibility. Models trained on this data could never achieve the accuracy needed for meaningful impact.

Rather than continuing to tweak algorithms, they paused AI development and focused on four specific data improvements:

1. Created a unified customer identifier that connected information across product lines
2. Established consistent risk classification standards across business units
3. Built automated data quality monitoring that identified issues before they affected models
4. Implemented a data access layer that simplified consumption while maintaining security

These changes required 10 months and $12M – substantial investments that faced scepticism. The breakthrough came when the team demonstrated the impact on a single use case: credit risk modelling for the commercial portfolio. The improved data foundation increased model accuracy by 28%, reducing false positives by 36% and false negatives by 22%.

This translated to approximately $14M in reduced credit losses in the first year – enough to justify the entire data investment. As they applied the same data improvements to other use cases, the cumulative value exceeded $50M within 2 years.

KEY INSIGHT: Rather than treating each AI use case as a separate initiative with its own data challenges, the team built foundational capabilities that enabled multiple applications. This created an exponential return on their data investment.

Healthcare: Patient Data Integration Driving Improved Outcomes

A regional healthcare system saw AI as critical to its clinical quality and cost management goals. It invested in several promising applications, including sepsis prediction, readmission risk modelling, and treatment optimization.

After multiple failed pilots, it recognized a pattern: its clinical data environment wasn't supporting AI success. Patient information was scattered across electronic health records, billing systems, pharmacy databases, and specialty department applications. Data collected at the bedside often took days to become accessible for analysis, and key relationships between events were lost.

Rather than continuing to build AI models that couldn't succeed, the organization developed a patient-centred data architecture with three key elements:

1. A unified patient timeline that synchronized events across all systems
2. Real-time data pipelines that made critical information available within minutes
3. Standardized clinical terminologies that enabled consistent interpretation

The first AI success came six months after implementing these changes: a sepsis prediction model that identified at-risk patients eight hours earlier than clinical protocols alone. This translated to reduced ICU time, lower treatment costs, and most importantly, improved survival rates.

Within two years, it had implemented seven AI applications on this foundation, creating measurable improvements in patient outcomes and operational efficiency. Leadership emphasized the patient outcome improvements as the true measure of success, with the data foundation enabling AI that augmented physician capabilities rather than creating additional burdens.

KEY INSIGHT: By creating a data architecture designed around clinical value rather than departmental boundaries, the organization enabled AI that augmented physician capabilities rather than creating additional burdens. This patient-centred approach created alignment across previously siloed departments.

RETAIL: CUSTOMER DATA PLATFORMS ENABLING PERSONALIZATION AT SCALE

A specialty retailer with both online and physical stores struggled to create personalized experiences despite years of collecting customer data. It had AI recommendation engines that performed poorly – and marketing personalization created awkward experiences that customers mocked on social media.

The root problem wasn't the approach to AI but customer data: fragmented across e-commerce, point-of-sale, marketing, and loyalty systems with no unified view. A customer who browsed online and purchased in-store appeared as two different people in their systems.

To remedy the situation, the company built a customer data platform with three primary capabilities:

1. Identity resolution that connected behaviours across channels
2. Preference management that respected privacy choices consistently
3. Real-time accessibility that enabled immediate response to customer actions

With this foundation, the previously ineffective recommendation engine suddenly performed exceptionally well – increasing average order value by 23% online and driving 8% incremental store visits when recommendations were included in marketing communications.

More importantly, customers noticed the difference. Net promoter scores increased by 17 points, with many specifically mentioning how the company "finally understands what I want" and "makes shopping easier."

The transformation required 14 months and significant investment, but the payback period was less than one year based on increased revenue alone.

KEY INSIGHT: Building a unified view of the customer wasn't just a technical challenge but a business transformation that changed how the entire organization thought about and responded to customer needs. The technology enabled new capabilities, but the organizational shift towards customer-centricity was equally important.

PUBLIC SECTOR: BREAKING DATA SILOS TO IMPROVE CITIZEN SERVICES

A state government agency responsible for social services struggled with program effectiveness and resource allocation. Citizens often navigated multiple departments with redundant paperwork and conflicting requirements, while administrators lacked visibility into overall service delivery.

An attempt to implement predictive models to identify at-risk populations and optimize interventions stalled because AI provided little value due to fundamental data limitations: each program operated as an independent silo with no visibility across services.

The breakthrough came when new leadership implemented a citizen-centred data strategy with appropriate privacy safeguards. Key elements included:

1. A secure citizen identity framework that connected services while protecting privacy
2. Standardized needs assessment methodology across programs
3. Outcome tracking that measured long-term effectiveness rather than just service delivery
4. Cross-program visibility with appropriate access controls

This foundation enabled AI applications that fundamentally changed service delivery. Predictive models identified citizens likely to need multiple services and facilitated proactive outreach. Resource allocation shifted from program-by-program decisions to holistic assessment of community needs.

The results included a 31% reduction in application processing time, 24% improvement in program completion rates, and 12% decrease in overall program costs. Most importantly, citizen satisfaction increased dramatically as they experienced more coordinated and responsive government services.

KEY INSIGHT: In public sector environments with inherent organizational boundaries, creating appropriate data connections can transform service delivery without requiring massive restructuring. In this case, a citizen-centred data approach created bridges between departments while respecting their distinct missions.

REFLECTION QUESTIONS: Which of these case studies most closely resembles your organization's situation? What specific approaches could you adapt from that example to improve your own data foundation?

THE DATA FOUNDATION FRAMEWORK: A PRACTICAL IMPLEMENTATION GUIDE

Based on the principles and cases we've explored, I've developed a practical framework to help you build the data foundation your AI initiatives need. This approach balances quick wins with sustainable capabilities, ensuring you create both immediate value and long-term advantage.

ACTION STEP: Use this four-phase implementation framework to build or strengthen your organization's data foundation for AI.

PHASE 1: ASSESS YOUR CURRENT STATE (2 TO 4 WEEKS)

Begin with a clear understanding of your starting point:

1. **Map Data Assets against Use Cases:** Identify which data sources are critical for your priority AI use cases, and assess their current state
2. **Evaluate Data Maturity:** Use the data maturity assessment Table 2.2 provided earlier to score your readiness
3. **Identify Critical Gaps:** Determine which limitations most significantly constrain your AI potential
4. **Quantify Impact:** Estimate the business value of addressing each gap

Key Outcome: A prioritized list of data improvements with estimated value impact

The Data Foundation

Phase 2: Create Quick Wins (1 to 3 Months)

Build momentum with targeted improvements that deliver immediate value:

1. **Select High-Value, Low-Effort Opportunities:** Focus on improvements that can create visible impact quickly
2. **Implement Tactical Solutions:** Address critical blocking issues that prevent AI progress
3. **Demonstrate Value:** Create clear before/after metrics that show the business impact
4. **Build Support:** Use early successes to generate enthusiasm for larger transformations

Key Outcome: Tangible improvements that demonstrate the value of data investment

Phase 3: Build Strategic Foundations (3 to 9 Months)

With credibility established, invest in more substantial capabilities:

1. **Establish Data Governance:** Create clear ownership, policies, and quality standards
2. **Build Integration Capabilities:** Implement the technical means to connect disparate sources
3. **Create Self-Service Access:** Enable appropriate data democratization with proper controls
4. **Develop Monitoring Systems:** Build automated quality and compliance tracking

Key Outcome: Sustainable data capabilities that enable multiple AI use cases

Phase 4: Scale and Optimize (Ongoing)

Continuously improve your data foundation as business needs evolve:

1. **Expand Data Sources:** Incorporate new data domains as AI applications grow
2. **Refine Governance:** Evolve policies based on emerging requirements and regulations
3. **Enhance Performance:** Optimize for scalability and real-time capabilities
4. **Measure Value Creation:** Track and communicate how data investments drive business outcomes

Key Outcome: An evolving data foundation that continuously expands AI possibilities

KEY INSIGHT: The most effective data transformations don't try to solve every problem at once. Instead, they balance quick wins that build credibility with strategic investments that create lasting capabilities. This balanced approach maintains momentum and support throughout the journey.

THE DATA FOUNDATION IMPERATIVE: FROM INSIGHTS TO ACTION

Before we move on, let me share a story that encapsulates everything we've discussed in this chapter.

A global insurance company had invested millions in separate AI initiatives across claims processing, customer service, and risk assessment departments. Each built their own data pipelines, created their own data cleaning procedures, and maintained their own versions of customer information. The result? Three incompatible AI systems working with inconsistent data, generating conflicting insights about the same customers.

The breakthrough came when a visionary data leader proposed a radical idea: "Let's build the data foundation first, then the AI applications." By creating a unified data infrastructure with standardized governance and cross-departmental access protocols, they transformed their approach. Teams that had previously competed for data resources now collaborated on data quality.

The resulting AI applications didn't just perform better technically, they worked together coherently, providing consistent customer experiences across touchpoints. This example demonstrates how data integration doesn't just support AI, it fundamentally enables cross-functional value creation that siloed approaches cannot achieve.

As Maureen Holland of AstraZeneca puts it:

> When you actually talk to stakeholders and you build these relationships, and even though they're doing different things with the data, 95% of the time, a lot of the data sources are the same. And so instead of everybody working independently, we need to start bringing these different teams together.

This simple truth – that most organizations are working with the same data in disconnected ways – represents both challenge and opportunity. By recognizing that data is an enterprise asset rather than departmental property, organizations can build foundations that enable AI success across boundaries.

FIVE KEY TAKEAWAYS

As we conclude this chapter, here are the most important points to remember:

1. **Your data foundation determines your AI value ceiling**. No algorithm, no matter how sophisticated, can overcome poor quality, inaccessible, or fragmented data.
2. **Data quality issues drain resources and delay value**. Most organizations drastically underestimate the time and cost of cleaning and integrating data for AI, with data scientists often spending 70%–80% of their time on data preparation.
3. **Cross-functional value requires cross-functional data**. AI's greatest potential often comes from connecting insights across traditional boundaries, which requires breaking down data silos.
4. **Data governance enables rather than constrains AI value**. Clear policies, ownership, and controls build trust and accelerate adoption while preventing costly compliance issues.
5. **Building the right foundation accelerates everything that follows**. Organizations that invest in data foundations before AI implementation achieve faster time-to-value and greater scale than those that rush to implementation.

ACTION STEP: Assess your organization's data foundation for AI readiness using the frameworks and tools provided in this chapter. Identify your most critical gaps and create a plan to address them before your next major AI investment.

REFLECTION QUESTIONS: What specific data foundation improvements would most significantly increase your organization's AI potential? How can you build support for investing in these improvements?

The path forward is clear: before rushing into AI implementation, assess your data foundation. Identify gaps, prioritize based on value, and invest in the capabilities that will enable sustained success. The time and resources spent building this foundation will return exponential value as AI initiatives move from isolated pilots to enterprise-wide transformation.

Remember: AI is only as good as the data it consumes. Your data foundation determines your AI value ceiling. By building on solid ground, you create the conditions for AI to deliver on its extraordinary promise.

In the next chapter, we'll explore why AI implementation often fails to deliver value – and how understanding these failure patterns can help you avoid common pitfalls and build initiatives that consistently deliver business impact.

3 Spotting Signals of AI Value Failure

Chapter Roadmap

In this chapter, we explore why so many AI initiatives fail to deliver real business value despite technical success. We examine implementation gaps, organizational misalignments, and value leakage points – all drawn from real-world cases. Each section highlights patterns, warning signs, and actionable strategies that help to avoid these common pitfalls.

THE REALITY OF AI IMPLEMENTATION CHALLENGES

"We've been working on this Artificial Intelligence (AI) project for 18 months. It works perfectly in the lab. The data science team is brilliant. But somehow, we still haven't seen any actual business impact."

The Chief Information Officer (CIO) of a global manufacturing company shared this frustration with me over dinner after a conference. His story wasn't unusual. In fact, it was surprisingly common.

Despite the hype and genuine potential, AI initiatives fail to deliver expected value with shocking frequency. Gartner research indicates 85% of AI projects fail. A RAND Corporation study interviewing 65 data scientists found that over 80% of AI projects fail. My experience working with organizations across sectors confirms this sobering reality.

REFLECTION QUESTIONS: Think about the AI initiatives in your own organization. How many have moved beyond the pilot phase to deliver measurable business value? What happened to the ones that didn't make it?

But here's what is interesting: these failures rarely stem from technical problems. The algorithms work; the models function; and the data science is sound.

Where these projects fall short is the space between technical success and business value: the messy middle where potential must transform into practice.

THE REALITY GAP

There's a persistent gap between AI's promise and its practical reality. This gap takes several forms:

- **The Technical-to-Business Translation Gap:** AI teams celebrate model accuracy improvements while business leaders wonder why customer satisfaction hasn't budged
- **The Lab-to-Field Performance Gap:** Models that perform brilliantly with test data struggle when exposed to real-world data and use
- **The Pilot-to-Scale Capability Gap:** Solutions that work perfectly for a single department collapse when deployed enterprise-wide
- **The Initial-to-Sustained Value Gap:** Systems that create impressive early results but deliver diminishing returns as conditions change

These gaps aren't due to technical failures – they are implementation failures that prevent technically sound solutions from delivering business value.

KEY INSIGHT: Technical success does not automatically translate to business value. The most sophisticated AI algorithm that doesn't solve a real business problem or can't be effectively deployed does ultimately fall short in creating value.

Pilot Purgatory and Scaling Challenges

One of the most common patterns I've observed is what I call "pilot purgatory" when promising AI initiatives get stuck in endless experimentation without ever reaching production scale.

A financial services client successfully piloted an AI credit scoring system that improved approval rates while reducing defaults. The pilot delivered clear value, yet three years later, the company was still running "extended pilots" in select markets without full deployment. The technology worked, but organizational, process, and change management challenges kept it trapped in purgatory.

N-iX's Yaroslav Mota reports that 90% of companies struggle to scale AI across their enterprises and close to half of AI projects fail. According to IBM's 2024 study, 43% of C-level technology executives express concern about their technology infrastructure and are now focused on upgrading it for scaling AI.

Scaling challenges typically emerge from:

Technical Limitations: Solutions designed for limited data volumes or transaction rates that cannot handle enterprise scale
Process Inconsistencies: Variations in workflows across departments that weren't apparent during limited pilots
Skills Gaps: Reliance on specialized expertise that cannot be replicated across the organization
Governance Hurdles: Compliance and risk management concerns that intensify at scale
Change Resistance: Cultural barriers and adoption challenges that multiply when moving beyond enthusiastic early adopters

WARNING SIGN: When your organization is running multiple "extended pilots" without clear paths to enterprise deployment, you're likely trapped in pilot purgatory.

Overcoming these challenges requires fundamentally different approaches than those that work for creating successful pilots.

WHY TRADITIONAL APPROACHES FALL SHORT

Many organizations apply the same implementation approaches to AI that served them well for traditional technology. This creates predictable failures because AI differs in crucial ways:

Iterative vs. Defined: Traditional technology typically follows a defined development path with clear requirements. AI requires continuous iteration and refinement as it learns from new data and user interactions.
Data Dependency: Traditional systems operate on structured inputs with predictable behaviour. For optimal performance, AI depends on data that is accurate, covers all scenarios, and reflects real-world conditions.
Explainability Challenges: Traditional systems follow explicit logic that can be traced and verified. Many AI approaches create "black box" decisions that require different validation approaches.
Ecosystem Impact: Traditional systems typically operate within defined boundaries. AI often creates ripple effects across processes, roles, and organizational structures.

A healthcare provider implemented an AI diagnostic support system using the same project methodology they had applied to electronic health record updates. The approach broke down completely when they couldn't define fixed requirements for a system designed to learn and evolve over time. Success only came when they adopted an entirely different implementation approach focused on continuous learning rather than fixed milestones.

ACTION STEP: Evaluate your current project methodology. If it relies on upfront requirements definition and fixed milestones, adapt it to allow for continuous iteration, learning, and adjustment before launching your next AI initiative.

VALUE LEAKAGE THROUGHOUT THE AI LIFECYCLE

Value doesn't typically disappear all at once in AI implementations. Instead, it leaks at multiple points throughout the lifecycle, with each leak reducing the ultimate impact.

Understanding where these leaks occur helps identify and address them before they drain the initiative's potential.

STRATEGY MISALIGNMENT AND UNCLEAR OBJECTIVES

Value leakage often begins at the very start – particularly due to strategic misalignment and poorly defined objectives.

Research, conducted by Ryseff and co-authors, identifies leadership misalignment as a primary cause of AI project failure. Their study shows that "AI projects often fail due to leadership misalignment and miscommunication regarding project objectives, resulting in premature abandonment or shifting priorities before realising value." This misalignment creates a cascade of issues throughout implementation.

Stuart King, Chief Technology Officer (CTO) of cybersecurity consulting firm AnzenSage, provides a compelling example of misaligned AI implementation. In conversations with organizations, he witnessed the following approach: "Here's this great new thing that we can use now, let's go out and find a use for it" rather than identifying problems first and then applying AI as a solution.

A retail client launched an "AI customer experience initiative" without specifying what aspects of experience they wanted to improve or how they would measure success. The data science team built impressive capabilities that ultimately delivered little value because the problems they solved weren't the ones customers cared about.

Common strategy leaks include:

- **Vague Value Propositions:** "Implementing AI for customer service" instead of "Reducing resolution time for high-volume customer issues by 30%"
- **Technical Objectives Masquerading as Business Goals:** "Deploying machine learning for inventory management" rather than "Reducing stockouts by 50% and decreasing inventory holding costs by 15%"
- **Misaligned Stakeholder Expectations:** Different leaders expecting entirely different outcomes from the same initiative
- **Shifting Priorities:** Objectives that change frequently without clear direction, creating confusion and wasted effort

QUICK WIN: Before starting any AI project, create a one-page value statement that clearly describes the specific business problem you're solving, how you'll measure success, and which stakeholders must agree on these outcomes.

Organizations that prevent strategy leakage maintain clarity about the specific business problems they're solving and how success will be measured.

Data Quality and Accessibility Limitations

As we explored in Chapter 2, data forms the foundation of AI success. Limitations here create significant value leakage.

Poor data quality costs organizations approximately $12.9 million annually according to Gartner. Rupert Brown, CTO and founder of Evidology Systems, identifies how legacy systems create data quality challenges that impede AI scaling: "Legacy systems that have limited input data fields or are forced to recycle account numbers are still prevalent in the fintech industry, which also give rise to corrections which AI cannot fathom." Brown emphasizes that "data quality is a problem that is going to limit the usefulness of AI technologies for the foreseeable future."

A manufacturing client built an advanced predictive maintenance system that performed brilliantly in the laboratory but failed in production. The culprit? Training data collected during normal operations didn't include enough examples of the edge cases that caused the most costly failures.

Common data leaks include:

Representativeness Issues: Training data that doesn't reflect the full range of real-world conditions
Quality Inconsistencies: The usage of data that is accurate enough for reports but not for algorithmic decision-making
Access Constraints: Essential data locked in systems that cannot support AI workloads or real-time needs
Context Loss: Critical relationships between data elements lost during processing
Feedback Limitations: Inability to capture outcomes to refine and improve models over time

PATTERN RECOGNITION: If your organization struggles with basic reporting reliability, you're likely to face even greater challenges with AI implementation. AI demands higher data quality and integration compared to traditional analytics.

Organizations that minimize data leakage build robust pipelines that deliver appropriate quality, representative training data, and continuous feedback for improvement.

Technical Implementation Challenges

Technical issues inevitably arise during implementation, creating another source of value leakage.

Matt Bostrom, VP of enterprise technology and strategy at Spirent Communications, encountered significant obstacles when attempting to integrate AI with existing systems: "We had integration tools at our company, but they were older, outdated tools. Achieving the large-scale integrations necessary for gen AI would have required significant and costly upgrades." This case illustrates how legacy integration tools can become a bottleneck for AI scaling, forcing companies to choose between expensive upgrades or limited AI implementation.

A financial services firm developed a fraud detection AI that worked flawlessly in testing but created unacceptable delays during actual transaction processing. The algorithm was accurate but too computationally intensive for the organization's production transaction volumes, forcing compromises that reduced its effectiveness.

Common technical leaks include:

Performance Constraints: Systems that cannot deliver results within required timeframes at production scale
Integration Complexity: Difficulties connecting AI capabilities with existing systems and workflows
Technical Debt Accumulation: Shortcuts taken during development that limit long-term sustainability

Monitoring Inadequacies: Insufficient visibility into how models perform in production, allowing degradation to go undetected

Maintenance Challenges: Difficulty updating models as conditions change, leading to declining performance over time

KEY INSIGHT: Many organizations focus exclusively on model accuracy during development while neglecting critical production requirements like latency, throughput, and integration complexity.

Organizations that reduce technical leakage balance immediate functionality with long-term sustainability and build robust monitoring and maintenance capabilities.

ORGANIZATIONAL MISALIGNMENTS THAT DEGRADE VALUE

Beyond specific leakage points, certain organizational misalignments consistently undermine AI value creation.

SILOED APPROACHES TO AI IMPLEMENTATION

Perhaps the most common organizational issue relates to siloed implementation, which fragments efforts and prevents enterprise value creation.

A global bank had 17 separate teams building customer-churn-prediction models – each for different products and regions. None could access data beyond their specific domain, severely limiting their effectiveness. Meanwhile, a comprehensive view across products would have revealed patterns invisible to any single team.

> Communicating in silos is a trap that you can sometimes fall into. I mean sometimes it's okay, it's something super tactical but you get my point right, like how do you keep people visible? Good visibility across this entire project as you work on it, and that's one that comes to mind in terms of potholes you can hit.
>
> **Jeremy Foster,**
> *Vice President, Cloud Infrastructure and Software Group, Cisco*

Siloed approaches create several problems:

Duplicated Effort: Multiple teams solving similar problems without sharing insights or resources

Fragmented Data: Critical information divided across organizational boundaries, preventing holistic analysis

Inconsistent Experiences: Customers or employees facing different AI-driven interactions across touchpoints

Limited Scale Economics: Inability to leverage investments across use cases

Success requires breaking down these silos through cross-functional teams, shared technical platforms, and coordinated prioritization processes.

ACTION STEP: Map your organization's AI initiatives across departments. Identify areas of potential duplication, opportunities for data sharing, and possibilities for creating shared infrastructure or expertise centres.

This disconnection from the bigger goal is precisely what undermines AI implementations. When departments focus solely on their immediate objectives without considering the broader organizational mission, even technically successful AI initiatives struggle to deliver meaningful value.

The Great Silo Dilemma

The challenge of silos runs deeper than just poor communication – it reflects how organizations naturally structure themselves around specialized functions and expertise. Elon Musk recognized this at Tesla and established a "no-silo rule," instructing managers to "work hard to ensure that they are not creating silos within the company that create an 'us vs. them' mentality," the problem is both structural and cultural.

The irony is striking. Multiple teams often work with the same underlying data, yet reach different conclusions and pursue divergent objectives because they aren't collaborating effectively.

Conflicting Incentives and Competing Priorities

Even when organizations attempt to collaborate, conflicting incentives can undermine success.

Research by Hanna Kleinings reveals how different departments prioritize fundamentally different AI applications: sales teams focus on "outbound email campaigns, demand forecasting, lead scoring"; marketing prioritizes "market research, competitor analysis, image generation, search engine optimisation"; operations emphasizes "inventory management"; human resources concentrates on "recruitment, organisation network analysis, and support hybrid and remote working"; and accounting uses AI to "automate tedious activities, improve accuracy and efficiency, and identify hidden trends."

This departmental misalignment plays out in practice in predictable ways and in my consulting work, I have seen this exact pattern unfold. For example, I witnessed this when a retailer's marketing team implemented AI-powered personalization to maximize immediate sales, while the customer experience team focused on long-term engagement. The competing objectives led to contradictory recommendations and confused customers who received mixed messages.

Common incentive conflicts include:

Short-Term vs. Long-Term Objectives: Teams measured on quarterly results vs. those building long-term capabilities

Efficiency vs. Quality Targets: Operations focused on cost reduction vs. service teams measured on customer satisfaction

Innovation vs. Risk Management: Technology teams pushing boundaries vs. compliance ensuring safety and regulatory adherence

Local vs. Enterprise Optimization: Department leaders optimizing their metrics vs. enterprise-wide outcomes

REFLECTION QUESTIONS: In your organization, how are different departments measured and rewarded? Do these metrics create natural alignment or conflict around AI initiatives?

Resolving these conflicts requires explicit discussion of trade-offs and creation of shared success metrics that align efforts towards common goals.

Unclear Ownership of Value Creation

When multiple functions contribute to AI initiatives, ownership of outcomes often becomes ambiguous.

A manufacturing company's quality prediction AI failed to deliver expected value, but responsibility was disputed: IT claimed that they delivered a functioning system, operations said that they weren't given adequate training, and the data team argued that process inconsistencies undermined their models. With everyone responsible for a piece, no one was accountable for the outcome.

Clear ownership requires:

End-to-End Accountability: Designated leaders responsible for value delivery, not just technical implementation

Cross-Functional Governance: Decision processes that balance specialized expertise with integrated outcomes
Shared Success Metrics: Performance measures that create mutual accountability across teams
Executive Sponsorship: Leadership that transcends organizational boundaries to drive integrated results

QUICK WIN: For your next AI initiative, create a matrix (responsible, accountable, consulted, and informed) that clearly defines roles across departments for each phase from planning through value realization.

Organizations that create clear ownership structures maintain focus on business outcomes rather than technical deliverables.

The Technical-Business Communication Gap

Perhaps the most persistent organizational challenge is the communication gap between technical and business stakeholders.

A retail banking client's data science team proudly presented a new credit risk model's impressive area under the curve (AUC) and F1 scores. Business leaders left the meeting confused and unconvinced, unable to connect these technical metrics to their concerns about approval rates, default risk, and portfolio profitability.

In my previous positions as chief data analytics and privacy officer – where I helped over 35 organizations deliver nine-figure transformations – I found three critical conversations that must happen to bridge this gap:

> The fear conversation focuses on the unspoken concerns keeping executives awake at night. The value conversation explores opportunities for maximum impact. The future conversation looks beyond immediate needs to long-term possibilities.

Without these conversations, AI initiatives struggle to connect technical capabilities with business priorities.

Common communication gaps include:

Technical vs. Business Language: Data scientists discussing model architecture, while business leaders care about business metrics
Different Risk Perspectives: Technical teams focused on false positives, while business leaders worry about regulatory compliance
Timeline Disconnects: Data science iterative development vs. business milestone expectations
Value Measurement Differences: Technical performance metrics vs. business outcome measures

KEY INSIGHT: The most successful AI initiatives have "translators" who can speak both technical and business languages, converting between model performance metrics and business outcomes.

Bridging this gap requires translators who understand both worlds and can convert between technical capabilities and business impacts.

THE TRUE COST OF AI VALUE FAILURE

When AI initiatives fail to deliver value, the costs extend far beyond wasted project expenses.

Direct Financial Impacts

The obvious costs include development expenses, technology investments, and specialized talent – often totalling millions for significant initiatives.

According to research by ESI ThoughtLab, organizations investing in AI face considerable costs. The 1,200 firms surveyed in 2020 invested an average of $38 million each in AI initiatives, which represented more than 0.75% of their revenue. One-third invested more than $50 million, while industry leaders invested over $99 million.

I came across a financial services firm, which invested $8.5 million in an AI-powered advice engine that was never deployed due to compliance concerns that could have been addressed during initial planning. Every dollar was lost, along with significant opportunity cost.

But direct costs are just the beginning.

Opportunity Costs and Competitive Implications

Perhaps even more significant are the opportunity costs. When organizations invest in unsuccessful AI, they fail to move forward while their competitors forge ahead.

A retailer delayed the launch of its personalization engine to perfect technical performance. Meanwhile, a competitor launched a "good enough" solution and captured market share through first-mover advantage. The lost revenue far exceeded the project budget.

In fast-moving markets, costs associated with missed opportunities often prove far greater than the investments for failed projects alone.

WARNING SIGN: When your AI initiative focus shifts from delivering business value to achieving technical perfection, you may be sacrificing competitive advantage for marginal technical improvements.

Organizational Momentum and Confidence Erosion

Failed AI initiatives create organizational drag that affects future efforts.

A healthcare provider's unsuccessful diagnostic AI created lasting scepticism among physicians. When the company later implemented a genuinely beneficial clinical decision support system, adoption was slow because the previous failure had damaged credibility.

This erosion of confidence creates increasing resistance to change, making each subsequent initiative more difficult regardless of its merits.

Regulatory and Reputational Risks

AI failures increasingly carry regulatory and reputational consequences beyond financial impacts.

Recent reports exposed how Microsoft's Tay chatbot rapidly learned toxic language from social media users and had to be shut down within hours of implementation, demonstrating the critical importance of data quality control and algorithmic safeguards. Similarly, Google's object recognition program showed higher error rates when identifying persons with darker skin, even misidentifying them as gorillas, highlighting the dangers of inadequate training data.

A financial institution deployed a credit scoring algorithm without adequate testing for bias. When analysis revealed disadvantages for certain demographic groups, the company faced regulatory scrutiny, negative press coverage, and erosion of customer trust. The combined impact far exceeded the implementation cost.

As AI regulation increases globally, the stakes for getting implementation right extend beyond business performance to legal compliance and brand integrity.

LEARNING FROM FAILURE: CASE STUDIES IN AI IMPLEMENTATION

Examining specific failure patterns provides valuable insights for avoiding similar outcomes.

RETAIL: HOW A PROMISING RECOMMENDATION ENGINE FAILED TO DELIVER

A specialty retailer invested heavily in an AI recommendation engine expected to increase basket size and customer retention. The technology worked perfectly in testing, consistently identifying relevant products based on purchase history.

Yet six months after deployment, the firm saw no significant improvement in key metrics. What went wrong?

Investigation revealed multiple implementation failures:

1. **Channel Disconnect:** While the AI made excellent recommendations, they were only displayed in the website footer where few customers noticed them
2. **Inventory Misalignment:** Many recommendations were for products frequently out of stock, creating frustration rather than sales
3. **Associate Exclusion:** Store associates weren't informed about the system and couldn't explain recommendations customers had seen online
4. **Measurement Flaws:** Success was measured by recommendation click rates rather than actual purchases, optimizing for interest rather than conversion

PATTERN RECOGNITION: Notice how none of these failures involved the AI algorithm itself. All were implementation and integration issues that prevented a technically sound solution from delivering business value.

The technology worked exactly as designed, but implementation failures prevented value realization. After addressing these issues – making recommendations more prominent, connecting with inventory systems, training associates, and refining metrics – the same core AI delivered a 14% increase in basket size.

FINANCIAL SERVICES: AUTOMATION NOT TRANSLATING INTO COST REDUCTION

A global bank implemented AI-powered process automation in their loan application workflow, expecting to reduce processing costs by 40% while improving accuracy. The technology successfully automated document classification and data extraction but costs barely decreased.

Analysis revealed several critical issues:

1. **Hidden Process Dependencies:** The automated components couldn't proceed without manual inputs from steps that hadn't been digitized
2. **Exception Proliferation:** The system required human review for too many cases, creating a new bottleneck
3. **Parallel Processes:** Compliance requirements mandated maintaining partial manual processes alongside automation, duplicating work rather than eliminating it
4. **Skill Transition Failure:** Staff weren't effectively redeployed to higher-value activities, limiting cost reduction despite reduced workload

David Kuder, Deloitte's US Market Offering Lead for AI Insights and Engagement, cautions that "AI doesn't just magically automate what hundreds of people are doing," highlighting the misconception that AI implementation automatically leads to cost-savings. He emphasizes that executives must "address the process complexity they are likely to find and be willing to reorganize to achieve the desired cost reductions."

After process redesign, exception handling optimization, compliance process integration, and staff role transformation, the same technology ultimately delivered a 35% cost reduction – close to the original target.

Healthcare: AI Diagnosis Tools That Never Reached Patients

A healthcare system developed an AI diagnostic assistant for identifying diabetic retinopathy from eye scans. Clinical validation showed it matched specialist accuracy while potentially expanding screening to underserved populations.

Yet two years after development, it remained unused in actual practice. Multiple implementation barriers prevented adoption:

1. **Workflow Integration Failure:** The tool required physicians to use a separate system, creating additional steps in already tight patient visits
2. **Liability Concerns:** Unclear responsibility for AI-assisted diagnoses created legal uncertainty that neither physicians nor administrators wanted to address
3. **Reimbursement Problems:** Insurance companies wouldn't recognize or pay for AI-assisted diagnoses, creating financial disincentives for adoption
4. **Training Gaps:** Physicians received minimal guidance on how to work with the AI assistant, creating uncertainty and distrust in its recommendations
5. **Patient Communication Challenges:** Doctors struggled to explain to patients how AI contributed to their diagnosis, leading many to simply ignore the system's input

ACTION STEP: For AI supporting clinical decision, create interdisciplinary implementation teams that include not just technical experts but also clinicians, administrative staff, legal/compliance representatives, and patient advocates to address the full spectrum of adoption barriers.

The technology was eventually adopted – after a comprehensive implementation approach that integrated the tool directly into the electronic health record workflow, established clear clinical guidelines for its use, negotiated reimbursement pathways with insurers, and created both physician and patient education programs.

Manufacturing: The Predictive Maintenance Project That Couldn't Scale

A global manufacturer successfully piloted an AI-powered predictive maintenance system that reduced unplanned downtime by 32% in its newest plant. Encouraged by these results, the company attempted to roll it out across all facilities – but implementation failed at older locations.

Analysis revealed several scaling barriers:

1. **Data Heterogeneity:** Older plants used different sensors, control systems, and data formats that the AI wasn't trained to handle
2. **Physical Variations:** Manufacturing lines that appeared identical on paper had subtle differences in configuration that affected reliability predictions
3. **Maintenance Practice Inconsistency:** Each plant followed slightly different maintenance protocols, making standardized recommendations ineffective
4. **Skills Distribution:** The pilot plant had engineers familiar with data-driven approaches, while other locations relied more heavily on experience-based maintenance
5. **Success Measurement Differences:** What counted as "improvement" varied across plants based on their baseline performance and priorities

The company eventually succeeded by adapting their approach: developing plant-specific models rather than a single global system, creating standardized data interfaces rather than requiring uniform systems, involving maintenance teams in model development, and customizing success metrics to each location's specific challenges.

KEY INSIGHT: Successful scaling often requires adaptation rather than standardization. It's important to recognize that conditions vary across an organization and build flexibility into the implementation approach.

TECH GIANTS AREN'T IMMUNE

You might think that tech giants like Microsoft and Google, with their vast resources and AI expertise, would be immune to such failures. But research reveals otherwise.

As mentioned earlier, in 2016, Microsoft launched Tay, a Twitter chatbot designed to learn from interactions with users. Within hours, Tay had learned toxic language from malicious users and began posting offensive content. Microsoft had to shut it down almost immediately after launch.

Google faced similar issues when its object recognition program showed higher error rates for people with darker skin, even misidentifying some as gorillas. The incident highlighted how even the most sophisticated AI can perpetuate harmful biases when not properly tested with diverse data.

These high-profile failures from tech leaders demonstrate a sobering reality: AI implementation is challenging for everyone, regardless of technical expertise or resources. The difference between success and failure lies not in technical brilliance but in thoughtful implementation that considers the full context in which AI operates.

REFLECTION QUESTIONS: If even Google and Microsoft experience AI implementation failures, what additional safeguards might your organization need to put in place?

THE AI ADOPTION PARADOX

A striking paradox emerges from these case studies: AI doesn't typically fail because of poor technology. It fails because of poor implementation.

Hugh Burgin, EY's Data and AI Leader for the Americas, captures this perfectly: "AI is successful everywhere. But my second point is, it's not always successful. Not every company is doing everything that they'd like to do around AI, not every company is seeing the Return on Investment (ROI) that they could achieve."

The stats confirm this paradox. While Gartner research indicates 85% of AI projects fail and a RAND Corporation study found over 80% of AI projects fail, organizations continue investing heavily – those surveyed by ESI ThoughtLab invested an average of $38 million each.

This disconnect between investment and results creates a critical question: how can organizations overcome these implementation challenges to realize AI's true potential?

FALSE STARTS

Ryan Kane, owner of IT managed services provider Soaring Towers, offers a revealing case about implementation issues, noting that he has witnessed companies deploying Microsoft Copilot "without any employee training concerning its use cases." This lack of preparation makes ROI determination nearly impossible. Kane bluntly states that he has "found very few companies who have found ROI with AI at all thus far," attributing this to most organizations "simply playing with the novelty of AI still."

Stuart King, CTO of cybersecurity consulting firm AnzenSage, adds: "I think back to the first discussions we had within the organizations that are working with [AI], and it was a case of, 'Here's this great new thing that we can use now, let's go out and find a use for it.'" King emphasizes that "What you really want to be doing is finding a problem to solve with it first."

This technology-first approach reverses the proper sequence for successful implementation. When organizations lead with technology rather than business challenges or opportunities, disappointing results are virtually guaranteed.

KEY INSIGHT: Success comes from starting with a clear business challenge, not from starting with AI technology and looking for applications.

THE VALUE-FIRST FRAMEWORK

To counter these common failure patterns, I've developed a simple framework that dramatically increases your chances of AI implementation success:

THE FIVE PRINCIPLES OF VALUE-FIRST AI IMPLEMENTATION

1. **Problem before Solution:** Clearly define specific business problems before considering AI solutions
2. **Value before Technology:** Establish clear value metrics before selecting specific technologies
3. **Integration before Innovation:** Ensure seamless integration with existing workflows before pursuing cutting-edge capabilities
4. **Adoption before Scale:** Focus on human adoption and behaviour change before scaling across the organization
5. **Iteration before Perfection:** Launch and learn rather than seeking perfect solutions from the start

This value-first framework serves as the foundation for the implementation approaches we'll explore in the remainder of this book.

ACTION STEP: Review a current or planned AI initiative using these five principles. Identify where your approach might be inverted (e.g., technology before value) and develop specific strategies to reorient towards a value-first approach.

IMPLEMENTATION CHECKLIST: PREVENTING AI VALUE FAILURE

Understanding why AI implementations fail provides the foundation for more successful approaches. The remainder of this book focuses on practical frameworks for avoiding these common pitfalls and creating sustainable AI value.

Before diving deeper into solutions, let's capture the key lessons from this chapter's exploration of failure patterns:

- **Start with clear business outcomes, not technical capabilities**. Define specific problems you're solving and how success will be measured before considering AI approaches.
- **Address data foundations before algorithm sophistication**. Even the most advanced AI cannot overcome fundamental data limitations.
- **Build cross-functional teams with end-to-end accountability**. Technical and business expertise must be integrated throughout the implementation process.
- **Design for human adoption from the beginning**. Consider workflow integration, trust building, and change management as core requirements, not afterthoughts.
- **Create comprehensive measurement approaches**. Track both technical performance and business outcomes, with clear attribution methodologies.

Spotting Signals of AI Value Failure

- **Plan for scale during pilot design**. Consider what will be required for enterprise deployment, not just initial proof of concept.
- **Anticipate and address regulatory and ethical implications**. These factors increasingly determine whether AI can be deployed regardless of technical performance.

Organizations that learn from common failure patterns position themselves to realize AI's potential rather than becoming another statistic in the discouraging implementation success rates.

FINAL REFLECTION QUESTIONS: Which of these failure patterns have you observed in your own organization? What change could you make to your current approach that would have the greatest impact on value realization?

4 The Value-First Approach to AI

Chapter Roadmap

In this chapter, we explore why technology-first thinking is the #1 trap in AI implementation – and how to adopt a value-first planning approach that ensures business impact. We cover practical tools, like the AI value mapping canvas for aligning stakeholders, how to build compelling business cases that secure sustained support, and strategies for setting realistic expectations that maintain momentum.

THE TECHNOLOGY-FIRST TRAP

"I need an AI strategy."

The CEO of a mid-sized insurance company sat across from me, coffee in hand, looking more uncertain than confident despite his direct statement.

"Why?" I asked.

He seemed startled by the question. "Well, everyone's doing it. Our competitors are talking about their AI initiatives in earnings calls. Board members keep asking what we're doing with AI. I just know we need to get moving before we're left behind."

Conversations like these happen daily in boardrooms and executive suites worldwide. Leaders feel immense pressure to "do AI," often without clarity about why or how it connects to their organization's actual goals.

This technology-first thinking represents the single biggest reason Artificial Intelligence (AI) initiatives fail to deliver value.

Where the Innovation-vs-Results Equation Creates Tension

PATTERN RECOGNITION: Every leader faces a fundamental tension with AI: the push to embrace cutting-edge technology versus the pull to deliver measurable business results.

On one side, there's legitimate fear of falling behind. AI capabilities are advancing rapidly, with competitors making bold claims about transformational results. The pressure to keep pace creates a powerful drive to implement something – anything – with "AI" in the name.

On the other hand, there's the reality that organizations ultimately succeed through business outcomes, not merely technological sophistication. If AI doesn't improve customers' experiences, reduce costs, increase sales, streamline operations, or create competitive differentiation, it can be just an expensive distraction.

While Dr. Andrew Ng advocates for starting with business problems rather than AI solutions, many organizations still approach implementation from the other direction. This tension between technology-first versus problem-first approaches to AI implementation affects organizations of all sizes and in all industries.

WARNING SIGN: This tension leads to a common but counterproductive pattern:

1. Leaders declare they need an AI strategy
2. Technical teams start exploring AI capabilities and use cases
3. Projects focus on technological feasibility rather than business value

The Value-First Approach to AI

4. Pilots demonstrate technical success but unclear business impact
5. Initiatives struggle to secure ongoing funding and support
6. AI gets branded as overhyped or impractical for "our business"

I've seen this cycle repeat across industries, from manufacturing to financial services to healthcare. The root cause is a lack of strategic thinking, resulting in a flawed approach.

WHY SUCCESSFUL AI RARELY STARTS WITH AI

KEY INSIGHT: The most valuable AI implementations typically don't begin as "AI projects" at all. They start as business initiatives targeting specific outcomes, with AI as an enabling capability rather than the central focus.

As Dr. Andrew Ng puts it when meeting with executive teams: "I don't want to hear about your AI problems, I want to hear about your business problems, and then it is my job to brainstorm with you whether there is an AI solution in terms of your key business problems."

Table 4.1 shows these contrasting approaches.

The difference isn't semantic, it's fundamental. Technology-first thinking focuses on capabilities; value-first thinking focuses on outcomes.

THE VALUE SHIFT: As the insurance CEO and I continued our conversation mentioned earlier, we shifted to his actual business challenges: rising claims costs, customer retention issues, and agency productivity. We identified specific, measurable outcomes where AI could create value. His "AI strategy" transformed into a business strategy enabled by AI, with clear success metrics tied to organizational priorities.

Six months later, his company had two AI initiatives in production: one reducing claims leakage and another optimizing customer retention efforts. Neither was branded as an "AI project" within the organization but both delivered measurable value through AI capabilities.

REFLECTION QUESTIONS: Think about a current or planned AI initiative in your organization. Does it start with technology ("implementing AI for X") or with a specific business outcome ("improving X through AI application)? How might reframing the initiative change stakeholder engagement and success metrics?

VALUE-FIRST AI PLANNING

Value-first planning reverses the typical AI implementation approach, starting with outcomes rather than technology. This approach dramatically increases success rates by ensuring alignment from the beginning.

TABLE 4.1
Technology-First Approach versus Value-First Approach

Technology-First Approach	Value-First Approach
"Let's implement AI for customer service"	"We need a significant reduction in resolution time for high-volume customer issues"
"We should use machine learning for inventory management"	"We need to reduce both stockouts and inventory costs"
"We need AI-powered fraud detection"	"We have to reduce fraud losses and decreasing false alerts"

STARTING WITH OUTCOMES, NOT TECHNOLOGY

Instead of asking "Where can we use AI?," value-first planning asks:

- "What specific outcomes would create the most value for our organization?"
- "Which of these outcomes might be enabled or accelerated by AI capabilities?"
- "How would we measure success in ways meaningful to the business?"

This shift sounds obvious but proves surprisingly difficult in practice. Technical teams naturally think in terms of capabilities and use cases. Business leaders often lack sufficient understanding of AI's possibilities to connect them to strategic priorities. Bridging this gap requires intentional processes that start with business outcomes and work backward to enabling technologies.

Case Study: TAG'S Data Pipeline Transformation

A real-world example comes from TAG, where shifting focus to outcomes rather than technology enabled the company to transform its data pipeline to make critical legacy processes not only up to 30 times faster but also much more resilient.

Rather than starting with a technical solution, the focus was on the business outcome – faster, more resilient data processing – which then informed the right approach.

ACTION STEP: For your next AI initiative, start by defining three to five specific, measurable business outcomes before exploring any technical solutions. Involve both business and technical stakeholders in this process to create shared understanding.

The value-first mindset creates several advantages:

- **Clearer Prioritization:** When multiple AI possibilities are explored, outcome-based planning provides a rational basis for selecting those with the greatest business impact
- **Better Alignment:** Everyone understands what success looks like in business terms rather than technical metrics
- **Stronger Executive Support:** Leaders more readily back initiatives tied to strategic priorities and measured by business outcomes
- **Faster Time-to-Value:** Focus remains on delivering business impact rather than technical perfection

IDENTIFYING HIGH-IMPACT USE CASES

Once focused on outcomes, the next step is identifying specific use cases where AI can create substantial value.

The key question shifts from "What can AI do?" to "Which AI applications would best advance our priority outcomes?"

This assessment considers multiple factors:

- **Value Potential:** The magnitude of impact if successful
- **Implementation Feasibility:** The organization's ability to execute effectively
- **Data Readiness:** The availability and quality of necessary data
- **Organizational Fit:** Alignment with existing processes and capabilities
- **Risk Profile:** Potential downsides and mitigation approaches

Case Study: A Gaming Company's Small-Scale Experiment

A gaming company took this approach by conducting a small-scale experiment that yielded 13% growth in just two weeks, providing the evidence needed to justify larger investments. By starting

The Value-First Approach to AI

with a focused use case, they quickly demonstrated value, which then served to build momentum for broader application.

QUICK WIN: Create a simple evaluation matrix for potential AI use cases. Score each opportunity on value potential (1-10), implementation feasibility (1-10), and data readiness (1-10). Prioritize those with the highest combined scores for initial implementation.

MAPPING AI CAPABILITIES TO STRATEGIC PRIORITIES

The most successful AI implementations directly connect to strategic priorities rather than existing as separate technology initiatives.

This mapping ensures that AI investment advances the organization's most important goals rather than creating interesting but ultimately peripheral capabilities.

Case Study: Commonwealth Bank of Australia's Document AI

Commonwealth Bank of Australia (CBA) exemplifies this approach with its implementation of Document AI for customer onboarding. Rather than pursuing AI for its own sake, the organization focused on specific strategic priorities: improving compliance with risk policies while enhancing customer experience.

By connecting AI capabilities directly to these strategic goals, they secured executive support and created measurable impact.

KEY INSIGHT: Outcomes are enhanced when organizations create explicit mapping between strategic priorities and AI initiatives, ensuring that every AI project directly advances core business goals.

Table 4.2 shows how a retail banking leader mapped AI applications to its strategic priorities.

This mapping created clear connections between AI capabilities and strategic outcomes that everyone in the organization could understand.

SETTING CLEAR SUCCESS METRICS BEFORE STARTING

The final element of value-first planning is establishing clear, measurable success metrics before development begins.

TABLE 4.2
AI Applications to Strategic Priorities Mapping

Strategic Priority	AI Applications	Expected Impact
Improve digital engagement	Personalized recommendation engine	22% increase in digital activity
	Conversational banking assistant	35% reduction in call centre volume
	Proactive financial insights	18% increase in mobile engagement
Reduce operational costs	Intelligent document processing	$12M annual savings
	Exception handling automation	40% faster processing times
	Process mining and optimization	25% reduction in manual work
Enhance risk management	Early warning credit signals	15% reduction in credit losses
	Fraud pattern detection	22% improvement in fraud detection
	Behavioural risk indicators	9% reduction in operational risk events
Maximize revenue per customer	Next best product prediction	14% increase in product adoption
	Life event anticipation	28% improvement in offer timing
	Retention risk identification	23% reduction in high-value attrition

These metrics should reflect business outcomes rather than technical performance, though technical metrics may serve as leading indicators.

WARNING SIGN: If you can't define clear success metrics for an AI initiative in business terms before starting development, you're likely not ready to proceed. Fuzzy objectives lead to fuzzy results.

A telecommunications company implemented an AI-powered customer churn prediction system with these predefined success metrics:

Primary Business Metrics:

- Reduce high-value customer churn by 20%
- Increase retention campaign Return on Investment (ROI) by 35%
- Improve net promoter score by 8 points

Leading Indicators:

- Prediction accuracy for high-risk customers (>80%)
- Timeliness of risk identification (30+ days before likely churn)
- Intervention effectiveness (>25% conversion of at-risk customers)

Implementation Metrics:

- Usage rates by customer service teams (>80%)
- Integration with campaign management systems
- User satisfaction with recommendations (>7/10)

This comprehensive measurement framework ensured everyone understood what success looked like, enabled early course correction, and created clear accountability for results.

ACTION STEP: For each AI initiative, create a three-tiered measurement framework that includes (1) primary business outcomes, (2) leading technical indicators, and (3) implementation metrics. Review and refine this framework with stakeholders before development begins.

When Less AI Means More Value

KEY INSIGHT: Organizations often create more value with simpler AI approaches applied to high-priority business problems than with sophisticated AI applied to peripheral issues.

A retail client spent millions building a cutting-edge computer vision system to analyse store traffic patterns. Meanwhile, a simple regression model analysing website cart abandonment delivered five times more revenue impact with one-tenth the investment.

Value-first planning prioritizes business impact over technical sophistication, sometimes leading to "less advanced" AI applications that create substantially more value.

Case Study: A Retail Company's Cautionary Tale

A world-class retail company requested an AI solution to predict which items would be returned before purchase, planning to offer customers a 10% discount to reduce return rates. Upon investigation, the team discovered that neither the historical average cost nor the overall cost of returns seemed to have been previously analysed, making it impossible to determine if the intervention would be profitable.

Further analysis revealed that the cumulative net effect of an intervention lacking in individual context sensitivity could actually lead to millions of dollars in negative investment returns. This stark illustration demonstrates the consequences of not taking a value-first approach.

REFLECTION QUESTIONS: In your organization, are resources flowing towards the most technically sophisticated AI projects or towards those with the clearest business impact? What might shift if you re-evaluated priorities based solely on expected business outcomes?

THE AI VALUE MAPPING CANVAS

To operationalize value-first planning, I've developed a practical tool: the AI value mapping canvas. It allows organizations to align AI initiatives with business objectives.

This canvas creates a shared visual understanding of how AI connects to value creation and serves as both a planning tool and a communication device.

CANVAS STRUCTURE AND COMPONENTS

The AI value mapping canvas has six interconnected sections:

1. **Strategic Objectives**
   - What specific business outcomes are we targeting?
   - How do these connect to organizational priorities?
   - What metrics will demonstrate success?
2. **Current State Assessment**
   - What challenges or limitations exist today?
   - What is the cost/impact of the status quo?
   - What constraints must be addressed?
3. **Stakeholder Value Expectations**
   - Who will be impacted by this initiative?
   - What value does each stakeholder expect?
   - Where might expectations conflict?
4. **AI Capability Mapping**
   - Which AI capabilities could address our objectives?
   - What data would these capabilities require?
   - What is our readiness for each capability?
5. **Implementation Requirements**
   - What organizational changes would be needed?
   - What skills and resources would be required?
   - What risks must be mitigated?
6. **Value Realization Plan**
   - How will impact be measured and attributed?
   - In what timeline can value delivery be expected?
   - How can sustained adoption be ensured?

The canvas creates a holistic view of the initiative, connecting business objectives to technical capabilities while addressing stakeholder needs and implementation requirements.

USING THE CANVAS IN PRACTICE

The power of the value mapping canvas comes from the collaborative process of completing it, which aligns diverse perspectives and surfaces potential issues early.

Case Study: Predictive Quality Initiative in Manufacturing

A manufacturing client used the canvas to plan a predictive quality initiative. The process revealed that while executives focused on scrap rate reduction, plant managers cared more about production uptime, and quality teams prioritized customer complaint reduction.

These seemingly aligned stakeholders actually had different primary value expectations. By recognizing these differences, the team designed a solution and measurement approach that addressed all three dimensions, securing broader support and ultimately delivering more comprehensive value.

PATTERN RECOGNITION: Successful AI implementations create value across multiple stakeholder groups, even when those groups have different primary objectives. The canvas helps to identify these different value dimensions and to ensure they're addressed in the solution design.

The canvas works best when completed in collaborative sessions including business leaders, technical teams, end users, and other key stakeholders. This diversity ensures that all perspectives are represented and creates shared ownership of the resulting plan.

ACTION STEP: Organize a two-hour workshop with diverse stakeholders to complete the AI value mapping canvas for your next AI initiative. Give participants a short preparation guide or a brief reflection exercise that helps them consider their value expectations and concerns before attending the session.

FROM CANVAS TO ACTION: NEXT STEPS

Once completed, the value mapping canvas guides several critical next steps:

1. **Go/No-Go Decision:** Based on the completed canvas, is this initiative worth pursuing compared to alternatives?
2. **Resource Allocation:** What specific investments in technology, data, and people will be required?
3. **Implementation Planning:** What sequence of activities will deliver value most effectively?
4. **Risk Mitigation:** What proactive steps can address identified challenges?
5. **Communication Planning:** How will we articulate the value story to different stakeholders?

The canvas doesn't replace detailed implementation planning but ensures that such planning occurs within a clear value context.

Note to Readers: We'll return to the AI value mapping canvas in Chapter 14, where we'll explore how to apply it within the data-to-value integration blueprint process. Chapter 16 includes a complete template along with practical guidance for facilitating team sessions. All templates will also be available in the book's online resource centre.

BUILDING THE BUSINESS CASE FOR AI VALUE

With a clear value map in place, organizations must still secure resources and support through effective business cases. AI initiatives present unique business case challenges that require specific approaches.

THE TENSION: TRADITIONAL ROI VS. CAPABILITY INVESTMENT

AI business cases often create tension between traditional ROI expectations and the reality of capability investments.

Traditional business cases assume relatively predictable inputs, outputs, and timeframes. AI initiatives involve more uncertainty, particularly for organizations early in their AI journey. Initial investments often build capabilities that enable multiple future applications rather than delivering standalone ROI.

Case Study: A Global Bank's Hybrid Business Case

A global bank struggled with this tension when building the business case for its data science platform. Traditional ROI calculations couldn't fully capture the platform's value since it was envisioned to enable dozens of future use cases. The team solved this by developing a hybrid business case that combined:

1. Direct ROI from initial applications
2. Option value of future capabilities
3. Risk reduction from improved compliance
4. Competitive positioning benefits

This comprehensive approach secured the necessary investment by addressing both traditional financial metrics and strategic capability development.

REFLECTION QUESTIONS: Does your organization's approach to AI business cases account for both direct ROI and capability development value? What might change if you expanded your business case framework to include both perspectives?

FINANCIAL FRAMEWORKS FOR AI INVESTMENTS

Effective AI business cases typically incorporate multiple financial frameworks:

Direct Return Models:

- Implementation costs (technology, data, and talent)
- Operating costs (ongoing maintenance and operation, and licensing)
- Direct benefits (cost reduction and revenue increase)
- Time-to-value and benefit distribution

Capability Valuation Approaches:

- Option value of new capabilities
- Future cost avoidance
- Competitive risk mitigation
- Strategic positioning impact

Risk-Adjusted Frameworks:

- Probability-weighted outcomes
- Scenario-based analysis
- Stage-gated investment approaches
- Portfolio risk balancing

Case Study: A Telecommunications Company's Stage-Gated Approach

A telecommunications company used a stage-gated investment approach for their customer experience AI. Initial funding covered only the first implementation phase with clear success criteria. Subsequent funding was unlocked when the company reached these milestones.

This approach balanced the need for accountability with the reality that AI projects often involve learning and adjustment during implementation.

QUICK WIN: Create a two-phased funding model for your next AI initiative. Phase 1 funding covers initial development with clear success criteria, while phase 2 funding depends on achieving those criteria. This approach reduces risk while maintaining momentum.

BALANCING SHORT-TERM AND LONG-TERM VALUE CONSIDERATIONS

Many AI initiatives create tension between short-term results and long-term transformation. Business cases must address both perspectives.

Case Study: A Retailer's Dual-Track Business Case

A retailer building an inventory optimization AI faced this challenge. The quick win would come from reducing safety stock in their current supply chain. The transformational value would come from enabling a new fulfilment model that wasn't possible without AI-driven prediction accuracy.

The resulting solution involved a dual-track business case:

- **Track 1:** Immediate value through inventory reduction (six-month payback)
- **Track 2:** Transformational value through new fulfilment capabilities (three-year horizon)

By securing short-term wins while building towards longer-term transformation, the company maintained support through an extended development process.

KEY INSIGHT: The most successful AI implementations deliver short-term value that funds long-term transformation. Structure your business case to highlight both immediate returns and future potential.

BUILDING FLEXIBILITY INTO VALUE PROJECTIONS

Given the level of unpredictability that is inherent in AI projects, effective business cases benefit from incorporating flexibility rather than clinging to false precision.

Case Study: The Healthcare Provider's Scenario-Based Approach

A healthcare provider developed a range-based business case for their clinical decision support AI (see Table 4.3).

Each scenario included specific indicators to track progress, allowing for course correction and expectation management as implementation progressed.

ACTION STEP: Create a scenario-based business case for your next AI initiative with at least three outcomes: minimal success, expected outcome, and transformational impact. Include specific indicators to track progress towards each scenario.

TABLE 4.3
Range-Based Business Case

Scenario	Patient Impact	Financial Impact	Timeframe
Minimal success	5% reduction in adverse events 3% reduction in length of stay	$4.2M annual savings	18 months
Expected outcome	12% reduction in adverse events 8% reduction in length of stay	$12.7M annual savings	12 months
Transformational	20% reduction in adverse events 15% reduction in length of stay	$22.3M annual savings	24 months

When Smaller Initial Targets Lead to Bigger Results

KEY INSIGHT: Many organizations instinctively choose AI projects that target their largest business problems. Counterintuitively, starting with smaller, well-defined challenges can lead to greater long-term value.

Case Study: A Global Manufacturer's Pivot

A global manufacturer initially planned to use AI to revolutionize its entire quality system. After struggling with scope challenges, they pivoted to a focused application in a single high-value production line. This limited scope enabled faster implementation, clearer results, and valuable learnings that ultimately supported broader transformation.

The lesson? Value often comes from starting small but in strategically chosen areas, proving results – and expanding based on demonstrated success rather than attempting comprehensive transformation in one leap.

WARNING SIGN: If your AI initiative aims to transform an entire business function rather than address a specific, well-defined challenge, you may be setting yourself up for stumbling blocks. Consider narrowing your initial focus to build momentum through demonstrated success.

STAKEHOLDER ALIGNMENT TECHNIQUES

Even the most compelling business case fails without stakeholder alignment. AI initiatives typically impact diverse stakeholders with different priorities, concerns, and success definitions.

Bringing Diverse Perspectives Together

Effective alignment starts with inclusive planning that brings together diverse perspectives early in the process.

Case Study: A Financial Services Firm's Cross-Functional Team

A financial services firm planning a credit-decisioning AI assembled a cross-functional team including:

- Credit operations (process efficiency and consistency)
- Risk management (default prediction accuracy)
- Sales (approval rates and turnaround time)
- Compliance (regulatory adherence and fairness)
- Customer experience (communication and transparency)
- Technology (implementation feasibility and security)

This diverse team identified value opportunities and potential conflicts that would have remained hidden in a more siloed approach.

PATTERN RECOGNITION: Research by McKinsey & Company identified "the presence of a tech-savvy C-level champion" who drives implementation throughout the organization as the top factor that differentiates winners from laggards in analytics adoption.

Reconciling Conflicting Priorities

AI initiatives inevitably surface conflicting priorities that must be addressed explicitly.

Case Study: A Healthcare Provider's Structured Prioritization

A healthcare provider implementing an AI triage system encountered tension between emergency department physicians who prioritized clinical accuracy and operations leaders focused on patient throughput. These weren't just different emphases but potentially contradictory optimization targets.

Rather than allowing this conflict to derail implementation, they used a structured prioritization process:

1. **Explicit Mapping:** Documenting each stakeholder's primary and secondary priorities
2. **Impact Analysis:** Assessing how different design choices affected each priority
3. **Principled Trade-Offs:** Establishing decision criteria for resolving conflicts
4. **Tiered Implementation:** Sequencing changes to address different priorities over time

This structured approach transformed potential conflict into productive collaboration around shared goals.

QUICK WIN: Create a simple "priority mapping" exercise for your next AI initiative. Ask each stakeholder to rank their top three priorities for the project. Compare results to identify alignments and potential conflicts before technical work begins.

BUILDING A SHARED UNDERSTANDING OF VALUE

Perhaps the most powerful alignment technique is developing a shared value narrative that connects to each stakeholder's priorities while maintaining a coherent overall direction.

Case Study: A Manufacturing Firm's Multi-Faceted Value Story

A manufacturing company that implemented predictive maintenance AI developed a multi-faceted value story (see Table 4.4).

This integrated narrative helped diverse stakeholders see how the initiative addressed their specific concerns while contributing to shared goals.

ACTION STEP: Create a stakeholder value matrix for your AI initiative that identifies each key stakeholder group, their primary value dimension, and a supporting narrative that speaks to their specific concerns and priorities.

CREATING ACCOUNTABILITY FOR VALUE REALIZATION

Alignment requires not just shared understanding but mutual accountability for results.

TABLE 4.4
Multi-Faceted Value Story

Stakeholder	Primary Value Dimension	Supporting Narrative
Operations leadership	Production uptime	"15% reduction in unplanned downtime worth $8.7M annually"
Maintenance teams	Work prioritization	"Focus on truly critical issues rather than scheduled maintenance for equipment running well"
Plant management	Cost management	"24% reduction in parts inventory and 18% decrease in emergency repair costs"
Quality department	Product consistency	"41% fewer quality deviations from equipment variation"
Executive leadership	Competitive positioning	"Industry-leading reliability enabling performance guarantees competitors can't match"

Case Study: A Telecommunications Company's Shared Metrics

A telecommunications company implementing an AI-powered network optimization system established shared performance metrics across departments:

- **Network Engineering:** System reliability and coverage improvements
- **Customer Service:** Reduction in service-related complaints
- **Marketing:** Net promoter score improvements in affected areas
- **Finance:** Reduced capital expenditure requirements

By connecting these metrics to team performance evaluations and bonus structures, they created mutual accountability that transcended organizational boundaries.

KEY INSIGHT: Shared accountability transforms AI from a technology initiative to a business initiative. When every stakeholder has skin in the game, the focus naturally shifts from technical implementation to value realization.

SETTING REALISTIC VALUE EXPECTATIONS

The final element of value-first planning is setting expectations that balance ambition with realistic expectations, and this can be particularly challenging in the hype-filled AI landscape.

MANAGING THE AI HYPE CYCLE

Every organization faces the challenge of navigating between AI enthusiasm and hype-driven disappointment.

Case Study: A Retailer's Expectations Management Failure

A retailer I worked with initially positioned their personalization AI as transforming the entire customer experience. When initial results showed incremental rather than revolutionary improvement, the initiative was labelled a "failure" despite delivering real value.

A major retail chain I worked with had just completed a successful pilot of their personalization AI system. The technology worked beautifully. Recommendation accuracy was high, customer engagement was up almost 20%, and average order values had increased by over 10%. By any technical measure, it was a success.

But the CEO was furious.

"This isn't what we promised the board," he said during our quarterly review meeting. "We told them this would transform the customer experience, that we'd see revolutionary changes in how customers shop with us. These numbers look like business as usual."

The problem wasn't the technology or the results. It was the story we'd told ourselves and our stakeholders. In our enthusiasm to secure funding, the marketing team had positioned the AI as a complete transformation of the customer journey. The board expected Netflix-level personalization that would fundamentally change shopping behaviour.

What we delivered was solid, incremental improvement that created real business value. The circa 10% increase in average order value translated to $8.4M in additional annual revenue. Customer satisfaction scores had improved. Repeat purchase rates were up. But because we'd set expectations for "transformation," incremental improvement felt like failure.

Six months later, despite continuing to deliver measurable value, the personalization initiative lost funding. The board redirected resources to what they called "more transformational" projects. A technically successful, financially valuable AI implementation died because of expectation mismanagement.

This dynamic reflects the tension observed by Katie Robbert, CEO of Trust Insights, who focuses on developing strategic focus rather than chasing every new technology. As highlighted in my research, Robbert poses critical questions about what CEOs actually need, asking,

> Do I need to be paying attention to all the little startups who have their own version of a skin on generative AI? Or do I need to be thinking bigger than that of, you know, what is this mean to bring a large learning model into my business period?

Managing expectations requires honesty about AI's capabilities and limitations. Effective approaches include:

- **Education:** Helping stakeholders understand what current AI can and cannot do
- **Analogies:** Connecting AI capabilities to familiar concepts and technologies
- **Case Examples:** Sharing realistic outcomes from similar implementations
- **Scenario Planning:** Presenting a range of possible outcomes rather than single-point predictions

WARNING SIGN: If your AI initiative is being presented as revolutionary transformation rather than being focused on value creation, you may be setting yourself up for perceived failure even when delivering real benefits.

Establishing Appropriate Timeframes for Value Delivery

AI value often follows different timing patterns than traditional technology implementations, creating tension when expectations aren't properly set.

A common pattern includes:

1. **Initial Capability Building:** Significant investment with limited direct return
2. **The Value Trough:** A period where costs are visible but benefits remain limited
3. **Value Acceleration:** Increasing returns as capabilities mature and applications multiply
4. **Sustained Advantage:** Compounding benefits as AI learning and organizational adaptation continue

Case Study: A Financial Services Firm's Proactive Timeframe Communication

A financial services firm implementing AI risk models experienced stakeholder frustration during the "value trough" phase. By proactively communicating this pattern and establishing realistic timeframes – eight months to initial deployment, 14 months to break-even, and 24 months to target ROI – they maintained support through the challenging middle period.

ACTION STEP: Create a value realization timeline for your AI initiative that explicitly acknowledges the "value trough" and set appropriate expectations for when different types of value will materialize.

Communicating Constraints and Limitations

Honesty about constraints and limitations builds credibility and prevents avoidable disappointment.

Case Study: A Healthcare Provider's Transparent Limitations

A healthcare provider implementing an AI diagnostic support system explicitly communicated key limitations:

- Initial focus on three specific conditions rather than comprehensive diagnosis
- Requirement for structured data input rather than natural language

- Advisory role supporting physician decisions rather than autonomous diagnosis
- Expected false positive rates and confidence thresholds

This transparency helped physicians develop appropriate trust in the system rather than either over-relying on its recommendations or dismissing them entirely.

KEY INSIGHT: Transparency about limitations builds more trust than exaggerated claims. When stakeholders understand what an AI system can and cannot do, they develop appropriate trust and usage patterns.

Building Credibility through Honest Assessment

Perhaps counterintuitively, acknowledging challenges and limitations often builds more support than overpromising.

Case Study: A Manufacturing Firm's Risk Transparency

A manufacturing client I worked with began its AI journey by sharing both opportunity and risk assessments with the board. Rather than undermining confidence, this honesty established credibility that helped maintain support when inevitable implementation challenges emerged.

The pattern holds true across industries: organizations that establish realistic expectations and demonstrate trustworthiness in their assessments gain more sustained support than those relying on hype and overpromising.

REFLECTION QUESTIONS: In your organization's AI communications, do you primarily emphasize opportunities or balance them with honest assessment of challenges and limitations? How might shifting towards greater transparency affect stakeholder trust and support?

CASE STUDIES: GETTING VALUE RIGHT FROM THE START

To illustrate the power of value-first planning, let's examine organizations that applied these principles successfully.

How a Global Consumer Products Company Saved Millions by Starting with Value Alignment

A global consumer products company had experienced three failed AI initiatives over two years, which had been technically interesting projects that never translated to business impact. For the fourth attempt, the company radically changed its approach.

Instead of starting with AI capabilities, it focused on the most pressing business challenge: trade promotion effectiveness. Its billion-dollar annual trade spend delivered inconsistent results and limited visibility into what worked.

Through a structured value mapping process, the organization:

1. **Defined Specific Success Metrics:** 12% improvement in promotion ROI within 18 months
2. **Engaged Key Stakeholders:** Sales, marketing, finance, and retail partners
3. **Assessed Data Readiness:** Promotional history, sales data, and competitive activity
4. **Designed a Phased Implementation:** Starting with a single product category in two markets
5. **Created a Balanced Measurement Approach:** Short-term sales lift and longer-term brand equity

The resulting AI-powered trade promotion optimization system delivered $24M in incremental value in the first year, more than all previous AI initiatives combined. By starting with a clear business problem rather than a technology solution, it created alignment that sustained momentum through implementation challenges.

Public Sector Example: Setting Citizen-Centric Value Objectives

A state government agency responsible for unemployment benefits faced surging demand during an economic downturn. The initial instinct was to implement generic "AI for government" solutions, but they instead took a value-first approach centred on citizen needs.

They defined primary success metrics from the citizen perspective:

- Reduce application processing time by 60%
- Decrease eligibility determination errors by 40%
- Enable 24/7 status checking and question answering

Working backwards from these objectives, they identified specific AI capabilities that would address each goal, focusing first on document processing automation and conversational AI for common questions.

The implementation delivered dramatic results: processing time dropped from 12 to three days, accuracy improved by 32%, and the virtual assistant handled over 70% of status inquiries without human intervention. More importantly, citizen satisfaction increased by 48% points during a challenging period.

By focusing on citizen-centred value rather than technology deployment, they created measurable impact that justified continued investment.

Commonwealth Bank of Australia's

CBA partnered with H2O.ai to implement Document AI for customer onboarding, improving compliance with risk policies while enhancing customer experience by extracting critical details from identifying documents. This implementation made customer onboarding quicker while maintaining regulatory compliance.

As Sonal Surana, General Manager at Commonwealth Bank of Australia, noted: "This really is just the beginning of where we can start to embed and re-look and radically reimagine our day-to-day operations and make lives better for our colleagues and our customers."

These results demonstrate the impact that can be achieved with a strategic focus on value and specific goals – improved compliance and customer experience – rather than generic technology implementation.

Startup Case Study: Value-Driven AI on a Limited Budget

A healthcare startup faced the classic constraint of limited resources with ambitious goals. Rather than trying to build comprehensive AI capabilities, it took a highly focused approach based on specific value targets.

After analysing potential impact areas, it identified medication adherence as its initial focus: a specific problem where even modest improvements would create meaningful clinical and financial outcomes for its target customers.

The company developed a tightly scoped AI application that:

- Predicted non-adherence risk using limited but available data
- Enabled personalized interventions for highest-risk patients
- Provided clear ROI measurement through refill rates

TABLE 4.5
Aligning AI Initiatives with Corporate Strategy

Strategic Priority	AI Initiative	Success Metrics
Customer-centred banking	Personalized financial insights	25% increase in mobile engagement 17% increase in product adoption
Operational excellence	Intelligent workflow automation	$45M reduction in processing costs 35% improvement in throughput time
Risk mitigation	Early warning monitoring	20% reduction in credit losses 30% improvement in fraud detection
Talent transformation	AI-enabled advisor tools	15% improvement in productivity 22% increase in employee satisfaction

This focused approach delivered a 22% adherence improvement in its initial pilot, creating clear value that attracted additional funding. By demonstrating measurable impact in a narrow domain rather than promising broad transformation, it built credibility that supported expansion to additional applications.

KEY INSIGHT: Resource-constrained organizations often achieve better results by focusing on a single high-value use case rather than spreading resources across multiple initiatives. This focused approach creates clearer results and stronger momentum.

GLOBAL FINANCIAL INSTITUTION: ALIGNING AI WITH CORPORATE STRATEGY

A global financial institution took a comprehensive approach to aligning AI initiatives with corporate strategy. Rather than treating AI as a separate technology function, they integrated it directly into their three-year strategic plan with specific connections to each priority (see Table 4.5).

By explicitly connecting each AI initiative to strategic priorities and defining success in business terms, the company maintained executive sponsorship and funding through implementation challenges. When budget constraints required prioritization, decisions were made based on strategic alignment rather than technical considerations.

QUICK WIN: Create a one-page strategic alignment document for your AI initiatives that explicitly maps each project to specific strategic priorities and includes measurable success metrics for each connection.

KEY TAKEAWAYS: IMPLEMENTING THE VALUE-FIRST APPROACH

As we conclude our exploration of value-first AI planning, let's recap the essential principles that any organization can apply.

VALUE-FIRST AI PLANNING FRAMEWORK

1. **Reverse the Planning Sequence:** Start with business objectives and work backwards to AI capabilities rather than starting with technology.
2. **Define Multidimensional Value:** Articulate how AI will create value across different stakeholders and timeframes to build comprehensive support.
3. **Create Explicit Connections to Strategy:** I've often stated: "Your technical expertise is valuable not just for what it can build, but for what it can prevent, accelerate, and enable." Directly link AI initiatives to strategic priorities rather than treating them as separate technology projects.

4. **Set Realistic Expectations:** Balance ambition with honesty about capabilities, limitations, and timeframes to build sustainable support.
5. **Use Structured Planning Tools:** Apply frameworks like the value mapping canvas to create alignment and surface potential issues early.
6. **Build Flexibility into Business Cases:** Acknowledge uncertainty and create adaptive approaches rather than false precision.
7. **Establish Comprehensive Metrics:** Measure both technical performance and business outcomes from the beginning.
8. **Start Strategically Small:** Build momentum through focused initial applications that demonstrate value rather than attempting comprehensive transformation.

LEADERSHIP REQUIREMENTS FOR VALUE-FIRST AI

Pedro Uria Recio, Chief Data Analytics and AI Officer at True Digital, emphasizes: "Generative AI is going to require new approaches to management and leadership, and the most important is determining who is going to be leading this change. The leader of AI in an organization has to be at the C-level."

REFLECTION QUESTIONS: In your organization, who is this tech-savvy C-level champion driving AI adoption? If this role isn't clearly defined, how might establishing it change your AI initiatives' chances of success?

YOUR "VALUE-FIRST" ACTION PLAN

ACTION STEP: Use these immediate steps to shift towards a value-first approach in your organization:

1. **Reframe Current Initiatives:** Review ongoing AI projects and restate their objectives in terms of specific business outcomes rather than technical capabilities
2. **Create a Value Map:** For your most important AI initiative, complete the AI value mapping canvas with key stakeholders
3. **Align Metrics:** Ensure every AI project has explicit success metrics tied to business outcomes, not just technical performance
4. **Build Executive Alignment:** Identify and engage a C-level champion who can drive implementation throughout the organization
5. **Plan for Transparency:** Develop a communication approach that acknowledges limitations and sets realistic timeframes for value delivery

The principles and frameworks highlighted in this chapter are designed to help organizations transform their approach to AI implementation – moving from interesting technology experiments to business initiatives that deliver measurable, sustainable value.

5 Creating a Cross-Organizational AI Strategy

Chapter Roadmap

In this chapter, we examine how to break down organizational silos to create AI strategies that deliver exponential value across the entire organization. We start with understanding why departmental thinking limits AI's potential, examine the CEO's critical role in driving integration, explore practical frameworks for cross-functional implementation, and finish with real-world case studies showing these principles in action. Along the way, I share counterintuitive insights that challenge conventional wisdom about organizational structure and AI governance.

BEYOND DEPARTMENTAL THINKING

"Our AI initiatives feel like a collection of random projects rather than a coherent strategy."

When the Chief Information Officer (CIO) of a global consumer goods company shared his frustration during a strategy review, I recognized a familiar pattern in his concerns. His company had dozens of Artificial Intelligence (AI) projects underway across marketing, supply chain, R&D, and customer service. Many of them showed promising results individually, yet something crucial was missing.

"Each department is pursuing its own priorities," he said. "We're solving local problems but missing the bigger opportunities that cut across functions. And we're rebuilding the same capabilities over and over in different parts of the organization."

Unfortunately, this isn't a rare scenario, it happens all too often. AI emerges in functional silos, creating pockets of value but falling far short of its transformational potential.

KEY INSIGHT: The fundamental limitation? Departmental thinking. Breaking free from this constraint represents one of the greatest opportunities for competitive differentiation through AI.

THE CROSS-FUNCTIONAL IMPERATIVE

AI's greatest value rarely comes from optimizing single functions. The real magic happens at the intersections of traditionally separate domains:

- When customer data informs product development
- When operational insights guide marketing decisions
- When supply chain signals drive financial planning
- When employee behaviour shapes customer experiences

Intersections like these represent AI's richest opportunities – but they tend to be invisible when narrow departmental lenses are applied.

A pharmaceutical company discovered this when it performed an analysis to find out why initially promising AI initiatives weren't delivering expected results. It found that each function had built capabilities around their specific needs:

- R&D used AI to accelerate compound screening
- Manufacturing applied it to process optimization

- Commercial teams leveraged it for market analysis
- Medical focused on treatment effectiveness patterns

Each initiative showed positive Return on Investment (ROI), yet the company wasn't seeing competitive differentiation or market share growth. The missing element? Integration across these domains to create insights and capabilities none could achieve alone.

REFLECTION QUESTIONS: In your organization, where do your most valuable customer or operational journeys cross departmental boundaries? These intersection points often represent the richest AI opportunities.

Tension between Centralized and Decentralized Approaches

Organizations typically oscillate between two AI organizational models, each with significant limitations:

- **Centralized AI:** A corporate function develops enterprise capabilities but struggles to understand specific business needs and secure adoption.
- **Decentralized AI:** Business units build function-specific solutions that address immediate needs but create redundancy, inconsistency, and missed opportunities for broader value.

Both models ultimately fall short because they reflect a false choice. Effective AI strategy benefits from a hybrid approach that balances enterprise coordination with business unit relevance.

A global retailer learned this lesson after two failed attempts. Their initial centralized AI team built impressive capabilities that business units largely ignored. When they pivoted to a fully decentralized model, each business unit created redundant capabilities at higher total cost.

The successful third approach combined:

- A central team developing shared data foundations and reusable capabilities
- Embedded specialists understanding specific business needs
- Cross-functional governance aligning efforts with enterprise strategy
- Flexible deployment models based on use case requirements

This hybrid model delivered both functional relevance and enterprise scale – a balance unachievable through either extreme.

QUICK WIN: Map your current AI initiatives on a centralization spectrum from fully decentralized to completely centralized. Look for clusters that suggest your organization might be stuck at one extreme.

Breaking Down Traditional Boundaries

Creating a cross-organizational AI strategy requires addressing boundaries that exist for legitimate historical reasons but limit AI's potential value.

These boundaries manifest in multiple ways:

- **Structural Boundaries:** Formal organizational divisions with separate leadership, budgets, and priorities
- **Data Boundaries:** Information silos where critical data remains trapped within functional but closed-off systems
- **Process Boundaries:** End-to-end workflows fragmented across multiple departments

- **Metric Boundaries:** Success measures that optimize departmental outcomes, potentially at the expense of enterprise results
- **Cultural Boundaries:** Tribal affiliations that create "us-vs.-them" dynamics between functions

A financial services institution recognized these boundaries were limiting their AI impact. Customer journeys spanned deposits, lending, investments, and insurance, yet each division built separate AI capabilities focused only on their specific products and services.

Their breakthrough came from reimagining these boundaries around customer journeys rather than products and services. By organizing data, processes, and AI capabilities around life events (home buying, retirement planning, and small business growth), they created integrated experiences that individual product teams couldn't achieve alone.

As described in *Making Data Work*, I've found that modern systems not only need to be able to contend with silos, they in fact need to actively and intentionally create them.

Not all boundaries are bad. Silos exist because they serve valuable functions. That's why the goal isn't eliminating silos entirely but connecting them effectively. Successful organizations create sufficient levels of integration while maintaining focus and efficiency alongside adaptability and connectedness.

CREATING SHARED OWNERSHIP OF AI OUTCOMES

Perhaps the most critical shift in cross-organizational AI strategy is moving from functional ownership to shared outcome ownership.

Traditional projects have clear owners. Cross-functional AI requires distributed responsibility with mutual accountability – a fundamentally different model that tends to create tension in most organizations.

A healthcare system implementing an AI-powered patient experience platform encountered this tension. The initiative required active participation from clinical, administrative, technical, and operational teams – none of whom had full responsibility or authority.

Their solution was creating a shared outcome ownership model with:

1. **Executive Sponsorship:** A C-suite leader with cross-functional authority
2. **Value Council:** Representatives from each impacted area with decision rights
3. **Shared Metrics:** Performance measures that reflected collective impact
4. **Joint Resource Commitment:** Staff and budget contributions from all participating functions
5. **Cross-Functional Recognition:** Visibility and rewards for contributions regardless of home department

This shared ownership model transformed what could have been a scattered initiative with ambiguous responsibility into a focused effort with clear accountability.

ACTION STEP: For your next cross-functional AI initiative, create a shared outcome agreement that explicitly defines how success will be measured, how resources will be contributed, and how decisions will be made across departmental boundaries.

BALANCING CENTRALIZED AND DECENTRALIZED APPROACHES

Highly effective AI strategies balance centralization and decentralization based on specific elements rather than choosing a single model.

I've found that certain elements typically benefit from centralization:

- Data governance and architecture
- Core AI/ML (machine learning) platforms and infrastructure

- Security and compliance frameworks
- Talent development and communities of practice
- Strategic prioritization and investment

Others require decentralization:

- Business problem definition
- Use case prioritization
- Solution design and customization
- Implementation and change management
- Value measurement and realization

A manufacturing conglomerate developed a balanced model they called "centralized foundation, distributed application." Central teams built enterprise capabilities, data platforms, and governance frameworks, while divisional teams applied these foundations to their specific business challenges.

This hybrid approach delivered the scale benefits of centralization with the relevance and adoption advantages of decentralization.

Breaking departmental boundaries for AI requires recognizing that significant value opportunities exist at the intersections of traditional functions. Success comes from a balanced approach that centrally coordinates foundations while enabling local relevance, with shared ownership of outcomes replacing traditional single-point accountability.

THE CEO'S ROLE IN AI VALUE CREATION

Cross-organizational AI strategy cannot succeed without active CEO engagement. While technical leadership typically drives implementation, only CEOs can create the conditions for enterprise-wide value creation.

WARNING SIGN: When CEOs view AI as "just another technology initiative" rather than a strategic transformation requiring their personal engagement, cross-organizational efforts will likely struggle to gain traction.

SETTING VALUE VISION AND PRIORITIES

CEOs must translate strategic priorities into clear direction for AI investment. Without this guidance, initiatives may scatter across disconnected opportunities without strategic coherence.

The CEO of a global telecommunications company exemplified this role when she established three enterprise AI priorities:

1. "Becoming the easiest provider to do business with" (customer experience focus)
2. "Creating the most efficient network operations in the industry" (operational excellence focus)
3. "Personalizing every customer relationship" (revenue growth focus)

These priorities provided clear direction without prescribing specific technologies or approaches. They focused the organization on outcomes rather than capabilities, creating a framework for evaluating potential AI investments.

COUNTERINTUITIVE INSIGHT: The most important CEO contribution often involves saying "no" to promising opportunities that don't align with strategic priorities, preventing resource dilution across too many initiatives.

Creating a Cross-Organizational AI Strategy

It's one thing to have the technology, but it's another to weave it into the fabric of your business strategy. This requires a vision that's shared across the executive team and an openness to iteratively refine your approach based on feedback from the ground.

Kirill Lazarev,
Founder and CEO of the design agency Lazarev

According to research cited by ProfileTree, CEOs play an essential role in setting clear direction for AI initiatives. A clear vision ensures that "AI applications align with business objectives, creating meaningful change and fostering competitive advantages." This alignment process is critical to moving beyond isolated projects to strategic transformation.

ACTION STEP: Work with your CEO to develop three to five outcome-focused AI priorities that explicitly connect to your organization's strategic objectives. These priorities should be broad enough to inspire creative approaches but specific enough to guide investment decisions.

Creating an Environment for AI Success

Beyond setting direction, CEOs must create organizational conditions where cross-functional AI can thrive.

This includes addressing five critical enablers:

1. **Resource Allocation:** Providing sufficient funding, talent, and time for AI initiatives to deliver results
2. **Organizational Structure:** Creating or modifying structural elements to enable cross-functional collaboration
3. **Incentive Alignment:** Ensuring performance measures and rewards to support collaborative behaviour
4. **Risk Tolerance:** Establishing appropriate appetite for experimentation and learning
5. **Cultural Signals:** Demonstrating through words and actions that AI transformation is a genuine priority

A healthcare CEO exemplified this role by establishing a "data and insights" function with cross-departmental authority, creating shared metrics tied to executive compensation, participating personally in AI governance meetings, and publicly celebrating early wins. These actions signalled that AI wasn't just another IT initiative but a strategic transformation with full executive commitment.

PATTERN RECOGNITION: Look for these signals to determine whether your CEO is creating conditions for cross-organizational AI success:

- Organizational changes that enable cross-functional collaboration
- Performance metrics and rewards that encourage integration
- Personal participation in key AI governance decisions
- Public recognition of cross-functional successes
- Consistent messaging about AI as a strategic priority

Driving Cross-Functional Collaboration

CEOs play a crucial role in breaking down silos that prevent collaborative AI initiatives.

The CEO of a financial services firm faced entrenched product-based silos that limited AI's potential impact on customer experience. Rather than attempting wholesale reorganization, she took targeted actions to enable cross-functional collaboration:

1. Established customer journey owners with authority across product lines
2. Created a portion of bonus compensation tied to enterprise metrics
3. Implemented rotation programs placing leaders in cross-functional roles
4. Personally reviewed cross-functional initiatives in quarterly business reviews
5. Celebrated and rewarded collaborative success stories

These interventions created spaces where cross-functional AI could succeed without disrupting the entire organizational structure.

QUICK WIN: Identify one high-value customer or operational journey that crosses departmental boundaries and establish clear ownership with authority to coordinate across functions.

Modelling Data-Driven Leadership

Perhaps the most powerful CEO contribution comes through personally modelling data-driven decision-making.

When leaders consistently ask for data, challenge assumptions, and visibly use insights to make decisions, they signal that the organization truly values these capabilities. When they make decisions based on intuition or hierarchy while espousing data-driven culture, the disconnect undermines transformation efforts.

A retail CEO demonstrated this commitment by transforming executive meetings. She required data-backed recommendations for all strategic decisions, asked probing questions about methodology and alternatives, and occasionally reversed her own initial positions based on compelling analysis. This behaviour cascaded through the organization, creating demand for AI-generated insights at every level.

REFLECTION QUESTIONS: How often does your CEO visibly change their position based on data? This behaviour sends a powerful signal about the organization's commitment to data-driven decision-making.

Leadership Strategies across Organizational Types

The CEO's role varies significantly across organization types, each with distinct challenges and approaches.

Commercial Organizations: CEOs in commercial settings tend to focus on competitive advantage, with AI strategy directly linked to market differentiation, revenue growth, and operational efficiency. Their primary tension involves balancing short-term performance demands with longer-term AI capability development.

A manufacturing CEO addressed this tension by explicitly dividing AI investments into two portfolios: 70% focused on near-term operational improvements with clear ROI, and 30% developing longer-term strategic capabilities. This balanced approach satisfied both quarterly performance expectations and future competitive positioning.

Government Organizations: Leaders in public sector organizations face different challenges, constrained budgets, strict procurement rules, public scrutiny, and complex stakeholder environments among them. Their AI strategy typically emphasizes service delivery, resource optimization, and policy effectiveness.

A government agency leader succeeded by framing AI not as a technology initiative but as a service improvement program with clear citizen benefits. By focusing on tangible outcomes like reduced wait times, improved access, and better decision consistency, she built support across political boundaries.

Non-Governmental Organizations (NGOs): NGO leaders navigate unique challenges including limited resources, complex impact measurement, and diverse stakeholder expectations. Their AI strategy often centres on mission advancement, donor engagement, and operational sustainability.

An NGO executive created a cross-organizational AI strategy by focusing on shared impact metrics that transcended departmental boundaries. By making beneficiary outcomes the central organizing principle, they aligned technology investments with mission advancement rather than functional optimization.

The United Nations provides a compelling example with their Global Initiative on Resilience to Natural Hazards through AI Solutions. This cross-organizational approach guides governments in leveraging AI for disaster management, optimizing sensor placements for damage detection after tropical storms and analysing satellite imagery to prioritize relief efforts. As Monique Kuglitsch, chair of the UN initiative's focus group, notes: "What I find exciting is, for one type of hazard, there are so many different ways that AI can be applied, and this creates a lot of opportunities."

ACTION STEP: Identify the specific contextual factors (regulatory environment, funding mechanisms, and stakeholder expectations) that shape AI leadership requirements in your organization and adapt your approach accordingly.

Differences in Approach across Geographic and Cultural Contexts

Cross-organizational AI strategy must also adapt to geographic and cultural differences, particularly in organizations operating globally or in multiple jurisdictions.

While core principles remain consistent, implementation approaches can vary significantly:

- **Decision Rights:** Some cultures emphasize consensus-building, while others accept more centralized decision-making
- **Risk Tolerance:** Attitudes towards experimentation and potential failure vary dramatically across regions
- **Collaboration Models:** Expectations around team structures and communication patterns differ substantially
- **Adoption Approaches:** Change management needs to adapt to local work practices and expectations

A global consumer products company recognized these differences when implementing their AI strategy across North America, Europe, and Asia. Rather than enforcing a single approach, they established common principles but allowed regional adaptation of implementation methods, governance structures, and adoption timelines.

This balanced regional sensitivity with enterprise consistency, creating a more effective cross-organizational approach than either extreme standardization or complete localization.

Addressing Conflicting Value Propositions between Departments

Cross-organizational AI inevitably surfaces conflicting value propositions that CEOs must actively address.

These conflicts typically manifest in several patterns:

- **Optimization Conflicts:** When AI improves one function at another's expense (e.g., inventory reduction hurting product availability)

- **Resource Conflicts:** When departments compete for limited AI talent, infrastructure, or implementation capacity
- **Priority Conflicts:** When functions have fundamentally different views on what matters most
- **Timeline Conflicts:** When some departments seek quick wins, while others prioritize long-term capability building

A retail CEO faced this challenge when her AI-powered pricing optimization created tension between merchandising (focused on margin) and marketing (focused on competitive positioning). Rather than allowing departments to pursue conflicting strategies, she:

1. Established shared success metrics incorporating both perspectives
2. Created a cross-functional team with joint accountability
3. Implemented a testing approach to validate impact on both dimensions
4. Personally reviewed results to reinforce the dual mandate

This intervention transformed a potential conflict into collaborative problem-solving that ultimately delivered better results.

PATTERN RECOGNITION: While departmental conflicts over AI initiatives can derail efforts, they also signal substantial value opportunities. When functions strongly disagree about an approach, they're usually recognizing different aspects of a significant opportunity that requires integration.

Securing Executive Buy-In for Long-Term AI Investments

Perhaps the most persistent CEO challenge involves building and maintaining executive support for AI investments that may not show immediate returns.
Several approaches prove consistently effective:

- **Value Staging:** Structuring initiatives to deliver incremental value while building towards larger transformation
- **Portfolio Balancing:** Combining quick-win projects with longer-term capability development
- **Milestone-Based Funding:** Releasing investment in stages tied to specific achievement criteria
- **Comparative Benchmarking:** Using competitor and industry comparisons to create urgency
- **Future Scenario Planning:** Illustrating potential consequences of investment delay or inaction

A telecommunications CEO secured board support for a major AI transformation by combining these approaches. The initiative delivered operational improvements from day one while building towards a comprehensive customer experience transformation. Quarterly reviews tracked progress against both immediate metrics and capability milestones, maintaining momentum through the inevitable challenges.

QUICK WIN: Structure your AI investment portfolio with explicit categorization of initiatives as "quick wins" (value within three to six months), "mid-term builders" (value within six to 18 months), and "transformational investments" (value beyond 18 months).

Leading through the "Value Trough" Period

Every significant AI transformation experiences what I call the "value trough" – a period when costs and disruption are highly visible while benefits remain limited. This challenging phase tests leadership commitment and organizational resolve.

CEOs play a crucial role in maintaining momentum through this difficult period by:

1. **Setting Realistic Expectations:** Preparing the organization for the trough before entering it
2. **Celebrating Meaningful Milestones:** Recognizing progress beyond financial metrics
3. **Maintaining Investment Consistency:** Preventing premature budget cuts that extend the trough or stall the project
4. **Visualization:** Connecting current efforts to future outcomes
5. **Personal Engagement:** Demonstrating continued commitment through direct involvement

A healthcare CEO successfully navigated this challenge during the organization's clinical AI implementation. When initial results fell short of expectations and clinician scepticism grew, he increased rather than decreased his personal involvement – attending implementation meetings, publicly acknowledging challenges while reaffirming commitment, and promoting a "learning mindset" that treated setbacks as valuable insights rather than failures.

This leadership approach maintained organizational commitment through the trough until the initiative began delivering clear value, ultimately transforming clinical outcomes across the system.

WARNING SIGN: When executives stop attending update meetings during the value trough, this signals eroding commitment that often precedes funding cuts. This disengagement makes the trough deeper and longer, sometimes threatening the entire initiative.

CEOs create the conditions for cross-organizational AI success by setting clear strategic priorities, allocating resources, aligning incentives, modelling data-driven leadership, and actively managing the inevitable conflicts and challenges. Their personal engagement is particularly critical during the "value trough," when costs are visible but benefits remain limited.

THE CROSS-FUNCTIONAL AI ROADMAP

With executive direction established, organizations need a structured approach for planning and implementing cross-functional AI initiatives. The cross-functional AI roadmap provides this framework.

Creating an integrated enterprise AI strategy serves to establish cross-functional ownership, align technical and business timelines, and manage dependencies across departments. The practical frameworks provided below can be applied immediately to support cross-functional initiatives.

BUILDING AN INTEGRATED ENTERPRISE AI STRATEGY

Effective AI roadmaps balance structure and adaptability, providing clear direction while allowing for learning and adjustment.

The roadmap development process includes five key phases:

1. **Value Opportunity Assessment**
   - Identifying cross-functional value opportunities
   - Prioritizing based on potential impact and feasibility
   - Mapping strategic objectives and priorities
2. **Capability Requirements Analysis**
   - Defining data, technology, and talent needs
   - Assessing current state and gaps
   - Determining build-vs.-buy decisions

3. **Dependency Mapping**
   - Identifying critical prerequisites and enablers
   - Establishing logical sequencing based on dependencies
   - Creating contingency paths for key risk factors
4. **Implementation Staging**
   - Defining phases with clear outcomes and milestones
   - Balancing quick wins and foundation-building
   - Creating feedback loops for continuous learning
5. **Resource Planning**
   - Allocating budget, talent, and technology resources
   - Defining shared-vs.-function-specific investments
   - Creating funding mechanisms that support cross-functional initiatives

A global manufacturer exemplified this approach when developing its enterprise AI roadmap. They began with cross-functional value opportunities, identified capability requirements and dependencies, and created a three-horizon implementation plan that balanced immediate impact with long-term transformation.

The resulting roadmap wasn't a static document but a living framework that evolved based on implementation learning and changing market conditions.

ACTION STEP: Catalogue your organization's current AI initiatives, then map them against strategic priorities to identify both alignment gaps (initiatives without clear strategic connections) and opportunity gaps (strategic priorities without supporting initiatives).

Cross-Functional Ownership of Outcomes

Roadmaps remain theoretical without clear ownership. Cross-functional AI requires rethinking traditional project accountability models.

A healthcare system implementing an AI-powered patient experience platform created a multi-level ownership structure:

- **Executive Sponsors:** C-suite leaders accountable for overall value realization
- **Initiative Owners:** Leaders responsible for specific cross-functional capabilities
- **Capability Teams:** Cross-functional groups implementing specific elements
- **Function Representatives:** Individuals ensuring alignment with departmental activities

This nested ownership model created clear accountability while maintaining cross-functional integration. Regular review sessions brought these groups together to assess progress, address barriers, and adapt plans based on learning.

COUNTERINTUITIVE INSIGHT: Sometimes less detailed roadmaps with clearer ownership create more value than comprehensive plans with ambiguous accountability.

> According to cross-functional collaboration research from Cisco,
>
> we still have a director [of] product being responsible for the product managers. We have a director [of] UX being responsible for the UX people in the team, and we have a director [of] engineering who typically leads the engineers. How do we change the style we lead?

This traditional structure creates what product leadership coach Tobias Freudenreich describes as "a cacophony of voices" where teams are "burdened" with translating between "all these different documents and stories."

Creating a Cross-Organizational AI Strategy

Jeremy Foster, VP of Cloud Infrastructure at Cisco, acknowledges the behavioural evolution required:

> These are behaviours that sometimes you may instinctively have, but they're also things that, as you grow in your career, you can get better and better at... how do I work cross-functionally, how are we going to solve a business objective? That might make sense to everybody, but it may not be easy to accomplish when everybody is involved.

Freudenreich recommends that leadership teams "speak one voice to the cross-functional teams so that they have clarity of direction in the end." This unified direction is essential for effective cross-functional AI implementation.

ALIGNING TECHNICAL AND BUSINESS TIMELINES

One of the most consistent challenges in cross-functional AI involves aligning technical and business timelines.

Technical teams typically prefer longer development cycles to build robust, scalable solutions. Business stakeholders want rapid results aligned with market opportunities and performance cycles. Without explicit alignment, this tension creates predictable conflict.

A financial services firm addressed this challenge through a dual-track roadmap approach:

- **Capability Track:** Longer-term development of foundational elements with technical milestones
- **Application Track:** Near-term implementation of specific use cases with business outcomes
- **Integration Points:** Explicit connections showing how each track enables the other

This approach gave technical teams the time needed for robust development while delivering business value at intervals that maintained stakeholder support.

QUICK WIN: Create a visual timeline that explicitly shows the relationship between foundational capability development and business value delivery, highlighting integration points where technical milestones enable business outcomes.

MANAGING DEPENDENCIES ACROSS DEPARTMENTS

Cross-functional AI initiatives typically involve complex dependencies across departmental boundaries. Without explicit management, these dependencies can become major failure points.

When a telecommunications company implemented an AI-powered customer experience platform, it faced dependencies across marketing, operations, IT, product, and customer service functions. Their dependency management approach included:

1. **Visual Dependency Mapping:** Creating clear visualization of cross-functional relationships
2. **Critical Path Analysis:** Identifying dependencies with highest impact on overall timeline
3. **Buffer Planning:** Building appropriate time reserves around high-risk dependencies
4. **Escalation Protocols:** Establishing clear processes for addressing dependency failures
5. **Regular Cross-Functional Reviews:** Joint sessions monitoring dependency status

This structured approach prevented the cascade failures that are common in complex cross-functional initiatives, where delays in one area create ripple effects across the entire roadmap.

WARNING SIGN: When cross-departmental dependencies are managed through informal relationships rather than structured processes, they typically become the primary cause of implementation delays and failure.

Effective cross-functional AI implementation requires integrated roadmaps with clear ownership models, deliberate alignment of technical and business timelines, and structured approaches to managing dependencies across organizational boundaries. Success comes not from perfect planning but from creating frameworks that enable continuous adaptation while maintaining strategic direction.

THE STAKEHOLDER VALUE ALIGNMENT MATRIX

Beyond roadmap planning, cross-functional AI requires a deeper understanding of how different stakeholders perceive and measure value. The stakeholder value alignment matrix provides a structured approach to this critical challenge.

The framework introduced below is a tool for mapping stakeholder objectives, identifying conflicts, creating win-win value propositions, and building effective coalitions. It represents a hands-on opportunity to transform stakeholder alignment from a political exercise into a value creation strategy.

Mapping Stakeholder Objectives and Concerns

The first step in creating stakeholder alignment involves comprehensively mapping objectives and concerns across all impacted groups.

A manufacturing company implementing an AI-powered supply chain optimization system identified these stakeholder perspectives (see Table 5.1).

This comprehensive mapping revealed potential conflicts and alignment opportunities that wouldn't have been visible through traditional planning approaches.

ACTION STEP: For your next cross-functional AI initiative, create a comprehensive stakeholder map that identifies both primary objectives (what each group wants to achieve) and key concerns (what they're worried about losing or risking).

Identifying Conflicts and Contradictions

The matrix analysis naturally reveals where stakeholder objectives potentially conflict – critical tension points that must be addressed proactively.

In the manufacturing example, several conflicts emerged:

- Supply chain's inventory reduction vs. sales product availability concerns
- Manufacturing's stability needs vs. rapid change implementation requirements
- System sophistication goals vs. IT resource constraints

TABLE 5.1
Stakeholder Map

Stakeholder	Primary Objectives	Key Concerns
Supply Chain	Inventory reduction	Complexity of new system
	Delivery reliability	Handling of exceptions
Manufacturing	Production stability	Rapid change implementation
	Resource utilization	Planning cycle adjustments
Sales	Product availability	Customer impact during transition
	Delivery commitments	Flexibility for key accounts
Finance	Working capital improvement	ROI timeline
	Cost reduction	Implementation costs
IT	System integration	Resource constraints
	Technical performance	Legacy system limitations

Creating a Cross-Organizational AI Strategy

Rather than proceeding with these conflicts unaddressed, the company created a structured resolution process that identified specific trade-offs and established decision principles for navigating them.

COUNTERINTUITIVE INSIGHT: Surfacing conflicts early actually accelerates implementation by preventing the passive resistance and workarounds that emerge when conflicts remain unacknowledged.

Research on cross-functional AI integration reveals that while organizations recognize the need for collaboration, many maintain traditional functional management where "a director [of] product [is] responsible for the product managers," with similar structures of responsibility for other departments as outlined earlier. This creates a "cacophony of voices," where teams struggle to translate between "all these different documents and stories." The conflict between functional leadership and cross-functional execution remains a significant obstacle in many organizations.

CREATING WIN-WIN VALUE PROPOSITIONS

The most powerful application of the alignment matrix involves identifying opportunities where multiple stakeholder objectives can be simultaneously addressed.

A healthcare organization implementing clinical decision support AI discovered several win-win opportunities (see Table 5.2).

By focusing initial implementation on these win-win opportunities, they built broad support that sustained momentum through more challenging aspects of the transformation.

This approach transforms stakeholder alignment from a political exercise to a value creation strategy – finding intersections where multiple objectives can be simultaneously advanced.

QUICK WIN: Identify at least one win-win value proposition for each critical stakeholder pairing in your initiative. These become your highest-priority implementation opportunities for building momentum and support.

BUILDING A COALITION FOR AI SUCCESS

The ultimate goal of stakeholder alignment is creating a coalition powerful enough to drive implementation success despite inevitable challenges.

A financial services organization used their alignment matrix to build a coalition strategy with three key elements:

1. **Core Coalition:** Stakeholders with highest impact and strongest support, who became active champions
2. **Support Network:** Groups with moderate impact who provided resources and passive support
3. **Engagement Priorities:** Stakeholders with significant impact but current scepticism, requiring focused attention

TABLE 5.2
Win-Win Propositions

Stakeholder Combination	Win-Win Value Proposition
Physicians + Administration	Reducing documentation burden while improving coding accuracy
Nurses + IT	Simplifying technology interaction while reducing support requirements
Patients + Finance	Improving care experience while reducing unnecessary utilization

This coalition approach concentrated relationship-building efforts where they would create the greatest implementation leverage rather than treating all stakeholders equally.

Successful implementations are often enabled when organizations actively build and maintain these coalitions throughout the AI lifecycle, recognizing that stakeholder alignment is not a one-time activity but an ongoing requirement for sustained success.

REFLECTION QUESTIONS: For your current AI initiatives, who constitutes your core coalition of active champions? If you can't identify at least three to five influential leaders actively supporting the effort, you likely have a coalition gap that needs addressing.

Effective stakeholder alignment requires mapping objectives and concerns across all impacted groups, proactively identifying and addressing conflicts, creating win-win value propositions that advance multiple objectives simultaneously, and building a coalition with sufficient power to drive implementation success. This structured approach transforms stakeholder alignment from a political challenge to a value creation opportunity.

AI GOVERNANCE FOR VALUE CREATION

Effective cross-organizational AI requires governance that enables rather than restricts – balancing innovation, risk management, and enterprise coordination.

We explore how to create governance structures that enable rather than restrict, balance innovation with risk management, establish clear decision rights and accountability frameworks, measure governance effectiveness, and adapt governance for emerging AI capabilities. These capabilities enhance the ability to transform governance from a perceived barrier to a valued enabler of success.

Governance Structures That Enable rather than Restrict

Traditional governance often focuses primarily on control and compliance. AI governance must balance these considerations with enabling speed, innovation, and value creation.

As shown in Table 5.3, a retail organization developed a tiered governance approach based on risk and impact.

This differentiated approach applied appropriate governance based on actual risk rather than treating all AI initiatives identically. High-risk applications involving customer data or business-critical processes received comprehensive oversight, while lower-risk applications followed streamlined paths.

TABLE 5.3
Risk Governance Matrix

Risk/Impact Level	Governance Approach	Key Elements
High	Comprehensive governance	Full review process
		Executive approval
		Detailed monitoring
Medium	Streamlined governance	Simplified reviews
		Self-certification
		Periodic auditing
Low	Guidelines and enablement	Clear standards
		Training and tools
		Post-implementation monitoring

The governance structure enabled rather than restricted by:

- Providing clear, documented paths appropriate to each risk level
- Offering pre-approved patterns for common use cases
- Focusing on outcomes and guardrails rather than prescriptive methods
- Including value creation as an explicit evaluation dimension

WARNING SIGN: While many C-level executives express confidence in their AI governance capabilities, a significant implementation gap exists. According to a 2024 Deloitte survey, 77% of C-level executives were confident in their workforce's ethical AI decision-making abilities, yet PwC's 2024 AI Business Survey found 75% of organizations lacked AI governance frameworks.

ACTION STEP: Assess your current AI governance approach against these enabling characteristics. Does it provide clear paths for different risk levels? Does it focus on outcomes rather than prescriptive methods? Does it explicitly include value creation as a key consideration?

Balancing Innovation with Risk Management

Perhaps the most challenging governance tension involves balancing innovation with appropriate risk management.

Organizations must navigate between two dangerous extremes:

- **Intense Focus on Risk:** Stifling innovation through excessive controls
- **Intense Focus on Innovation:** Creating unmanaged risks through inadequate oversight

A healthcare organization navigated this balance through a portfolio approach to AI governance:

1. **Innovation Zone:** Designated space for experimentation with appropriate safeguards but maximum flexibility
2. **Transition Zone:** Structured process moving promising innovations towards production
3. **Production Zone:** Comprehensive governance for operational AI systems

This zoned approach allowed different governance models appropriate to each phase of the AI lifecycle rather than applying production standards to early experimentation or maintaining loose governance for operational systems.

Steve Jarrett, Senior Vice President of Data and AI at Orange Innovation, recognizes this balance in stating that Orange has established "a group Data and AI Ethics council – and per country local AI ethics referent to adapt methodologies and tools," while emphasizing that "AI ethics is not negotiable, it is the foundation of our AI strategy." This case illustrates how Orange prioritizes ethics as fundamental to its AI strategy rather than treating it as an afterthought, with governance structures that operate at both global and local levels to ensure contextual implementation.

Decision Rights and Accountability Frameworks

Effective governance requires clear decision rights, particularly for cross-functional initiatives where traditional hierarchies may not apply.

A financial services organization implemented a structured framework that I will call RACI+ for AI initiatives:

R – Responsible: Who performs the work?
A – Accountable: Who makes decisions and has ultimate ownership?

C – Consulted: Who provides input to decisions?
I – Informed: Who receives communication about decisions?
+ – Veto: Who can block decisions based on specific risk concerns?

They applied this framework to key decision points throughout the AI lifecycle, from initial approval through design, implementation, and ongoing operation. The clarity prevented the decision paralysis and confusion that had plagued earlier cross-functional initiatives.

COUNTERINTUITIVE INSIGHT: Effective governance doesn't necessarily mean centralized decision-making. Instead, it provides clear decision rights at appropriate organizational levels with transparency about who decides what and how.

QUICK WIN: For your current AI initiatives, create a simple RACI+ chart that explicitly defines decision rights for key governance actions like use case approval, model validation, and production deployment.

Measuring Governance Effectiveness

Governance itself requires measurement to ensure it enables rather than hinders value creation.
Leading organizations track governance effectiveness through metrics such as:

Value Enablement Measures:

- Time from concept to implementation
- Resource efficiency in governance processes
- Innovation adoption rates
- Value realization from governed initiatives

Risk Management Measures:

- Compliance with relevant regulations
- Quality and security incidents
- Data protection effectiveness
- Bias and fairness outcomes

Balanced Scorecard Approach:

- Combined metrics assessing both enablement and protection
- Regular review with governance stakeholders
- Continuous improvement based on feedback

A global bank implemented quarterly governance effectiveness reviews that drove continuous improvement in the AI oversight approach. When initial reviews showed excessive implementation delays, they streamlined processes for lower-risk applications while maintaining appropriate controls for high-risk use cases.

PATTERN RECOGNITION: Organizations with effective AI governance typically evolve from compliance-focused approaches to value-enabling frameworks through a predictable maturity journey. Look for these signs of evolution:

- Tiered governance based on risk rather than a one-size-fits-all approach
- Active focus on reducing the time to implementation
- Regular review and refinement based on value metrics
- Increasing emphasis on enablement alongside control

How Governance Can Enhance rather than Hinder Value Realization

When designed effectively, governance actively contributes to value creation rather than simply controlling risk.

A manufacturing organization transformed its AI governance from a compliance function to a value enabler through several targeted changes:

1. **Pattern Recognition:** Identifying successful approaches that could be replicated
2. **Knowledge Sharing:** Creating visibility across initiatives to prevent redundant work
3. **Resource Prioritization:** Aligning investments with highest-value opportunities
4. **Capability Building:** Developing reusable components and accelerators
5. **Cross-Functional Coordination:** Facilitating collaboration across departmental boundaries

This evolved approach moved governance from a perceived barrier to a valued enabler of successful implementation, maintaining appropriate controls while actively contributing to value creation.

According to Beena Ammanath, Executive Director of the Global Deloitte AI Institute and Trustworthy AI leader at Deloitte LLP, "as leaders look to strike a balance between innovation and regulation, ethically designed governance structures are important to hold both leaders and employees accountable in the responsible use of this technology." She states that successful cross-organizational AI governance involves "recruiting and upskilling to build a prepared talent pool, providing employee trainings and establishing structures of leadership" as tactics "that have emerged to drive AI innovation with an ethical focus."

QUICK WIN: Evaluate your AI governance processes against these value-enabling criteria. Identify one specific opportunity to enhance value creation through governance improvements in pattern recognition, knowledge sharing, resource prioritization, capability building, or cross-functional coordination.

Adapting Governance for Agentic AI and Autonomous Systems

As AI evolves towards more autonomous and agentic systems, governance must adapt to new challenges and opportunities.

Traditional governance focuses on static models with defined behaviours. Agentic AI requires governance for systems that learn, adapt, and potentially make independent decisions within defined boundaries.

A financial institution implementing agentic AI for customer service adapted their governance approach to address these unique considerations:

1. **Boundary Definition:** Clearly specifying the limits of agent autonomy
2. **Runtime Monitoring:** Implementing continuous oversight of agent behaviour
3. **Explainability Requirements:** Ensuring agent decisions can be understood and justified
4. **Human Oversight Models:** Defining when and how humans remain in the loop
5. **Feedback Integration:** Creating mechanisms to incorporate learning and adjustments without compromising safety

This adapted governance recognized that traditional "check-once-and-approve" approaches aren't sufficient for systems that evolve over time. Instead, governance became a continuous process throughout the agent lifecycle.

REFLECTION QUESTIONS: As your organization implements more autonomous AI capabilities, how are you adapting your governance approach? Have you explicitly defined boundaries for AI autonomy and established continuous monitoring processes beyond traditional pre-deployment approval?

ADDRESSING GOVERNANCE GAPS IN REAL-TIME DECISION-MAKING SYSTEMS

AI systems making real-time decisions create particular governance challenges, especially when human review isn't practical before actions are taken.

Organizations implementing such systems need governance approaches that provide appropriate oversight without creating impractical operational constraints.

A telecommunications company deploying AI for network management created a layered governance approach:

1. **Design-Time Governance:** Comprehensive review of system design, training approach, and potential risks
2. **Bounded Autonomy:** Clearly defined parameters within which the system could operate independently
3. **Exception Escalation:** Automated triggers for human review of unusual situations
4. **Outcome Monitoring:** Continuous assessment of decision patterns and impacts
5. **Periodic Recertification:** Regular comprehensive reviews of system performance

This approach provided appropriate governance for decisions that needed to occur in milliseconds without direct human oversight, balancing operational requirements with risk management.

Saudi Telecom Company provides a practical example with its field surveyor AI solution that employs a "human-in-the-loop (HITL)" methodology. The system allows surveyors to capture images of buildings and other objects through an app with built-in computer vision models that automatically suggest descriptions. The surveyor maintains oversight by verifying AI outputs – accepting accurate descriptions or rejecting and amending incorrect ones – demonstrating a practical implementation of maintaining human oversight in AI systems while improving efficiency.

ACTION STEP: For AI systems making real-time decisions, implement a bounded autonomy approach by explicitly defining the parameters within which the system can operate independently versus when human review is required.

BALANCING ETHICAL COMPLIANCE WITH BUSINESS VALUE REALIZATION

The final governance challenge involves balancing ethical considerations with business objectives, particularly as AI ethics receives increasing attention from regulators, customers, and the public.

Organizations must navigate potential tensions between business optimization and ethical principles like fairness, transparency, and human autonomy.

A retail organization implementing AI-powered pricing found that pure algorithmic optimization would disproportionately disadvantage vulnerable customer segments. Rather than choosing between business performance and ethical concerns, they developed a balanced approach:

1. **Value-Ethics Mapping:** Explicitly identifying where values and ethics aligned or conflicted
2. **Principled Trade-Off Framework:** Establishing decision rules for navigating conflicts
3. **Stakeholder Participation:** Including diverse perspectives in ethical review processes
4. **Transparent Communication:** Clearly explaining how principles were applied
5. **Continuous Monitoring:** Tracking both business and ethical outcomes over time

This integrated approach prevented ethics from becoming either an afterthought or an absolute constraint divorced from business reality. Instead, ethical considerations became part of a balanced value creation approach.

WARNING SIGN: Research exposes a troubling disconnect in AI ethics implementation. Ethics was identified as one of the biggest challenges facing AI in a 2018 Deloitte survey of 1,400 executives, with 32% listing ethical concerns as one of the top three hazards. Conversely, McKinsey Global Institute Partner Michael Chui noted that "across all of these different risks ... less than 50% of our respondents said [ethics] was even relevant" and "even a smaller percentage of them had done anything to mitigate against those risks."

GLOBAL REGULATORY TRENDS AND THEIR IMPACT ON GOVERNANCE FRAMEWORKS

Cross-organizational AI must increasingly navigate complex and evolving regulatory requirements across jurisdictions.

Organizations operating globally face particular challenges with regulations that may have contradictory requirements or different implementation timelines.

A financial services institution developed a regulatory adaptation layer in their governance framework:

1. **Regulatory Monitoring:** Tracking emerging requirements across all operating jurisdictions
2. **Common Controls Framework:** Identifying governance elements that satisfied multiple regulatory regimes
3. **Jurisdictional Adaptations:** Creating specific additions for local requirements
4. **Forward-Looking Design:** Building governance to accommodate anticipated future regulations
5. **Regulatory Engagement:** Actively participating in policy development where appropriate

This approach created efficiency by identifying common elements across regulations while maintaining flexibility for jurisdiction-specific requirements. It also positioned the organization to adapt more quickly as regulatory requirements continued to evolve.

The global regulatory landscape for AI is diverse and evolving rapidly. According to recent research, the European Union (EU) has embraced a risk-based framework with clear prohibitions through the AI Act. The United States, however, favours incremental approaches that build on existing laws. China implements regulations focused on social control, requiring clear labelling of AI-generated content. Meanwhile, Gulf nations are developing tailored frameworks, with Bahrain enacting the Gulf's first comprehensive artificial intelligence law and the UAE establishing the Artificial Intelligence and Advanced Technology Council.

This patchwork of regulations creates both opportunities and challenges for organizations implementing cross-organizational AI strategies.

The EU's approach categorizes AI systems into four risk levels: unacceptable, high, limited, and minimal risk. High-risk AI systems face stricter obligations and enforcement, while limited, low, and minimal risk systems face less stringent or no obligations. This risk-based framework provides a potential model for internal governance approaches that balance enablement with appropriate controls.

As Alek Tarkowski, Director of Strategy at the Open Future Foundation, notes: "I am happy that Europe is thinking about banning, for instance, social scoring."

PATTERN RECOGNITION: Organizations with effective global AI governance typically adopt a "highest common denominator" approach – identifying the most stringent requirements across jurisdictions and building a common framework that satisfies these requirements, with jurisdiction-specific adaptations where necessary.

Case Studies in Strategic Integration

To illustrate these principles in action, let's examine how organizations have successfully implemented cross-organizational AI strategies. We explore real-world examples from Apple, public sector organizations, healthcare, and financial services that demonstrate cross-organizational AI strategy in action. These cases highlight the different approaches to integration that the organizations took, the challenges they encountered, and the transformational results that were achieved.

How Apple Transformed Its Business Model through AI

Unlike the siloed approach common in many organizations, Apple demonstrates a remarkably integrated approach to AI implementation that transcends traditional departmental boundaries.

According to research on cross-functional collaboration, Apple "abandoned the traditional autocratic leadership model" in favour of empowering team members to "collectively refine their thinking for optimal solutions." Jason, a Wireless Software Engineering Manager at Apple, notes the pride in seeing "all the cross-functional work and the relationships we've forged come together in one product."

The transformation began with leadership establishing clear strategic priorities centred on customer experience. The implementation approach included:

Organizational Changes:

- Creating cross-functional teams focused on user experiences rather than product features
- Establishing shared leadership models that transcended functional boundaries
- Developing integrated work processes that connected technical capabilities to customer benefits

Technology Approach:

- Building unified platforms that supported multiple products and services
- Implementing consistent AI experiences across devices and services
- Creating seamless integration between hardware, software, and services

Value Realization:

- Measuring success through customer experience metrics rather than functional key performance indicators (KPIs)
- Balancing innovation with reliability and usability
- Communicating progress through compelling user stories

The results proved transformational, creating a seamless ecosystem where AI capabilities enhance customer experiences across products and services in ways that functionally organized competitors struggle to match.

The key success factor wasn't any single AI application but the cross-organizational integration that transformed isolated capabilities into a coherent customer experience.

COUNTERINTUITIVE INSIGHT: Apple's success stems not from superior technology alone but from organizational structures that deliberately break down functional barriers and create what the Cascade Team describes as a culture that enables teams to "collectively refine their thinking for optimal solutions."

Public Sector Example: Cross-Agency AI Implementation

Government organizations face unique challenges implementing cross-organizational AI due to rigid agency boundaries, complex procurement rules, and limited technology resources.

A state government overcame these barriers when addressing a citizen services challenge. Previously, multiple agencies interacted with the same citizens, but without coordination, creating fragmented experiences and missed intervention opportunities.

The cross-agency approach included several innovative elements:

Governance Innovation:

- Creating a multi-agency oversight council with shared decision authority
- Establishing data-sharing agreements with appropriate privacy protections
- Developing common standards while allowing agency-specific implementations

Technical Approach:

- Building a secure citizen data platform with appropriate access controls
- Implementing a shared services model for AI capabilities
- Creating flexible integration points respecting agency system constraints

Organizational Model:

- Forming virtual teams across agency boundaries
- Embedding technical specialists within agencies
- Creating a central capability team supporting agency implementations

Funding Structure:

- Pooling resources across agencies for shared capabilities
- Developing a chargeback model for agency-specific applications
- Securing executive budget for enterprise foundation elements

The initiative delivered significant improvements in both operational efficiency and citizen experience, including a 42% faster service delivery, a 35% reduction in redundant requests, and dramatically improved citizen satisfaction scores.

The success demonstrated that cross-organizational AI is possible even in environments with strong structural boundaries, provided governance and technical approaches are appropriately adapted.

Another compelling example comes from the United Nations' Global Initiative on Resilience to Natural Hazards through AI Solutions, mentioned previously, which shows that even large, complex governmental organizations can implement cross-functional AI strategies when they organize around shared outcomes rather than departmental structures.

ACTION STEP: For public sector organizations, identify common user journeys that cross agency boundaries as potential opportunities for cross-functional AI implementation. These journeys often represent high-value integration points.

HEALTHCARE: CLINICAL AND OPERATIONAL AI INTEGRATION

Healthcare organizations face particular challenges integrating clinical and operational domains, which typically operate with different priorities, workflows, and technical systems.

A regional healthcare system overcame these barriers with a novel approach to cross-organizational AI. Facing rising costs, declining reimbursements, and increasing patient expectations, they recognized that optimizing either clinical or operational domains in isolation would yield limited value.

The cross-domain integration strategy included several innovative elements:

Value Definition:

- Creating shared metrics connecting clinical quality and operational efficiency
- Developing a unified framework for prioritizing initiatives across domains
- Establishing joint accountability for both clinical and financial outcomes

Organizational Approach:

- Forming integrated teams with both clinical and operational expertise
- Creating a dedicated translation function to bridge domain languages
- Implementing rotation programs to build cross-domain understanding

Technology Integration:

- Building a unified data platform combining clinical and operational sources
- Implementing AI solutions that spanned traditional boundaries
- Creating common user experiences regardless of underlying systems

The results transformed both clinical and operational performance. Readmission rates decreased by 23%, while hospital length of stay was reduced by 1.2 days. Provider satisfaction improved significantly as AI reduced administrative burden on clinicians, allowing them to focus more time on patient care.

COUNTERINTUITIVE INSIGHT: The key insight? Neither clinical nor operational excellence alone creates transformational value. The real opportunity emerges at their intersection, for example, when clinical insights inform operational decisions and operational capabilities enable clinical innovation.

FINANCIAL SERVICES: ENTERPRISE-WIDE AI TRANSFORMATION

A global bank provides our final case study in cross-organizational AI strategy. Facing fintech disruption and changing customer expectations, the organization implemented a comprehensive transformation integrating previously siloed AI initiatives.

The approach centred on three organizing principles:

1. **Customer Journeys as Organizing Units:** Restructuring around key journeys (home buying, wealth building, and business growth) rather than product lines
2. **Value Streams as Funding Mechanisms:** Creating end-to-end funding models that transcended departmental budgets
3. **Capability Platforms as Shared Foundations:** Building enterprise resources that supported multiple business applications

This integrated approach delivered several breakthrough capabilities:

Customer Intelligence Platform:

- 360-degree view across previously separate products
- Life event prediction enabling proactive outreach
- Next-best-action recommendation optimizing for lifetime value

Intelligent Operations System:

- End-to-end process optimization across departmental boundaries
- Predictive resource allocation preventing bottlenecks
- Exception management prioritizing highest-impact cases

Risk Management Ecosystem:

- Holistic risk assessment across product boundaries
- Early warning indicators enabling proactive intervention
- Scenario modelling supporting strategic decision-making

The transformation delivered substantial business impact within three years:

- Customer acquisition costs decreased by 32%
- Share of wallet for existing customers increased by 24%
- Operational costs reduced by 28% for targeted processes
- Risk losses decreased by 17%, while approval rates improved

The critical success factor was integration: creating connections across previously separate domains that transformed isolated efficiencies into customer experiences competitors couldn't easily replicate.

QUICK WIN: Identify key customer journeys that cross traditional product or service boundaries in your organization. These journeys can serve as organizing principles for cross-functional AI initiatives that deliver integrated experiences.

These case studies demonstrate how organizations across sectors –from technology and financial services to healthcare and government – have successfully implemented cross-organizational AI strategies. Despite different contexts and challenges, common patterns emerge: organizing around customer journeys rather than functional departments, creating shared technology foundations while enabling local customization, establishing cross-functional governance with clear accountability, and measuring success through integrated outcome metrics.

KEY LESSONS IN CROSS-ORGANIZATIONAL AI STRATEGY

As we conclude our exploration of cross-organizational AI strategy, several key principles emerge.

STRATEGIC INTEGRATION CREATES EXPONENTIAL VALUE

The most powerful AI opportunities rarely exist within functional silos. They emerge at the intersection of traditionally separate domains, creating value that isolated applications cannot achieve.

Organizations that recognize this principle move beyond departmental optimization to enterprise transformation, focusing on cross-functional value opportunities rather than localized improvements.

PATTERN RECOGNITION: Look for "seams" between departments that separate customer experiences or operational processes. These friction points often represent the richest opportunities for cross-organizational AI.

STRUCTURAL CHANGES MUST SUPPORT INTEGRATION

Organizational structures designed for industrial-era efficiency often inhibit the cross-functional collaboration AI requires. Successful transformations adapt these structures without necessarily replacing them entirely.

Whether through formal reorganization, virtual teams, or matrix approaches, leading organizations create structural elements that enable collaboration while respecting necessary functional expertise.

Cisco transformed its structure "from a command-and-control system to a collaborative and organic work environment," according to Sean Worthington, Vice President of IT, Operational Excellence and Service Enablement at Cisco. The company "created a bridge between siloed architecture experts and customer-facing staff in diverse groups across the company," enabling teams to collaborate on planning and technology sharing.

Governance Becomes an Enabler, Not Just a Control

Traditional governance focuses primarily on risk management and compliance. Cross-organizational AI requires governance that actively enables value creation while providing appropriate oversight.

The most effective governance approaches balance innovation and risk management, with differentiated models based on specific use case characteristics rather than one-size-fits-all processes.

Leadership Actively Manages Inevitable Tensions

Cross-organizational AI inevitably creates tensions between functional priorities, short vs. long-term objectives, and innovation vs. stability. These tensions aren't flaws to be eliminated but realities to be actively managed.

Leaders in successful organizations directly address these tensions, creating processes for balancing competing priorities rather than allowing them to derail transformation efforts.

Shared Foundations Enable Scaled Success

While application specifics may vary across functions, certain foundational elements benefit from enterprise coordination: data architecture, AI/ML platforms, talent development, and governance frameworks.

Organizations that build these shared foundations create economies of scale and accelerate implementation while still allowing for function-specific customization where appropriate.

REFLECTION QUESTIONS: Which of these five principles represents the greatest opportunity – or challenge – for your organization's cross-organizational AI strategy? Where would focused attention create the most significant impact?

A clear understanding of cross-organizational AI strategy is important for addressing one of the most challenging aspects of implementation: moving from successful pilots to enterprise-wide scale.

Most AI initiatives struggle at this critical juncture, and in the following chapter, we explore the reasons as well as provide practical frameworks for navigating the transition successfully. We'll examine technical scaling considerations, organizational readiness requirements, and measurement approaches that sustain momentum through the challenging middle phase of transformation.

KEY TAKEAWAY: Creating a cross-organizational AI strategy isn't about finding the perfect organizational model or governance framework. It's about building adaptive systems that balance enterprise coordination with local relevance, establish clear accountability while enabling collaboration, and manage risks while accelerating innovation. Organizations that master these balancing acts create sustainable competitive advantage through their ability to implement AI at the intersections of traditional boundaries, where the greatest value opportunities lie.

6 From Pilot to Scale
Building Sustainable AI Value

Chapter Roadmap

In this chapter, we explore how to bridge the critical gap between successful AI pilots and scaled implementations that create lasting business value. Among the topics we cover are the hidden success trap, the invisible middle zone where AI value is made or lost, the four pillars of scaling success, and the three horizons of AI scaling. I share practical frameworks for assessing scale readiness, and tips for building technical and organizational foundations for scale and for overcoming common scaling obstacles, all based on insights from real-world case studies.

THE $2 MILLION WAKE-UP CALL

Let's begin with personal stories that changed everything about how I approach Artificial Intelligence (AI) scaling.

I had spent months leading data transformation efforts at a global financial services firm, and was ready for a big presentation to showcase our successful AI pilot.

The meeting was going perfectly, at least that's what I thought. The technical metrics were flawless: 95% accuracy, 60% faster processing, and significant error reduction across the board.

My slides were polished. The demo worked beautifully.

Then came the comment that tripped me up. "This all looks impressive," the CEO said, setting down his coffee cup. "But I have just one question: What does this mean for our market position in three years?"

The room went silent. I had every technical detail perfected, but I couldn't answer the one question that actually mattered.

This wasn't just an isolated experience. When I was leading transformations at AXA, a similar high-pressure moment left me speechless.

These experiences taught me something that would transform my entire approach to scaling AI: technical excellence alone captures only about 20% of potential value. The other 80% – where the real opportunity lives – comes from something more subtle: the ability to see and seize strategic opportunities that are easily missed.

KEY INSIGHT: The most significant part of AI value comes not from technical excellence but from strategic integration with business objectives and opportunities.

REFLECTION QUESTIONS: Think about your recent AI projects. What percentage of your focus was on technical details versus strategic business alignment? How might this balance need to shift?

Since these key wake-up calls, I've focused considerable attention on how to bridge the critical gap from successful pilots to scaled AI implementations that create lasting business value. Here, I share the detailed frameworks and approaches refined through years of practice in dozens of organizations.

What you'll learn isn't just theory. It's everything I wish someone had told me before the questions posed by these CEOs left me speechless.

THE HIDDEN SUCCESS TRAP

Most AI initiatives follow a predictable pattern: they succeed as pilots but fail to scale. The statistics tell a sobering story.

According to research by Accenture, the pattern is stark – only about 15% to 20% of organizations reach the "Strategically Scaling" stage with their AI initiatives, while fewer than 5% achieve full "Industrialization for Growth." The Data Analysis Bureau reports that this gap creates a substantial difference in returns – an average of $110 million between companies stuck in pilot phases and those successfully scaling AI.

WARNING SIGN: If your organization has multiple successful AI pilots but few scaled implementations, you're likely caught in the hidden success trap – focusing on technical success rather than business transformation.

This isn't because the technology doesn't work. It's because scaling AI requires something fundamentally different than what makes pilots successful.

Think about it like this: pilot projects are about proving technical feasibility, such as, can the AI model accurately predict customer churn? Can it efficiently categorize documents? Can it meaningfully enhance product recommendations?

But scaling is about creating sustainable business value, for example, how does this capability transform our customer experience? How does it reshape our operational model? How does it drive revenue growth or cost reduction at scale?

PATTERN RECOGNITION: Successful pilots focus on "Can we do this?," while successful scaling focuses on "How does this transform our business?"

The trap is thinking that scaling is simply doing more of what worked in the pilot. It's not. It's a fundamentally different challenge requiring different approaches, different skills, and different measurements.

THE TALE OF TWO TRANSFORMATIONS

Here is a comparison that illustrates this perfectly.

Two global financial institutions – let's call them Bank A and Bank B – both launched AI initiatives around the same time. Both had successful pilots, strong technical teams, and significant executive support.

Three years later, Bank A had successfully deployed AI at scale across multiple business functions, creating over $300 million in documented value. Bank B was still running pilots, with minimal business impact to show for their investment.

What made the difference?

Bank A recognized from day one that scaling AI wasn't just about the technology – it was about business transformation. They established clear value targets tied to strategic priorities. They redesigned processes to leverage AI capabilities. They built cross-functional teams that combined technical and business expertise. Most importantly, they measured success in business terms, not technical metrics.

Bank B took a different approach. They viewed AI as a technical project – something to be perfected before being handed off to the business. They focused on model accuracy rather than business outcomes. They built technical capabilities but didn't redesign processes or develop the business capabilities needed to leverage them effectively.

The contrast couldn't be clearer: Bank A created sustainable business value, while Bank B created technically impressive but strategically limited point solutions.

From Pilot to Scale

This pattern repeats across industries and organizations. The companies that successfully scale AI understand something fundamental: scaling isn't primarily a technical challenge – it's a transformation challenge with technology at its core.

ACTION STEP: Evaluate your current AI initiatives and identify which ones are being treated as technical projects versus business transformation opportunities. Reframe the technical projects by identifying their potential strategic impact.

THE INVISIBLE MIDDLE: WHERE AI VALUES ARE MADE OR MISSED

Between the gleaming vision of AI transformation and the nuts and bolts of implementation lies what we call "the invisible middle," a zone where many promising AI initiatives end up dying. This middle zone contains four key friction points that can derail scaling efforts:

1. **Strategy-to-Execution Misfire:** The original strategic intent gets lost in technical implementation
2. **Product vs. Policy Drift:** The AI solution and organizational policies diverge over time
3. **Cross-Functional Disconnect:** Different departments have conflicting priorities and approaches
4. **Speed vs. Integrity Trade-Offs:** Pressure to deliver quickly compromises solution quality and integrity

BMW's Cross-Functional AI Integration

Here is an example of how BMW successfully navigated this invisible middle.

BMW recognized the challenges of scaling AI early in their transformation journey. Rather than launching disconnected AI projects across various departments, they established a centralized "Project AI" with clear strategic oversight while simultaneously ensuring deep domain expertise within each functional area.

When their production team developed AI for quality control, they didn't just focus on the technical aspects – they carefully mapped how these new capabilities would integrate with existing workflows, what training staff would need, and how success would be measured.

This deliberate attention to the "invisible middle" is why BMW now boasts one of the most sophisticated cross-functional AI implementations in the automotive industry, successfully integrating AI across seven distinct functional areas:

1. R&D (in-vehicle energy management)
2. Production (automated image recognition, nameplate checks, and dust particle analysis)
3. Supply chain (robotics applications for object identification)
4. After-sales support (customer service issue resolution)
5. Building management (energy optimization)
6. Administration (focus groups and client reviews)
7. Customer-facing functions (driver assistance systems)

I've previously highlighted BMW's people-centric perspective, which includes a focus on both employees and customers.

A common element of tried-and-true approaches is that they go beyond the mere technical – to address other key dimensions through attention to the connections between technology, people, and processes.

KEY INSIGHT: Success in scaling AI requires deliberate attention to the "invisible middle" between strategy and execution, where technical and organizational elements must be carefully integrated.

QUICK WIN: Create a visual map of your AI initiative that explicitly connects strategic objectives to technical implementation details and organizational changes, highlighting potential disconnect points.

THE SCALING CHALLENGE: LESS ABOUT DOING MORE, MORE ABOUT CREATING VALUE

Scaling isn't about doing more – it is about creating more value with the same or less effort. This counterintuitive insight comes from hard-won experience, including what I call my "$2 million scaling mistake."

After several successful engagements with more demand than I could handle, I took on everything thinking I would figure it out later. The result was nearly destroying two seven-figure client relationships by trying to be everywhere at once, amounting to what nearly became a very costly lesson.

The scaling challenge is particularly acute with AI because it touches so many aspects of an organization. As Partha Gopalakrishnan, Partner and President at Brane Group, observes:

> Executives are rushing – they don't want to be seen as lagging behind, and investors and boards are hungry for AI transformation due to the promised efficiency gains – but this means that they are less focused on putting in place the necessary foundations and guardrails.

This rush often leads to what I call the "technical success-business failure" pattern: AI initiatives that work perfectly from a technical standpoint but fail to deliver meaningful business impact.

Tesla's Mission-Driven Approach

Tesla provides a compelling contrast to this common pattern. As documented by the AI Expert Network, Tesla has embedded AI capabilities across manufacturing (robots with AI performing assembly tasks), operations (streamlining processes), R&D (autonomous driving and humanoid robots), and IT (machine learning systems and real-time data analysis).

This success stems from embedding AI capabilities at the core of their organizational mission rather than treating AI as separate initiatives. This integrated approach enables the organization to overcome the typical scaling challenges that prevent many AI projects from moving beyond pilot stages.

PATTERN RECOGNITION: Organizations that struggle with scaling AI often treat AI initiatives as technology projects separate from core business operations, while those that succeed embed AI directly into their organizational mission and operational processes.

REFLECTION QUESTIONS: Does your organization treat AI as a separate technology initiative or as an integral part of your mission and operations? What would need to change to create deeper integration?

But what does this actually mean in practice? Let me break down the difference between scaling approaches that fail and those that succeed.

The "More Is Better" Trap

Most organizations approach AI scaling by thinking bigger: more data, more models, more use cases, more teams, and more resources. This linear thinking leads to predictable problems:

- **Resource Exhaustion:** Teams spread thin across multiple initiatives, none receiving adequate attention or resources to succeed.

- **Quality Degradation:** As initiatives multiply, the quality of implementation, governance, and value realization inevitably suffers.
- **Coordination Complexity:** Managing multiple parallel AI efforts creates exponential coordination challenges that consume more energy than they create value.
- **Value Dilution:** Attention gets divided among so many initiatives that none achieves transformative impact.

This is exactly what happened in my scaling mistake. I accepted five major engagements simultaneously, thinking I could manage them all. Instead of creating 5× the value, I created chaos. Client A received fragmented attention during their critical board presentation period. Client B's AI implementation stalled because I couldn't provide the strategic guidance they needed at key decision points. What should have been transformative partnerships became damaged relationships.

The "Value Multiplication" Approach

Successful scaling takes a fundamentally different approach. Instead of doing more things, it focuses on creating exponentially more value from focused efforts. Here's how this works:

- **Strategic Concentration:** Rather than spreading resources across many initiatives, successful scaling concentrates on fewer, higher-impact opportunities that can create multiplicative value.
- **System Building:** Instead of creating isolated solutions, it builds systems and capabilities that can generate value across multiple contexts with minimal additional effort.
- **Leverage Creation:** Rather than adding more people to do more work, it creates mechanisms that allow existing resources to create exponentially greater impact.
- **Compounding Benefits:** Instead of seeking immediate returns from every initiative, it builds capabilities that create increasing returns over time.

Real-World Example: The European Steel Manufacturer's People-First Scale

Yet, such insights are not entirely limited to my individual client engagements. According to McKinsey Digital research, a leading European steel manufacturer approached AI scaling with a fundamentally different mindset. Tim Fountaine, McKinsey Senior Partner, explains that the CEO viewed scaling AI as "a people problem" rather than a technology challenge, asking: "How will my people deliver AI? What kinds of skills do they need to have? How do I fit this into our culture?"

Instead of launching multiple AI initiatives across different departments, they concentrated on building human capabilities first. The company created an academy for analytics that trained 400 of their 9,000 workers in the first year.

The results demonstrate the power of value multiplication: this focused investment in people capabilities led to 40 initiatives with a 15% EBITDA (Earnings Before Interest, Taxes, Depreciation, and Amortization) improvement within 18 months. By investing in systems that multiplied human effectiveness rather than simply adding more AI projects, they achieved exponentially greater impact.

This wasn't about doing more AI – it was about creating an organizational capability that could generate AI value across multiple contexts simultaneously.

The CNH Industrial Approach: Hypothesis-Driven Value Creation

Marc Kermisch, Global Chief Digital and Information Officer at CNH Industrial, takes a rigorous approach to scaling that exemplifies value multiplication thinking. Rather than pursuing every possible AI opportunity, CNH focuses on clearly defined hypotheses about value creation.

As Kermisch explains: "If I believe I can reduce 10,000 man-hours, that's a hypothesis. What experiments am I going to run to prove that or disprove that?"

CNH has deployed AI across multiple fronts: equipment health monitoring that alerts owners about maintenance needs, code development support, autonomous driving technology enhancement, and an AI-powered chatbot providing service technicians with instant access to equipment manuals and repair details.

But here's the key insight: each of these initiatives is designed to test specific value hypotheses that can inform and enhance the others. The equipment health monitoring data, for example, simultaneously improves maintenance predictions, informs autonomous driving safety protocols, and enhances the chatbot's diagnostic capabilities.

This systematic approach to value creation means that each AI investment creates compounding returns across multiple use cases rather than isolated benefits in single applications.

THE "SCALE-OR-FAIL" REALITY CHECK

The research reveals a sobering reality about AI scaling. According to CGI's investigation across leaders globally in different industries, as Diane Gutiw, Vice President and Global AI Research Lead, explains: "79% of organizations are experimenting with generative AI, but only 26% are actually moving into production."

This massive gap between experimentation and value realization illustrates the fundamental scaling challenge. The organizations stuck at 79% are caught in the "more is better" trap – by running more experiments, testing more models, and exploring more use cases. The 26% who reach production understand something different: scaling requires moving from experimentation to value multiplication.

As Gutiw emphasizes: "The key to getting a return on investment is focusing on something with the intention of going into production, with the intention of getting that actual benefit."

THE INFRASTRUCTURE INVESTMENT PARADOX

Alibaba's massive $53 billion investment in AI and cloud infrastructure – the largest private tech investment in China – seems to contradict the "less is more" principle. But CEO Eddie Wu's explanation reveals the value multiplication logic: "This will be the most intensive period of infrastructure development in the company's history."

This isn't about doing more AI projects – it's about building foundational capabilities that can multiply value creation across every business function and customer interaction. Such infrastructure investments create what economists call "increasing returns to scale" – where each additional AI application becomes more valuable because it can leverage the shared infrastructure and data effects.

Wu expects this approach will "improve efficiency and create more value across its business and consumer platforms" – classic value multiplication rather than simple addition.

THE PRACTICAL FRAMEWORK: VALUE MULTIPLICATION PRINCIPLES

Based on these examples and additional research insights, here are the practical principles I recommend for scaling through value multiplication rather than activity multiplication:

1. **Identify Core Value Drivers:** Instead of pursuing multiple opportunities simultaneously, identify two to three core mechanisms through which AI can create the most value for your organization.
2. **Build Reusable Capabilities:** Rather than creating custom solutions for each use case, build AI capabilities that can be applied across multiple contexts with minimal additional investment.

From Pilot to Scale

3. **Create System Effects:** Look for opportunities where AI improvements in one area can automatically enhance value creation in other areas through shared data, processes, or capabilities.
4. **Focus on Leverage Points:** Identify the minimal set of changes that can create maximum value across the organization rather than trying to optimize everything simultaneously.
5. **Measure Value Density:** Track not just total value created, but value created per unit of resource invested – optimizing for efficiency and impact rather than scale alone.

KEY INSIGHT: The most successful AI scaling efforts create exponentially more value with the same or fewer resources by building systems and capabilities that multiply impact rather than simply adding more activities or initiatives.

PATTERN RECOGNITION: Organizations that struggle with AI scaling typically have many pilots and few scaled successes (like the 79% experimenting but not producing). Organizations that succeed have fewer initiatives but each one creates multiplicative value across multiple areas of the business.

ACTION STEP: Review your current AI portfolio and identify opportunities to consolidate initiatives around core value drivers that can create multiplicative effects across your organization. Look for ways to build reusable capabilities rather than custom solutions for each use case.

This fundamental shift – from scaling through addition to scaling through multiplication – is what separates organizations that achieve transformative AI impact from those that remain stuck in pilot purgatory with impressive technology but limited business results.

THE FOUR PILLARS OF SCALING SUCCESS

Through helping dozens of organizations move from pilot to scale, I've identified four critical elements that determine success. Think of them as the foundation upon which sustainable AI value is built.

STRATEGIC ALIGNMENT

The most successful AI initiatives are deeply connected to strategic priorities. They don't just solve problems – they advance the organization's most important goals.

At AXA, this approach led to a nine-fold increase in time to market for AI initiatives. How? Through strategically aligning every technical decision with business priorities, ensuring that AI capabilities directly supported strategic goals rather than existing as interesting technical experiments.

A common misconception here is that strategic alignment means abandoning technical excellence. The reality is quite different: it means connecting technical excellence to business value.

When I worked with a leading European retailer, we didn't just build impressive machine learning models, we built capabilities that directly enhanced customer personalization, a core strategic priority. The result? A 34% increase in conversion rates (the percentage of website visitors who actually made a purchase) and a 28% increase in average order value.

ACTION STEP: For each AI initiative, explicitly document how it connects to at least one top-level strategic priority. If you can't make a clear connection, reconsider the initiative or reframe it to create strategic alignment.

VALUE TRANSLATION

This is the ability to express technical concepts in business terms – to connect AI capabilities directly to financial and operational outcomes.

When I collaborated with the data science team at a large insurance company, we didn't just talk about model accuracy, we discussed how improved predictions impact underwriting risk (where we achieved an annual reduction of $4.2 million) and decision-making (where we achieved a 76% acceleration).

This translation isn't just about communication, it is about how you design and deploy AI systems. Every feature, every capability, and every aspect of the implementation benefit from a direct connection to value creation.

KEY INSIGHT: The ability to translate technical capabilities into business outcomes is perhaps the single most important skill for scaling AI successfully. It bridges the gap between what AI can do and what the business needs.

ORGANIZATIONAL READINESS

AI at scale doesn't exist in isolation, it operates within complex organizational systems. Success requires preparing those systems to absorb and leverage new capabilities.

This means addressing:

- **Process Redesign:** How will existing workflows change to leverage AI?
- **Skill Development:** What new capabilities do teams need?
- **Change Management:** How will you help people adapt to new ways of working?
- **Governance Structures:** How will you manage AI systems over time?

At Lloyds Banking Group, this preparation made all the difference. Before scaling their customer service AI, they redesigned the entire service process, developed new skills among customer service representatives, created clear change management plans, and established governance structures to monitor and improve the system over time.

The result? A smooth transition that delivered immediate value rather than the disruption and resistance that often accompanies technology deployments.

SUSTAINABLE IMPLEMENTATION

The final pillar focuses on building systems that create enduring value. This isn't about quick wins, it's about establishing the foundation for long-term success.

Key elements include:

- Technical architecture that can evolve over time
- Data pipelines that maintain quality at scale
- Monitoring systems that ensure ongoing performance
- Feedback loops that drive continuous improvement

At a large Middle Eastern Insurance Group, this approach transformed the organization's fraud detection capabilities. Rather than focusing solely on the initial implementation, they built comprehensive monitoring systems, established clear feedback loops with fraud investigators, and created a technical architecture that could continuously evolve as fraud patterns changed.

Three years later, their system wasn't just still functional – it was significantly more effective than at launch, having evolved through hundreds of iterations driven by real-world experience.

QUICK WIN: Create a simple feedback loop for an existing AI system by establishing a regular review process that brings together technical and business stakeholders to evaluate performance and identify improvement opportunities.

THE PEOPLE FACTOR: A SURPRISING PERSPECTIVE ON SCALING SUCCESS

A counterintuitive insight from our research is that successful AI scaling is often more about people than technology. This means successful AI scaling requires attention to – and investment in – human capital alongside technological infrastructure.

Awareness of this important dimension, as illustrated by the example of the European steel manufacturer mentioned earlier, tends to lead to stronger outcomes.

A people-centric approach is also embedded in BMW's philosophy, where Michael Würtenberger emphasizes that "the focus remains on people" in their AI transformation.

KEY INSIGHT: The most successful AI scaling efforts prioritize people and organizational change alongside technological implementation, recognizing that technology alone cannot create value without the right human capabilities and cultural context.

EARLY WINS VS. FOUNDATION BUILDING: THE STRATEGIC TENSION

One of the most challenging decisions in scaling AI relates to balancing quick wins against building the right foundations for long-term success. This tension is perfectly illustrated in two contrasting approaches revealed in our research.

We've seen the previously mentioned commitment to long-term Return on Investment (ROI) by Alibaba, with a massive upfront investment – balanced against the perspective of Diane Gutiw, who is looking for evidence of "actual benefit."

Hugh Burgin, EY's Data and AI Leader for the Americas, argues:

> AI is proven overall: it does drive benefits, does drive actions and decisions. So, the focus should not be on, can we prove that AI works, the question should be, how can we adopt AI and make sure that the insights and the value that it generates actually gets integrated into our business in a sustainable way.

This contradicts the previously highlighted view of Marc Kermisch, who is looking for "experiments ... to run to prove that or disprove" specific hypotheses.

PATTERN RECOGNITION: Organizations that successfully scale AI find ways to create early wins that build momentum while simultaneously investing in the foundations necessary for long-term success. They don't treat these as either/or choices but as complementary strategies.

PwC demonstrates this balance with the Project AIR for the financial services sector. The proof-of-concept successfully demonstrated that AI could ingest English requests, interpret them accurately, and produce code to extract the correct data and fulfil reporting requirements, delivering immediate value while establishing foundations for broader implementation.

ACTION STEP: Review your AI portfolio and ensure that you have a balanced mix of initiatives, some focused on quick wins to build momentum and credibility and others focused on building the technical and organizational foundations for long-term success.

THE SCALE-READY ASSESSMENT FRAMEWORK

How do you know if your AI initiative is ready to scale? The framework below helps you assess your readiness across the four pillars we've discussed.

For each dimension, rate your current state on a scale from 1 to 5:

Strategic Alignment Assessment:

- **Connection to Priorities:** How directly does the AI initiative support strategic goals?

- **Executive Understanding:** How clearly do executives understand the strategic value?
- **Value Quantification:** How well have you quantified the business impact?
- **Resource Commitment:** How appropriate are the resources allocated relative to the strategic opportunity?

Value Translation Assessment:

- **Business Case Clarity:** How clearly can you articulate business outcomes?
- **Stakeholder Understanding:** How well do key stakeholders understand the value proposition?
- **Measurement Approach:** How effectively can you measure business impact (not just technical performance)?
- **Value Capture Mechanisms:** How clearly defined are the processes to realize and document value?

Organizational Readiness Assessment:

- **Process Integration:** How effectively have you redesigned processes to leverage AI?
- **Skill Development:** How prepared are teams to work with new AI capabilities?
- **Change Management:** How robust is your approach to helping people adapt?
- **Governance Structures:** How clearly defined are your governance mechanisms?

Sustainable Implementation Assessment:

- **Technical Architecture:** How well does your architecture support evolution over time?
- **Data Infrastructure:** How prepared are your data systems to operate at scale?
- **Monitoring Capabilities:** How effectively can you track performance in production?
- **Feedback Mechanisms:** How robust are your systems for capturing and acting on feedback?

A total score below 40 indicates significant scaling risks. Between 40 and 60 suggests partial readiness with specific gaps to address. Above 60 indicates strong scaling readiness.

WARNING SIGN: The most dangerous situation is not a low score overall but significant imbalances across the four pillars. If you score high on technical implementation but low on organizational readiness, for example, you're likely headed for the "great technology, poor adoption" trap.

But the real value isn't in the score, it's in identifying specific areas where you need to focus before scaling. No organization scores perfectly across all dimensions, but successful scaling requires addressing the most critical gaps before moving forward.

ACTION STEP: Complete the scale-ready assessment for your highest-priority AI initiative and identify the three most critical gaps to address before scaling. Create a specific action plan for each gap.

THE THREE HORIZONS OF SCALING

Scaling isn't a single event, it's a journey that unfolds across three distinct horizons, each with different focus areas and challenges.

Horizon 1: Value Realization (0 to 6 Months)

The first horizon focuses on capturing immediate value from your AI capabilities. This isn't about trying to solve every problem at once. Instead, it is about finding specific, high-impact opportunities where you can create measurable business outcomes quickly.

Key activities include:

- Identifying high-value, low-complexity use cases
- Implementing focused solutions with clear business impact
- Measuring and documenting value creation
- Building credibility through demonstrated results

When I worked with a global pharmaceutical company, we started the scaling journey by focusing on a single high-impact use case: optimizing clinical trial site selection. Within four months, we had reduced site selection time by 62% and improved trial enrolment rates by 28%, creating immediate, measurable value while building momentum for broader initiatives.

Horizon 2: Capability Building (3 to 12 Months)

With initial value established, the second horizon focuses on building the capabilities needed for broader impact. This includes technical foundations, organizational capabilities, and governance structures that support sustainable scaling.

Key activities include:

- Developing reusable technical components
- Building cross-functional teams that combine technical and business expertise
- Establishing governance frameworks for managing AI at scale
- Creating training programs to develop needed skills

At AXA, this horizon transformed how the company approached data and AI. They built reusable data pipelines, established cross-functional "AI pods" that combined technical and business talent, created comprehensive governance structures, and developed training programs that reached thousands of employees, creating the foundation for enterprise-wide impact.

Horizon 3: Transformation (6 to 24 Months)

The final horizon leverages the value and capabilities established earlier to drive fundamental business transformation. This goes beyond incremental improvements – to reshape how the organization operates and competes.

Key activities include:

- Redesigning core business processes around AI capabilities
- Creating new products and services enabled by AI
- Transforming customer and employee experiences
- Establishing AI as a core strategic capability

At a global telecommunications company, this horizon led to a complete reinvention of their customer experience. They integrated AI capabilities across every customer touchpoint, created new AI-powered services, and fundamentally changed how they acquired and served customers. The result? A 42% increase in customer satisfaction and a 17% reduction in churn.

The power of this approach comes from the way the horizons build on each other. Initial value creation builds credibility and momentum. Capability building enables broader impact. And transformation creates a sustainable competitive advantage.

KEY INSIGHT: Most organizations try to skip directly to horizon 3 (transformation) without building the value credibility (horizon 1) and organizational capabilities (horizon 2) needed for success. This leads to ambitious but ultimately unsuccessful transformation efforts.

REFLECTION QUESTIONS: Which horizon is your organization currently focused on? Are you trying to achieve transformation without having built sufficient value credibility and organizational capabilities?

BUILDING THE TECHNICAL FOUNDATIONS FOR SCALE

Technical foundations for scaling AI need to balance flexibility with stability, allowing for evolution while maintaining performance. Here's where many organizations face the "scale-or-fail" dilemma.

THE E-COMMERCE SCALE-OR-FAIL EXAMPLE

In *Value-Driven Data*, I documented an e-commerce company facing explosive digital growth that was overwhelming their legacy data systems. As their online channel scaled to over 20 million transactions monthly in just its second year, their infrastructure couldn't keep pace.

Despite significant investment in a centralized data platform, the existing infrastructure was projected to reach capacity limits within 12 months, threatening business continuity. The company's chief technology officer acknowledged that current platforms would need upgrading and they would outgrow server capacity within a year, creating an urgent need to develop a more adaptable and scalable technology stack.

This example illustrates a common challenge: the very success of AI initiatives can create scaling challenges if the technical foundations aren't built with growth in mind.

THE GAMING GURU'S INTENTIONAL SILOS

The Gaming Guru case from *Making Data Work* provides another instructive example. The company struggled with scaling its data platform due to tightly coupled monolithic architectures. By applying "the principle of intentional silos" – breaking down their systems into smaller autonomous components – they were able to achieve significant performance improvements without disrupting their 24/7 operations.

KEY INSIGHT: Sometimes creating strategic, intentional silos can actually facilitate scaling better than forced integration of everything into monolithic systems. The key is to ensure that these intentional silos have clear interfaces and integration points.

Key technical foundations for scale include:

1. **Modular architecture** that allows components to evolve independently
2. **Scalable data pipelines** that can handle increasing volumes and sources
3. **Robust monitoring and feedback systems** that ensure performance at scale
4. **Versioning and lifecycle management** capabilities for both models and data
5. **Integration frameworks** that connect AI capabilities with existing systems

These foundations enable what Tesla has achieved through a skilled approach to AI implementation. As documented by the AI Expert Network, Tesla has embedded AI capabilities across manufacturing, operations, R&D, and IT departments. This approach treats AI not as separate initiatives but as core to their organizational mission, enabling them to overcome the typical scaling challenges that prevent many AI projects from moving beyond pilot stages.

QUICK WIN: Conduct a technical scaling assessment of your current AI infrastructure, focusing specifically on identifying potential bottlenecks that could limit growth. Prioritize addressing the most critical constraints.

THE THREE LEVELS OF AI ADVISORY SCALE

Through helping dozens of technical professionals build successful practices, I've identified three distinct levels of scale for AI initiatives. Each requires different systems and creates different types of value.

Level 1: Personal Leverage

This is where most technical professionals start. You're doing the work yourself but using systems to multiply your impact.

At this level, focus on:

- Methodology documentation
- Value creation frameworks
- Knowledge management systems
- Client engagement tools

Examples from my work with clients at large scale organizations as significant as Barclays and Lloyds: by creating a systematic approach to value assessment, I reduced the time needed for initial client engagements by 60% while increasing identified value by three times.

Level 2: Team Amplification

This level involves building a small team that can multiply your impact without sacrificing quality.

Key elements include:

- Knowledge transfer systems
- Quality control frameworks
- Team development processes
- Value delivery standards

When I led the data transformation at AXA, we scaled from a five-person team to managing 300+ professionals using these exact systems.

Level 3: Practice Scale

This is where individual success becomes organizational impact. You're now creating value through multiple teams and partnerships.

Focus areas:

- Partnership networks
- Innovation systems
- Market positioning
- Value multiplication

At a hugely successful tech scale up company, this approach helped us scale from individual consulting to multi-million dollar engagements across multiple industries.

PATTERN RECOGNITION: Most scaling efforts stall at level 1 or make a disorganized leap to level 3 without building the critical systems and capabilities at level 2. The key is to progress deliberately through each level, building the right foundations before moving to the next.

THREE SUCCESS PATTERNS FROM LEADERS

Organizations that successfully scale AI follow consistent patterns that transcend industry, size, and specific technologies. By understanding and applying these patterns, you can significantly increase your chances of success.

Pattern 1: The Value-First Implementation

Leading organizations start with value rather than technology. They identify specific business outcomes, quantify their impact, and work backwards to the AI capabilities needed to deliver them.

When Barclays Group launched their customer service transformation, they didn't start by selecting AI technologies. They began by identifying specific customer experience improvements that would drive loyalty and retention, quantified the value of these improvements, and then determined what AI capabilities were needed to deliver them.

This approach ensured that technology decisions were driven by business needs rather than the other way around, creating direct alignment between AI capabilities and value creation.

Pattern 2: Ecosystem Thinking

Rather than viewing AI projects as standalone initiatives, leaders understand them as part of broader business ecosystems. They consider how AI systems interact with people, processes, data, and other technologies to create holistic capabilities.

AIG applied this thinking to their underwriting transformation. They mapped the entire ecosystem including underwriters, business rules, data sources, existing systems, and new AI capabilities. This comprehensive view enabled them to identify critical integration points, potential friction areas, and opportunities for unexpected value, creating a solution that worked within the complex reality of their business rather than an idealized environment.

Pattern 3: Capability Building in Parallel

While many organizations focus exclusively on technical implementation, leaders build the organizational capabilities needed for success in parallel with technology deployment.

A global retailer applied this approach to their inventory optimization initiative. While developing their AI solution, they simultaneously redesigned inventory management processes, developed new skills among merchandise planners, created change management programs, and established new performance metrics – ensuring that the organization was ready to leverage new capabilities as soon as they were available.

By following these patterns, organizations create the conditions for successful scaling rather than hoping success will somehow emerge from technology implementation alone.

ACTION STEP: Review your current scaling approach against these three patterns. Identify specific actions you can take to strengthen each pattern in your organization.

THE SUSTAINABLE VALUE FRAMEWORK

One of the biggest challenges in scaling AI is maintaining value creation over time. Unlike traditional technology implementations that deliver value through stable operations, AI systems require continuous evolution to maintain and enhance their impact.

The sustainable value framework addresses this challenge through four interconnected elements.

Value Measurement Systems

You can't manage what you don't measure. Sustainable value requires robust systems for tracking impact across multiple dimensions.
Key components:

- **Direct Impact Metrics:** Revenue generated, costs reduced, and time saved
- **Indirect Impact Metrics:** Customer satisfaction, employee experience, and risk reduction
- **Leading Indicators:** Early signals of potential issues or opportunities
- **Value Realization Tracking:** Mechanisms to ensure anticipated benefits materialize

At a leading retailer, this approach transformed how they managed their AI portfolio. Rather than focusing solely on model performance, they tracked direct financial impact, customer satisfaction improvements, and leading indicators of potential issues. This enabled them to continuously optimize for business value rather than technical metrics.

Feedback Integration Mechanisms

AI systems improve through continuous learning, but this requires structured approaches to capturing and acting on feedback.
Key components:

- **User Feedback Systems:** Mechanisms for capturing insights from those using the system
- **Performance Monitoring:** Automated tracking of key metrics
- **Anomaly Detection:** Systems for identifying unexpected behaviours
- **Improvement Cycles:** Structured processes for implementing enhancements

A global insurance company implemented this approach for their claims-processing AI. They created dedicated feedback channels for claims adjusters, established comprehensive performance monitoring, built anomaly detection that identified emerging issues, and implemented bi-weekly improvement cycles. The resulting system continuously improved rather than degrading over time.

Organizational Learning Loops

Sustainable value isn't just about improving individual AI systems, it's about enhancing the organization's ability to create value with AI over time.
Key components:

- **Knowledge-Sharing Mechanisms:** Ways to distribute insights across teams
- **Best Practice Identification:** Processes for identifying what works
- **Capability Development:** Ongoing skill-building based on emerging needs
- **Cross-Functional Collaboration:** Structures that combine diverse perspectives

At Barclays Group, these learning loops transformed how the teams approached AI development. They established regular knowledge-sharing sessions, documented best practices from successful implementations, created continuous learning programs based on project experiences, and built cross-functional communities of practice. This enabled the organization to become more effective with each AI initiative.

EVOLUTION PLANNING

Finally, sustainable value requires planning for how AI systems will evolve over time in response to changing business needs and technological capabilities.

Key components:

- **Technology Refresh Cycles:** Planned updates to technical components
- **Business Alignment Reviews:** Regular reassessment of business needs
- **Capability Roadmaps:** Forward-looking plans for system evolution
- **Resource Planning:** Ensuring appropriate support for ongoing development

A global pharmaceutical company implemented this approach for a research AI platform. They established regular technology refresh cycles, conducted quarterly business alignment reviews, maintained a three-year capability roadmap, and implemented rolling resource planning, ensuring their system continued to create value as business needs and technological capabilities evolved.

Together, these four elements create a framework for sustainable value creation, enabling AI implementations to deliver increasing rather than diminishing returns over time.

KEY INSIGHT: The most successful AI implementations get better over time rather than degrading, but this doesn't happen by chance. It requires deliberate systems and processes designed to capture feedback, facilitate learning, and enable evolution.

OVERCOMING FIVE COMMON SCALING OBSTACLES

Even with the right frameworks and approaches, companies inevitably encounter obstacles in scaling AI. Let's examine the five most common challenges and proven strategies for overcoming them.

OBSTACLE 1: THE DATA QUALITY GAP

Many AI initiatives succeed as pilots using carefully curated data, only to struggle when faced with the messier reality of enterprise data at scale.

Solution: Implement a Data Readiness Program

Rather than assuming data quality or trying to fix everything at once, successful organizations implement structured programs that:

- Assess data quality for specific use cases rather than more broadly
- Prioritize improvements based on business impact
- Implement "good enough" quality standards for initial deployment
- Create continuous improvement mechanisms for ongoing enhancement

At a global manufacturer, this approach transformed their predictive maintenance initiative. Rather than waiting for perfect data, they defined minimum quality standards, prioritized improvements for high-value assets, and implemented feedback loops that continuously enhanced quality. This enabled them to start creating value quickly while building towards comprehensive coverage.

OBSTACLE 2: THE SKILL MISMATCH

Many organizations find they lack the combination of technical and business skills needed to scale AI effectively.

Solution: Develop Hybrid Talent Strategies

Successful organizations address this challenge through multifaceted approaches:

- Create cross-functional teams that combine technical and business expertise
- Develop training programs that build AI literacy in business roles
- Implement "translation" roles that bridge technical and business domains
- Establish communities of practice that share knowledge across domains

A financial services company implemented this strategy by creating dedicated "AI translator" roles, developing an AI literacy program that reached thousands of employees, establishing cross-functional "AI pods" for key initiatives, and building a community of practice that shared knowledge across the organization. This created the hybrid skills needed for sustainable scaling.

OBSTACLE 3: THE PROCESS DISCONNECT

AI capabilities often fail to deliver value because existing processes aren't designed to leverage them effectively.

Solution: Implement Process-First Design

Rather than treating AI as a bolt-on to existing processes, successful organizations:

- Start with process redesign before implementing technology
- Involve process experts in AI solution design
- Create feedback loops between AI outputs and process execution
- Measure impact on process outcomes rather than just AI performance

A healthcare organization applied this approach to their diagnostic AI implementation. They began by redesigning clinical workflows, involved physicians and nurses in solution design, created tight feedback loops between AI recommendations and clinical decisions, and measured impact on patient outcomes rather than just model accuracy. The result was a seamless integration that delivered value immediately.

OBSTACLE 4: THE GOVERNANCE GAP

As AI deployments grow, many organizations struggle with managing them effectively at scale.

Solution: Establish Balanced Governance Frameworks

Successful organizations create governance frameworks that:

- Balance innovation and control
- Focus on risk management rather than just compliance
- Establish clear accountability for AI systems
- Create mechanisms for continuous monitoring and improvement

A global bank implemented this approach for their AI portfolio. They established a tiered governance model that applied appropriate controls based on risk, created clear accountability for each AI system, implemented comprehensive monitoring capabilities, and established regular review cycles. It enabled them to manage hundreds of AI deployments with appropriate oversight without stifling innovation.

OBSTACLE 5: THE VALUE DILUTION CHALLENGE

Many organizations find that anticipated value gets diluted as initiatives move from concept to implementation.

Solution: Implement Value-Assurance Processes

Leading organizations protect value through structured approaches:

- Document value drivers during initial planning
- Track value-critical decisions throughout implementation
- Establish clear value realization responsibilities
- Create feedback loops that connect actual outcomes to expected outcomes

A telecommunications company took this approach with a customer experience AI. They documented specific value mechanisms during planning, tracked decisions that impacted these mechanisms, assigned clear accountability for value realization, and implemented weekly reviews that compared actual to expected outcomes. This helped ensure that anticipated value materialized rather than dissipating during implementation.

By proactively addressing these common obstacles, organizations can navigate the challenges of scaling AI and maintain momentum towards transformative impact.

WARNING SIGN: If your implementation team can't clearly articulate how design decisions impact value creation, you're likely experiencing value dilution. Create explicit connections between technical choices and business outcomes to prevent this.

PRACTICAL SUCCESS: 3B-FIBREGLASS AND REAL-WORLD VALUE

To bring these principles to life, let's look at how a manufacturing company successfully scaled AI to solve a tangible problem.

A PwC publication introducing the implementation of AI in manufacturing details how 3B-Fibreglass implemented AI to solve fibre breakage issues in their manufacturing process. By setting up cameras to monitor fibre flow and using deep learning computer vision to analyse the data, they identified breaking points and predicted failures in advance, demonstrating successful AI scaling from problem identification to predictive maintenance.

What makes this example powerful is the clear progression from a specific operational challenge (fibre breakage) to a scaled AI solution that created measurable value (predictive maintenance). They didn't start with AI technology, they started with a business problem that had a significant impact on production quality and efficiency. Then they worked backwards to the AI capabilities needed to solve it.

This "problem-back" approach – rather than a "technology-forward" approach – is a common pattern among many organizations that successfully scale AI.

PATTERN RECOGNITION: Successful AI scaling often follows a "problem-back" approach (starting with business problems and working backwards to technology) rather than a "technology-forward" approach (starting with technology and looking for applications).

FROM RESEARCH TO REALITY: GAME-CHANGING PUBLIC SECTOR IMPACT IN GLASGOW

While private sector examples are plentiful, public sector organizations also demonstrate powerful AI scaling success. As documented in the CommBox case study within our research, Glasgow successfully scaled AI-driven surveillance technology to improve public safety.

By leveraging "powerful analytics and predictive tools to gain real-time actionable intelligence," the city was able to identify and address potential threats promptly. The implementation resulted in a 10% reduced crime rate over two years.

This case illustrates how the same principles of successful scaling apply in the public sector, with clear outcomes that benefit citizens directly.

A similar pattern emerges in the World Bank's work in Edo State, Nigeria. In collaboration with Data Science Nigeria, an AI pilot project for citizen feedback focused on monitoring project progress in select locations using a mobile app called DataCrowd, based on AI technology. The four-week pilot in 2020 collected citizens' feedback on the State Employment and Expenditure for Results (SEEFOR) project.

Following positive initial outcomes, the project was scaled up for implementation in three additional states, demonstrating how a well-designed pilot can lead to broader adoption and impact.

KEY INSIGHT: Successful AI scaling follows similar patterns across public and private sectors. The key is starting with clear problems to solve, building appropriate organizational capabilities, and measuring impact in terms that matter to stakeholders.

NEXT STEPS: THE 30-DAY SCALING ACTION PLAN

Translating these concepts into action starts with specific steps you can take over a one-month period to accelerate your scaling journey. Here's a concrete plan to move forward:

Week 1: Opportunity Assessment:

- Map current AI initiatives against strategic priorities
- Identify three to five high-potential scaling opportunities
- Assess readiness using the scale-ready framework
- Select an initial focus area for scaling

Week 2: Value Framework Development:

- Define specific value metrics for your focus area
- Establish baseline measurements for comparison
- Create a value-tracking mechanism
- Align key stakeholders around value expectations

Week 3: Obstacle Analysis:

- Identify potential scaling obstacles for your focus area
- Develop mitigation strategies for each obstacle
- Assess organizational readiness for implementation
- Create a risk-management plan

Week 4: Implementation Planning:

- Develop a 90-day implementation roadmap
- Define clear roles and responsibilities
- Identify required resources and support
- Establish success metrics and review cycles

This structured approach converts concepts into concrete action, creating momentum towards sustainable scaling rather than just theoretical understanding.

QUICK WIN: Start with a single high-potential AI use case that has clear strategic alignment and create a detailed value framework for it, including specific metrics and measurement approaches. This creates a foundation for successful scaling even if you're not ready to implement the full action plan.

THE VALUE ALIGNMENT LOOP: A PRACTICAL FRAMEWORK

To help bridge the gap between strategic intent and practical execution, I've developed a value alignment loop framework. This simple but powerful approach helps organizations maintain focus on value throughout the scaling process:

1. **Declare:** Clearly articulate the specific value you intend to create
2. **Design:** Build technical solutions explicitly aligned with that value
3. **Deploy:** Implement the solution with value realization mechanisms
4. **Diagnose:** Measure actual versus expected value and identify gaps
5. **Repeat:** Use insights to refine value targets and improve solutions

This loop creates a continuous cycle of value creation and refinement, ensuring that AI initiatives stay connected to business outcomes throughout their lifecycle.

ACTION STEP: Apply the value alignment loop to an existing AI initiative by explicitly documenting your current state for each step. Identify where the loop is broken or weak – and create specific actions to strengthen those areas.

LOOKING AHEAD: USING THE AI VALUE MAPPING CANVAS

As you move forward on your scaling journey, the AI value mapping canvas (available in the downloadable templates mentioned at the end of the book) provides a powerful tool for aligning AI initiatives with business value. This visual framework helps you connect technical capabilities directly to strategic outcomes, creating a shared understanding among technical and business stakeholders.

The canvas includes sections for:

- Strategic priorities and objectives
- Key stakeholders and their needs
- Value creation mechanisms
- Technical capabilities and requirements
- Success metrics and measurement approaches
- Implementation considerations and dependencies

By completing this canvas for each scaling initiative, you create a foundation for successful implementation that maintains focus on value creation throughout the process.

REFLECTION QUESTIONS: What tools or frameworks do you currently use to align AI initiatives with business value? How might a structured approach like the AI value mapping canvas enhance your ability to create and maintain this alignment?

CONCLUSION: BEYOND TECHNOLOGY TO TRANSFORMATION

As we conclude this chapter, I want to return to the story I shared at the beginning, to the moment when I couldn't answer the CEO's question about strategic impact. It's a moment many technical leaders experience, and it highlights a fundamental truth about scaling AI successfully.

From Pilot to Scale

The gap between pilot success and scaled impact isn't primarily technical, it is strategic. It's not about building better models or more sophisticated algorithms. It's about connecting technical capabilities directly to business transformation.

The frameworks, approaches, and examples in this chapter provide a roadmap for bridging this gap, for moving from impressive pilots to transformative impact. They represent lessons learned through both success and failure across dozens of organizations in multiple industries.

But perhaps the most important insight is this: scaling AI successfully requires more than just good technology or even good processes. It requires a fundamental shift in how we think about the relationship between technology and business value: a shift from viewing AI as a technical project to understanding it as a transformative capability that can reshape how organizations operate and compete.

When we make this shift, the question is no longer, "How do we scale our AI pilots?" It now is, "How do we leverage AI to transform our business?" And answering that question opens up possibilities far beyond what most organizations initially imagine when they begin their AI journey.

As Tesla has demonstrated through their mission-driven AI implementation approach, when AI capabilities are embedded into the core of an organization's mission rather than treated as separate initiatives, scaling becomes less about technical challenges and more about strategic integration. This mindset shift allows companies to overcome what my research identifies as "the invisible middle" – the dangerous zone between vision and execution where promising AI initiatives often fail.

Your journey from pilot to scale is unique, but the principles, frameworks, and lessons in this chapter provide a road-tested map for navigating this complex terrain successfully. Whether you're leading a global enterprise or a small team, these approaches can help you create sustainable AI value that drives true business transformation. The difference between organizations that struggle with scaling and those that succeed isn't technical sophistication, it is strategic clarity, organizational alignment, and disciplined execution across the full spectrum of people, process, and technology.

KEY INSIGHT: The most successful AI scaling efforts start with the question "How can AI transform our business?" rather than "How can we implement AI?" This shift in perspective changes everything about how organizations approach scaling and dramatically increases the chances of success.

In the next chapter, we'll explore how to secure executive buy-in and board support for AI initiatives, building on the value-focused approach we've discussed here. We'll examine specific techniques for communicating AI value in ways that resonate with senior leaders and drive organizational alignment around your vision.

7 Securing Executive Buy-In and Board Support

Chapter Roadmap

This chapter explores the art and science of securing executive support for AI initiatives. To that end, it will cover understanding the unique priorities of different C-suite roles, creating compelling AI narratives that resonate with leadership, building rigorous business cases tailored to executive audiences and demonstrating early value to build and maintain credibility. I also highlight the benefits of developing effective board-level dashboards for oversight.

THE EXECUTIVE BUY-IN PARADOX

Let's begin by exploring the fundamental challenge that derails many promising AI initiatives: a disconnect that happens when technical teams build impressive AI capabilities but struggle to translate their work into terms that resonate with executive leaders and board members.

Time and time again, I've seen examples of the following scenario play out in countless boardrooms.

The room fell silent as the CEO's question hung in the air.

"Your AI project sounds impressive. But how does it help us achieve our strategic objectives?"

The data science team exchanged nervous glances. For 20 minutes, they had meticulously explained their sophisticated approach to customer segmentation – the algorithms, the accuracy metrics and the technical innovations. Yet somehow, they had completely missed a crucial element: the outcomes the executive team actually cared about.

This happens daily in organizations worldwide – and with serious results, including limited sponsorship, insufficient resources, and ultimately, unrealized potential.

KEY INSIGHT: Understanding what executives and board members care about – and not only building AI initiatives with such outcomes in mind but also framing them in such terms – is perhaps the most underappreciated success factor in enterprise AI.

UNDERSTANDING EXECUTIVE AND BOARD PRIORITIES

THE C-SUITE LENS: WHAT DIFFERENT LEADERS REALLY CARE ABOUT

PATTERN RECOGNITION: Each C-suite role evaluates AI through a distinct lens. Recognizing these perspectives allows you to pay attention to appropriate capabilities and tailor your communications effectively.

The CEO Perspective

CEOs evaluate AI through strategic and market-focused lenses:

- How does this strengthen our competitive position?
- Does this create sustainable differentiation or just a temporary advantage?
- How will this affect our overall business model and growth trajectory?

Securing Executive Buy-In and Board Support

- What strategic risks does this address or potentially create?
- How does this align with our organizational purpose and vision?

Real-World Example: A technology company CEO was initially unimpressed by technical descriptions of a new AI system. When the team reframed the discussion around how the capability would create a 12-to-18-month competitive advantage in a rapidly evolving market, his interest and support increased dramatically.

The CFO Perspective

Chief Finance Officer (CFOs) bring financial discipline and investment rigour to AI discussions:

- What specific financial returns can we expect and when?
- How reliable are these projections compared to other investments?
- What is the total cost beyond the initial implementation?
- How will this affect our financial metrics and Key Performance Indicators (KPIs)?
- What financial risks does this create or mitigate?

Real-World Example: At Microsoft, CFO Amy Hood demonstrated how financial leadership influences AI initiatives. As Tom Dotan's research reveals, Hood balanced ambitious AI investments with fiscal responsibility, emphasizing "meticulous financial oversight to ensure AI initiatives align with the company's strategic objectives while maintaining financial discipline."

REFLECTION QUESTIONS: How are you currently framing your AI initiatives for financial leaders? Are you speaking the language of ROI, risk management, and financial discipline?

The CIO/CTO Perspective

Technology leaders focus on integration, architecture, and capability building:

- How does this fit with our overall technology strategy?
- What implications does this have for our technical architecture?
- How will we support and maintain this capability over the long term?
- What skills and resources are required for implementation and operation?
- How does this affect our security, compliance, and resilience posture?

Real-World Example: A retail Chief Information Officer (CIO) initially resisted an AI pricing initiative that seemed technically elegant but would create significant integration challenges with existing systems. When the team revised their approach to better align the solution with the organization's technical architecture, his opposition transformed into active championship.

The COO Perspective

Operations leaders evaluate AI through a focus on execution and efficiency.

- How will this affect our core operational processes?
- What implementation challenges and risks should we anticipate?
- How will this impact operational metrics and performance?
- What organizational changes will be required?
- How will this integrate with our current operational systems?

Real-World Example: A logistics Chief Operating Officer (COO) became an enthusiastic supporter of an AI route optimization system only after the team demonstrated how it would integrate with existing driver workflows and operational processes. The technical capabilities were less important to her than the practical implementation considerations.

The CMO Perspective

Marketing leaders assess AI through customer and market impact lenses:

- How will this affect customer experience and engagement?
- What competitive advantage does this create with customers?
- How does this support brand positioning and promises?
- What market insights or capabilities does this enable?
- How will this impact marketing effectiveness and efficiency?

Real-World Example: A financial services Chief Marketing Officer (CMO) initially dismissed an AI customer analytics initiative until the team reframed it around improving personalized offers and reducing customer attrition. This connection to her priority concerns transformed her from sceptic to champion.

The CHRO Perspective

Human resource leaders focus on talent implications and organizational impact:

- How will this affect roles, skills, and workforce planning?
- What talent acquisition or development needs will this create?
- How will employees react, and what change management is required?
- What implications does this have for our culture and values?
- How should performance management evolve to support this initiative?

Real-World Example: A healthcare Chief Human Resource Officer (CHRO) became a crucial ally for a clinical decision support AI system only after the team directly addressed physician reactions, workflow changes, and training implications. His support proved essential in navigating the complex human dimensions of implementation.

The CRO/CLO Perspective

Risk and legal leaders evaluate AI through compliance and governance lenses:

- What regulatory or compliance implications does this create?
- How will we ensure appropriate oversight and control?
- What new risks might this introduce to the organization?
- How will we protect against bias, privacy, or ethical concerns?
- What documentation and evidence will we maintain?

Real-World Case: Air Canada's chatbot liability case provides a cautionary tale of inadequate AI governance. When their AI-powered virtual assistant provided incorrect fare information to a passenger, the airline initially claimed it couldn't be held liable. However, a Canadian tribunal ruled against the airline, determining "that the company did not take reasonable care to ensure the accuracy of its chatbot." This case demonstrates the real financial and reputational consequences of inadequate AI governance and oversight.

WARNING SIGN: If your AI initiative lacks clear answers to the compliance and risk concerns of legal and risk officers, you're likely to face significant resistance.

BOARD-LEVEL CONCERNS AND EXPECTATIONS

Beyond the C-suite, board members bring additional perspectives that must be understood and addressed.

Board members typically focus on five key dimensions:

Strategic Alignment and Value

- How does this initiative advance our strategic objectives?
- Is this creating sustainable competitive advantage?
- Does this position us appropriately for industry and societal evolution?
- What strategic alternatives were considered?
- How does this compare to competitor approaches?

Risk Oversight and Governance

- What new risks does this create and how are they managed?
- Is appropriate governance in place for responsible implementation?
- How are we addressing regulatory and compliance considerations?
- What reputational implications should we consider?
- How does this affect our overall risk profile?

Financial Performance and Investment

- What is the expected return and investment requirement?
- How reliable are these projections and what validation exists?
- How does this affect our short- and long-term financial outlook?
- What is the opportunity cost of this investment?
- How will we measure and track financial impact?

Organizational Capability and Readiness

- Do we have the right leadership and talent for successful implementation?
- What organizational changes or challenges should we anticipate?
- How does this affect our culture and operating model?
- What new capabilities will this create or require?
- Are we properly resourced for successful execution?

Market and Customer Impact

- How will customers and markets respond to this initiative?
- What competitive implications should we anticipate?
- Does this strengthen our market position and how?
- What external validation supports our approach?
- How does this affect our customer relationships and brand?

Real-World Example: A global retailer successfully secured board support for their AI transformation by explicitly addressing these five dimensions rather than focusing primarily on technological innovation. Taking a comprehensive approach, the team anticipated and addressed board concerns before they were raised, building confidence that enabled appropriate risk-taking.

The Geopolitical Tension in AI Governance

Multinational organizations seeking executive and board support face a significant complication: fundamental differences in AI governance approaches among major global powers. This tension creates strategic uncertainties that executives must navigate.

The United States and the United Kingdom prioritize economic growth and technological leadership with the assumption that strict regulations might impede innovation. Countries like France,

China, and India, on the other hand, advocate for more robust ethical standards and inclusive governance that emphasize societal implications.

As Dr. Joerg Storm, founder of Digital Storm Weekly, explains: "On the one hand, [regulation] can help AI and promote public trust, transparency, and also support the ethical development of AI. On the other hand, it can hinder AI due to strict regulations, and this could slow down innovation and also stifle development in certain regions."

These perspectives came to a clash in February 2025 when the Global AI Action Summit in Paris saw 58 countries sign a joint declaration on inclusive AI governance – with the United States and the United Kingdom notably absent. The United States expressed concerns that "the agreement did not sufficiently address national security issues," while the United Kingdom cited "the need for clearer global governance frameworks."

For executives, these divergent approaches complicate the establishment of consistent internal AI standards that are valid across the jurisdictions in which their organizations operate. Differences in rules and regulations – both those currently in effect and those in development – must be factored into strategic decision-making.

KEY INSIGHT: Research reveals significant differences in regulatory approaches. The EU AI Act can result in penalties of up to 35 million euros or 7% of a company's worldwide annual turnover for non-compliance with Article 5, while American approaches vary significantly by state.

Aligning AI Initiatives with Executive Incentives

Perhaps the most overlooked aspect of executive buy-in involves understanding – and alignment with – incentive structures. Since incentive structures signal executives' personal objectives and compensation targets, AI initiatives advancing these goals naturally receive stronger support than those that appear disconnected from these priorities.

ACTION STEP: Create an executive incentive alignment map to connect your AI initiative to the personal objectives of each key leader.

Real-World Example: A pharmaceutical company struggled to gain executive sponsorship for an R&D AI initiative until they explicitly mapped how it would impact the executive team's annual objectives (see Table 7.1).

This explicit alignment with existing incentives transformed executive engagement from polite interest to active championship. The initiative was no longer competing with executives' priorities but directly advancing them.

Karen Stroup, Chief Digital Officer at WEX, reinforces this point, sharing how the team secured executive buy-in by creating shared goals around AI across the entire executive leadership team. Stroup explains that this alignment was "incredibly important" because "we win together or we lose together," emphasizing how cross-functional executive support was essential to AI success.

TABLE 7.1
Mapping of Executive Team's Objectives

Executive	Annual Objective	AI Initiative Projected Impact
CEO	15% revenue growth	Accelerated product pipeline contributing 3% to growth target
CFO	2% margin improvement	$15M annual cost reduction through more efficient trials
CIO	Digital transformation milestones	Key capability building for overall transformation
R&D Head	Pipeline advancement targets	30% faster candidate evaluation
Commercial Lead	Launch effectiveness	Earlier market insights improving positioning

SPEAKING THE LANGUAGE OF THE BOARDROOM

The final element of understanding executive and board priorities involves communicating in a language that resonates with leadership audiences.

Technical teams often default to expertise-centred communication, focusing on aspects they find most interesting or innovative. This approach typically fails with executive audiences who need business-centred communication focused on outcomes rather than methods.

Peter Altabef, Chair and CEO of Unisys, provides valuable insight on this communication challenge: "The CIOs now have stuff in their domain that is of great interest to the CEOs and CMOs and the CCOs and the CFOs. So, this is the time for CIOs to shine. This is the time not to be defensive. This is the time to show your stuff." He continues, "One of the great skills of a superior CIO is not to get trapped in engineering speak and not to get trapped in functions, but actually to communicate in a way that a non-expert, the rest of the C-levels, can understand the value and opportunity you are providing."

Real-World Transformation: A healthcare AI team transformed their executive communications by shifting from technical to business language (see Table 7.2)

This translation wasn't about simplifying complex concepts but reframing them around business impact rather than technical implementation. The shift dramatically improved executive engagement and support.

QUICK WIN: Review your last AI presentation and identify three technical statements that could be reframed as business outcomes.

THE AI NARRATIVE FRAMEWORK

Beyond understanding priorities, securing executive buy-in requires a compelling narrative that positions AI initiatives as critical business opportunities and solutions rather than merely interesting technology projects.

FRAMEWORK: The Four-Part AI Narrative

A compelling AI narrative for executives contains four essential elements:

1. **The Tension:** A compelling problem or opportunity that demands attention
2. **The Shift:** How AI changes the equation in ways traditional approaches cannot

---

TABLE 7.2
Shifting from Technical to Business Language

Technical Language	Business Language
"Our model achieved 94% accuracy with significantly improved AUC compared to baseline approaches."	"This capability reduces misdiagnosis by 47%, avoiding approximately $42M in unnecessary treatments annually."
"We've implemented a novel attention-based architecture with reinforcement learning components."	"The AI system learns continuously from clinical outcomes, consistently improving its diagnostic recommendations as more cases are processed."
"The system processes structured and unstructured data through our proprietary pipeline."	"The system integrates insights from patient records, clinical notes, and research literature that would take physicians hours to review manually."
"We've achieved substantial dimensionality reduction while maintaining feature importance."	"We've simplified the decision process while maintaining transparency about key factors."

3. **The Impact:** Tangible outcomes and metrics relevant to executive priorities
4. **The Ask:** Clear next steps for decision-makers

Let's explore each component in detail.

THE TENSION: FRAMING A COMPELLING PROBLEM

Effective narratives begin by establishing tension, ideally around a compelling problem or opportunity that demands attention. This tension creates the "why now" urgency that moves AI initiatives from interesting options to strategic imperatives.

> **Real-World Example:** A financial services organization struggled to gain executive support for an AI risk management initiative before taking an approach focused on tension.
> **Weak Tension:** "We should implement AI for risk management."
> **Strong Tension:** "Our current risk identification approach misses 27% of emerging credit risks until they've already impacted performance. Meanwhile, competitors that use AI-enhanced approaches are identifying these same risks 60–90 days earlier, allowing them to mitigate exposure while we're still discovering the problem."

This tension-centred opening immediately established both problem significance and competitive implications, creating urgency that their previous technology-centred approach lacked.

Andrew McKinshnie, Senior NLP Engineer, identifies why tension is so important: "The biggest challenge is usually getting buy-in from all relevant parties" due to "distrust in AI, upfront costs," and "lack of education around AI." Your narrative must address these sources of resistance directly.

Effective tension statements typically include:

- Specific business impact of the current situation
- Clear metrics establishing significance
- Competitive or market context creating urgency
- Implicit cost of inaction
- Connection to strategic priorities or pain points

The most powerful tension statements aren't generic industry trends but specific organizational challenges that executives already recognize but haven't connected to AI potential.

REFLECTION QUESTIONS: What specific business problem or opportunity creates urgency for your AI initiative? How can you quantify its significance?

THE SHIFT: HOW AI CHANGES THE EQUATION

Once tension establishes why action is necessary, the narrative must explain how AI specifically changes the equation. This isn't about technology capabilities in isolation but how these capabilities address the established tension in ways traditional approaches cannot.

> **Real-World Example:** A manufacturing company effectively illustrated this shift in their predictive maintenance initiative.
> **Weak Shift:** "AI can analyse equipment data and predict potential failures."
> **Strong Shift:** "Traditional maintenance approaches face fundamental limitations because they rely on either fixed schedules that waste resources on healthy equipment or reactive responses that create expensive downtime. AI fundamentally changes this equation by identifying specific failure patterns days or weeks before they cause operational impact, creating a new option: precisely targeted intervention exactly when needed."

This explanation focused not on AI capabilities themselves but on how they created new possibilities unavailable through traditional approaches. It positioned AI as a strategy enabler rather than merely a technology implementation.

Effective shift explanations typically include:

- Limitations of current approaches
- Specific mechanisms through which AI creates new possibilities
- Practical examples illustrating the difference
- Clear contrast with traditional alternatives
- Connection to established tension points

The most compelling shift explanations help executives see AI not as incremental improvement but as fundamental transformation of what's possible.

THE IMPACT: TANGIBLE OUTCOMES AND METRICS

With tension establishing urgency and shift explaining the solution approach, the narrative must then quantify expected impact in terms directly relevant to executive priorities.

> **Real-World Example:** A retail organization transformed executive support by shifting from capability-focused to impact-focused communication:
>
> **Weak Impact:** "Our AI system will provide personalized product recommendations with high accuracy."
>
> **Strong Impact:** "Based on pilot results and industry benchmarks, this capability is projected to deliver three specific outcomes: average basket size improvements of 15–20%, inventory efficiency gains of 20–25%, and net promoter score (NPS) increases of 3–5 points by creating more satisfying customer experiences. These projections reflect conservative estimates based on similar implementations in our industry."

This impact statement connected directly to revenue growth, operational efficiency, and customer experience – core executive priorities – rather than focusing on technical performance metrics.

Meta provides a compelling case study in communicating AI ROI to executives: "Meta's successful communication of AI ROI represents a noteworthy example. The company experienced robust growth in advertising sales due to AI-powered technology, with earnings exceeding market expectations and driving a 7% increase in shares (with shares up 34% year to date). Additionally, Meta achieved 22% revenue growth through its digital advertising income, where AI algorithms optimize ad targeting, bidding, and content personalization to maximize advertiser ROI. By focusing on these concrete financial outcomes, Meta successfully demonstrated the ROI of AI initiatives to shareholders and market analysts.

Effective impact statements typically include:

- Multiple value dimensions relevant to different stakeholders
- Specific, measurable outcomes targets with timeframes
- Connection to established strategic metrics and KPIs
- Reasonable ranges reflecting implementation variables
- Comparison to relevant benchmarks or alternatives

The most compelling impact statements balance ambition with credibility, providing specific projections supported by relevant evidence rather than vague assertions or unrealistic promises.

The Ask: Clear Next Steps for Decision-Makers

The final narrative element provides clear direction on what executives should do next. Many presentations fail at this critical juncture, ending with implicit rather than explicit requests or asking for decisions without providing necessary context.

> **Real-World Example:** A healthcare organization effectively concluded their AI narrative with a structured ask.
> **Weak Ask:** "We'd like your support for this initiative."
> **Strong Ask:** "We're seeking three specific commitments to move this initiative forward:
> 1. Executive sponsorship from the chief medical officer to signal clinical leadership
> 2. $1.2M initial funding for phase one implementation in cardiology and oncology
> 3. Formation of a clinical governance committee to oversee implementation and expansion

We've prepared detailed planning for each element and would appreciate your decision by March 15th to maintain our implementation timeline."

This structured approach clarified exactly what support was needed, by whom, and when – enabling clear decision-making rather than vague endorsement.

Effective asks typically include:

- Specific, actionable requests with clear ownership
- Resource requirements with appropriate justification
- Decision timelines connected to implementation needs
- Governance or oversight recommendations
- Next-step options for different decision outcomes

The most successful asks make it easy for executives to say yes by providing complete information for decision-making rather than requiring additional clarification or elaboration.

ACTION STEP: Download the AI narrative framework template from the book's online resource centre to create your own compelling executive presentation.

Fill-in-the-Blank Narrative Builder Template

To help create effective executive narratives, use this structured template that guides the development of each narrative element:

Tension Statement

[Specific business challenge] is creating [quantified negative impact]. This situation [competitive or market context creating urgency], while our [strategic priority] requires [desired outcome]. The gap between our current approach and what's needed represents [cost of inaction].

Shift Explanation

Traditional approaches to this challenge face [specific limitations] because [root cause explanation]. AI changes this equation by [key capability difference], enabling [new possibility]. Unlike [traditional alternative], this approach [key differentiating factor] while [addressing potential concern].

Impact Projection

This initiative will deliver [timeframe] impact across multiple dimensions:

- [First value dimension]: [Specific, quantified outcome] leading to [business result]

Securing Executive Buy-In and Board Support

- [Second value dimension]: [Specific, quantified outcome] leading to [business result]
- [Third value dimension]: [Specific, quantified outcome] leading to [business result]

These projections are based on [evidence source] and compare favourably to [relevant benchmark].

Structured Ask

To move this initiative forward, we're seeking:

1. [Specific request] from [owner] by [deadline] to [purpose]
2. [Specific request] from [owner] by [deadline] to [purpose]
3. [Specific request] from [owner] by [deadline] to [purpose]

The timeline for these decisions aligns with [implementation consideration] and enables [desired outcome].

Organizations that systematically develop each narrative element create compelling executive communications that transform support from passive endorsement to active championship.

BEYOND FILLING IN THE BLANKS

While the four-part structure (tension, shift, impact, ask) provides a foundation, effective executive narratives require contextual adaptation rather than template completion.

NARRATIVE ADAPTATION GUIDELINES

For Different Organizational Contexts:

- **Startup Environments:** Emphasize speed and competitive advantage; minimize governance complexity
- **Large Enterprises:** Address integration challenges and risk management; emphasize staged approaches
- **Public Sector:** Focus on citizen impact and transparency; address political sustainability
- **Regulated Industries:** Lead with compliance and risk mitigation; demonstrate governance maturity

For Different Executive Audiences:

- **Innovation-Focused Leaders:** Emphasize competitive differentiation and market opportunity
- **Risk-Averse Leaders:** Lead with proven approaches and staged implementation
- **Financially Driven Leaders:** Quantify returns with conservative projections and clear attribution
- **Operationally-Focused Leaders:** Demonstrate workflow integration and change management

CONTEXT-SENSITIVE FRAMEWORK QUESTIONS

Rather than a fill-in-the-blank template, consider these diagnostic questions:

Tension Development:

- What keeps your specific executives awake at night?
- Which competitors are making moves that concern your leadership?

- What regulatory or market pressures create urgency for your organization?
- Where are current approaches failing in ways that resonate with leadership priorities?

Shift Articulation:

- What unique constraints does your organization face that AI specifically addresses?
- How does your competitive position influence the urgency of AI adoption?
- What organizational capabilities make certain AI approaches more viable for you?
- Which traditional approaches have failed specifically in your context?

Impact Calibration:

- What evidence sources will your executives find most credible?
- Which metrics already drive executive attention and decision-making?
- How conservative should projections be given your organization's track record with technology initiatives?
- What secondary benefits might matter more than primary financial returns?

Ask Contextualization:

- What decision-making processes and timelines characterize your organization?
- Which governance structures need to be engaged for different types of requests?
- What resource constraints or budget cycles influence timing and sizing of requests?
- Which executives have the authority and motivation to champion different aspects of your initiative?

Dynamic Narrative Examples

Conservative Financial Services Approach: "While our risk management processes have served us well, emerging regulatory requirements and competitive pressures from fintech companies are exposing gaps that traditional approaches cannot address within acceptable timeframes or resource constraints…"

Aggressive Technology Startup Approach: "Our current customer acquisition costs are 40% higher than our main competitor, who deployed AI-powered targeting six months ago. Every month we delay gives them additional market share that becomes exponentially harder to recapture…"

Risk-Aware Healthcare Context: "Patient safety incidents related to diagnostic delays have increased 15% over two years, while our peer institutions implementing clinical decision support have seen 23% reductions. The clinical evidence base now supports AI augmentation as a patient safety imperative rather than an experimental technology…"

Iteration and Refinement Process

Rather than completing a template once, effective narratives emerge through:

1. **Stakeholder Feedback Loops:** Test narrative elements with trusted advisors
2. **Pilot Presentations:** Refine based on executive questions and engagement
3. **Competitive Intelligence:** Adjust based on market developments
4. **Implementation Learning:** Update projections based on early results

KEY INSIGHT: The most compelling executive narratives feel custom-crafted for the specific organization, audience, and moment rather than adapted from a generic template.

Securing Executive Buy-In and Board Support

The principle: Use the four-part structure as a diagnostic framework to ensure completeness, but develop each element through deep organizational context rather than template completion.

VISUAL: BOARDROOM STORYTELLING STACK

Beyond the narrative structure, visual presentation plays a crucial role in executive communication. The "storytelling stack" approach creates visually compelling executive presentations:

Level 1: The One-Page Summary

Every executive presentation should begin with a single slide capturing the complete narrative in condensed form – the essential information executives need even if they read nothing else. This includes tension, shift, impact, and ask in highly condensed form.

Level 2: The Executive Narrative

The core presentation expands each narrative element with supporting evidence, examples, and context. This level typically includes 7–10 slides that collectively tell a complete and compelling story.

Level 3: Supporting Detail

Additional slides provide deeper information on specific aspects: technical approach, implementation plan, risk assessment, etc. These aren't presented sequentially but are available as needed to address specific questions or concerns.

Level 4: Technical Appendix

Detailed technical information supports the presentation but remains in the appendix unless specifically requested. This level satisfies technical diligence without overwhelming the core narrative.

This progressive disclosure approach ensures executives can engage at appropriate depth based on their interests and needs.

A pharmaceutical company using this approach found their AI initiatives receiving more substantive board discussion and more decisive support compared to traditional presentations that mixed narrative and technical detail without clear structure.

QUICK WIN: Create a one-page executive summary that captures your entire AI initiative in a single view using the four-part narrative framework.

BUILDING THE EXECUTIVE BUSINESS CASE

While a compelling narrative creates engagement, rigorous business cases create confidence. Executive and board support ultimately depends on demonstrating that AI investments represent sound business decisions rather than merely interesting technical opportunities.

FINANCIAL FRAMING FOR DIFFERENT LEADERSHIP ROLES

PATTERN RECOGNITION: Different leaders evaluate impact through different lenses. Tailoring your financial presentation to each audience dramatically increases engagement.

Effective business cases reflect the recognition that different leaders evaluate impact through different lenses. Rather than creating a single financial perspective, sophisticated approaches address multiple dimensions.

Real-World Example: A manufacturing organization successfully secured cross-functional executive support by presenting their AI quality initiative through multiple financial frames:

For the CFO: ROI and P&L Impact

- Implementation costs: $3.2M over 18 months
- Annual cost savings: $5.7M from reduced scrap and rework
- Margin improvement: 0.8 percentage points on affected product lines
- Payback period: 11 months
- Three-year ROI: 437%

For the CEO: Strategic Value and Market Position

- Revenue protection: $14M annually from reduced customer defections
- Premium positioning: Supporting 12% price premium through quality differentiation
- Market share opportunity: 2.3 percentage point increase in targeted segments
- Competitive necessity: Matching capabilities already deployed by two major competitors

For the COO: Operational Efficiency and Performance

- Throughput improvement: 9% increase through reduced rework cycles
- Resource reallocation: 34 FTEs shifted from inspection to value-added activities
- Process stability: 47% reduction in quality-related disruptions
- Capability building: Foundation for broader operational excellence initiative

For the CMO: Customer and Brand Impact

- Customer satisfaction: 18-point NPS improvement with priority accounts
- Quality perception: Supporting brand premium positioning
- Complaint reduction: 73% decrease in quality-related issues
- Relationship enhancement: Proactive quality communication capabilities

This multi-dimensional approach ensured each leader could connect the initiative to their specific priorities and evaluation frameworks.

ROI Models for Different Types of AI Initiatives

Research shows that different AI applications require fundamentally different ROI approaches tailored to their specific value-creation mechanisms.

For Automation Initiatives

Automation-focused AI typically delivers value through efficiency, consistency, and scale. Appropriate ROI models include:

- Cost displacement analysis: Quantifying reduced labour and operational costs
- Throughput enhancement: Measuring increased processing capacity and speed
- Error reduction valuation: Calculating savings from decreased exception-handling
- Scalability economics: Demonstrating cost advantages due to increased volumes

Real-World Example: A financial services organization effectively used cost displacement analysis for their document processing AI, demonstrating 74% reduction in processing costs with specific department-level impacts that resonated with affected executives.

For Augmentation Initiatives

Augmentation-focused AI delivers value by enhancing human capabilities and decision quality. Appropriate ROI models include:

Securing Executive Buy-In and Board Support

- Decision quality impact: Measuring improved outcomes from enhanced decisions
- Productivity enhancement: Quantifying increased output per employee
- Effectiveness improvement: Calculating impact of better resource targeting
- Opportunity capture: Measuring previously inaccessible opportunities

Real-World Example: A pharmaceutical company used decision quality impact modelling for research augmentation AI, demonstrating 23% improvement in candidate selection accuracy, worth approximately $42M annually in reduced failed trials.

For Insight-Generating Initiatives

Insight-focused AI creates value through revealing previously hidden patterns and opportunities. Appropriate ROI models include:

- Revenue enhancement: Quantifying increased sales and services from better targeting
- Risk reduction: Calculating avoided losses from detecting potential issues earlier
- Opportunity identification: Measuring new revenue streams or increased market reach
- Speed advantage: Valuing accelerated market or operational responses

Real-World Example: A telecommunications company used risk reduction modelling for network optimization AI, demonstrating $27M annual savings from predicted and prevented outages that would otherwise impact both costs and customer experience.

For Personalization Initiatives

Personalization-focused AI delivers value through improved customer experiences and relationships. Appropriate ROI models include:

- Conversion improvement: Measuring increased purchase completion
- Basket enhancement: Quantifying increased transaction value
- Retention impact: Calculating reduced churn and extended lifetime value
- Premium pricing: Measuring willingness to pay for personalized experiences

Real-World Example: When a retail organization used basket enhancement and retention impact modelling for a recommendation engine, the demonstrated results included a 14% increase in average order value and 8% improvement in customer retention with specific financial projections.

ACTION STEP: Identify which of these four AI types best characterizes your initiative, then select the most appropriate ROI model from that category.

BALANCING SHORT- AND LONG-TERM VALUE CONSIDERATIONS

One of the most challenging aspects of AI business cases involves balancing immediate returns with longer-term strategic value. Executives need both perspectives to make informed decisions.

Real-World Example: A healthcare organization effectively balanced these timeframes for a clinical AI business case:

Short-Term Value (0–12 months)

- Operational efficiency: $4.2M annual savings through reduced administrative burden
- Quality improvement: 27% reduction in documentation errors
- Utilization optimization: 9% reduction in unnecessary diagnostic testing

Medium-Term Value (1–2 years)

- Clinical outcomes: 12% improvement in treatment plan adherence
- Care coordination: 31% reduction in care gaps for chronic conditions
- Provider satisfaction: 18% improvement in physician experience scores

Long-Term Value (2+ years)

- Care model transformation: Enabling proactive intervention models
- Precision medicine: Foundation for personalized treatment protocols
- Competitive differentiation: Clinical excellence positioning in market

Strategic Optionality

- Capability building for value-based care transitions
- Platform for additional AI applications in clinical workflows
- Data asset creation supporting research and innovation

This balanced perspective enabled executives to see both immediate returns justifying investment – and strategic value creating sustained competitive advantage.

WARNING SIGN: If your AI business case only addresses immediate financial returns without considering longer-term strategic value, you're likely undervaluing your initiative and making it vulnerable to short-term cost-cutting.

BUILDING FLEXIBILITY INTO VALUE PROJECTIONS

AI initiatives involve inherent uncertainty that traditional business cases often inadequately address. Effective approaches build flexibility rather than false precision into value projections.

Real-World Example: A retail organization implemented three specific approaches in their customer analytics business case:

Scenario-Based Projections

Rather than single-point estimates, they provided three scenarios with clear drivers (see Table 7.3).

This approach acknowledged uncertainty while providing structured understanding of potential outcomes and their drivers.

TABLE 7.3
Scenarios, Drivers, and Probabilities

Scenario	Revenue Impact	Key Drivers	Probability
Conservative	+4.8%	Basic personalization adoption Minimal customer behaviour change Limited cross-selling impact	30%
Base Case	+8.3%	Moderate personalization adoption Measurable behaviour changes Effective cross-selling to receptive segments	50%
Accelerated	+12.7%	Strong personalization adoption Significant behaviour changes Comprehensive cross-selling impact	20%

Staged Investment Approach

Rather than pursuing a single, large investment decision, the team structured a phased approach with defined stage gates:

- **Phase 1:** $1.2M investment in foundation capabilities and initial use cases
- **Decision Point 1:** Evaluate metrics against targets before proceeding
- **Phase 2:** $2.7M investment in expanded capabilities based on phase 1 results
- **Decision Point 2:** Assess broader business impact before final expansion
- **Phase 3:** $3.5M investment in enterprise deployment and advanced features

This staging created flexibility to adjust – or potentially discontinue – investment based on actual results rather than committing the full amount upfront.

Option Value Recognition

Beyond direct financial returns, the team explicitly valued strategic options created by the investment:

- **Platform Optionality:** Ability to rapidly deploy additional AI use cases
- **Market Response Optionality:** Capability to adapt to competitive moves
- **Business Model Optionality:** Foundation for potential subscription offerings

This option value recognition captured important benefits beyond direct financial projections, creating a more complete picture of potential returns.

Together, these approaches created a business case that acknowledged uncertainty while providing executives with the structured information needed for confident decision-making.

Keith McCormick, Executive Data Scientist in Residence at Pandata, reinforces why flexible approaches are essential: "In order for an AI project to be profitable, it can't be open-ended. You've got to structure it to have a specific revenue goal from the onset."

KEY INSIGHT: Organizations are making substantial investments in AI, with companies surveyed by ESI ThoughtLab in 2020 investing an average of $38 million each (more than 0.75% of their revenue). One-third invested more than $50 million, while industry leaders invested over $99 million.

DEMONSTRATING EARLY VALUE TO BUILD CREDIBILITY

Beyond compelling narratives and rigorous business cases, building and maintaining executive support requires demonstrating tangible value throughout the implementation journey. This ongoing validation builds the credibility essential for sustained backing.

QUICK-WIN STRATEGIES THAT BUILD MOMENTUM

Successful organizations deliberately build AI roadmaps that include "quick win" opportunities, targeted applications that deliver visible value early in the implementation journey.

Real-World Example: A financial services institution exemplified this approach with a customer analytics initiative. Rather than pursuing the ultimate goal of comprehensive customer journey optimization immediately, the team deliberately sequenced three quick wins:

1. **Attrition Prediction (Month 3)**
   - Focused on high-value customers with clear retention signals
   - Generated $3.7M by preventing 22% of predicted defections
   - Required minimal process changes and limited data integration

2. **Next-Best-Action Recommendation (Month 5)**
   - Deployed with contact-centre agents for inbound calls only
   - Increased conversion rates by 34% in targeted interactions
   - Built on existing systems with streamlined implementation
3. **Marketing Campaign Optimization (Month 7)**
   - Applied to email campaigns for a single product line
   - Improved response rates by 41% through improved targeting
   - Required limited technical integration with campaign tools

Each success served to build credibility for the broader transformation while delivering tangible business impact that maintained executive support. By month 9, when they began implementing the more ambitious cross-channel orchestration, the team had already delivered $8.2M in documented value, thereby creating momentum that sustained the initiative through more complex implementation challenges.

Effective quick-win strategies typically demonstrate several characteristics:

- **Clear Business Impact:** Focusing on use cases with measurable outcomes
- **Implementation Simplicity:** Minimizing technical and organizational complexity
- **Stakeholder Visibility:** Creating results visible to key decision-makers
- **Foundation Building:** Contributing to longer-term capability development
- **Risk Minimization:** Selecting applications with high probability of success

The most successful approaches don't treat quick wins as disconnected side projects but as deliberate stepping stones toward broader transformation.

PATTERN RECOGNITION: Research shows that only 20% of AI projects reach wide deployment across organizations. Among underperforming AI implementers, the deployment rate is half the average. Quick wins are essential for avoiding this scaling trap.

ACTION STEP: Identify three potential quick-win applications for your AI initiative that could deliver measurable value within 90 days.

Meaningful Metrics for Early-Stage Reporting

Maintaining executive support requires demonstrating progress before full implementation is complete. Effective early-stage metrics balance leading indicators with tangible business outcomes.

Real-World Example: A healthcare organization implementing clinical decision support AI developed a balanced measurement framework:

Technical Foundation Metrics

- Data integration completion percentage
- Model development milestones achieved
- Testing coverage and accuracy results
- Infrastructure deployment progress

Adoption Indicators

- Physician engagement by department
- Training completion rates

Securing Executive Buy-In and Board Support

- Usage frequency in available modules
- Feedback sentiment analysis

Initial Impact Measures

- Documentation of quality improvement
- Time savings for participating clinicians
- Alert response rate changes
- Initial clinical outcome indicators

Business Outcome Tracking

- Pilot department efficiency gains
- Quality measure improvements
- Patient satisfaction changes
- Initial financial impact indicators

This balanced approach demonstrated meaningful progress across multiple dimensions before enterprise-wide deployment was complete. Crucially, they didn't rely solely on technical metrics but included business impacts even during early implementation phases.

REFLECTION QUESTIONS: How balanced is your current measurement approach? Does it include technical foundations, adoption indicators, initial impacts, and business outcomes?

BALANCING SHORT- AND LONG-TERM VALUE STORIES

Value narratives are more impactful when they connect immediate results with longer-term transformation. This balanced communication maintains executive patience for strategic outcomes while demonstrating tangible progress.

Real-World Example: A manufacturing company implementing predictive maintenance AI effectively balanced these timeframes in executive communications:

Current Value Demonstration

- Specific equipment failures prevented with quantified impact
- Maintenance cost reduction in pilot facilities
- Technician time reallocation to preventive activities
- Initial parts inventory optimization results

Building Block Demonstration

- Data foundation progress enabling future applications
- Skill development across maintenance organization
- Process redesign preparing for broader transformation
- Infrastructure deployment supporting enterprise scale

Future Value Projection

- Roadmap connecting current results to future capabilities
- Implementation milestones with expected value delivery
- Capability evolution from reactive to predictive to prescriptive
- Competitive positioning trajectory relative to industry peers

This balanced communication maintained executive confidence by connecting visible current results with longer-term strategic outcomes, preventing the common pattern where initial enthusiasm fades before full value realization.

Managing through the "Value Trough" Period

As previously mentioned, every significant AI implementation experiences the "value trough" – a challenging middle period when costs and disruption are highly visible while benefits remain limited. Since this phase tests executive commitment, it requires deliberate management approaches.

WARNING SIGN: If your AI initiative lacks specific strategies for maintaining executive support during the implementation dip, you're at high risk of losing momentum precisely when you need it most.

Real-World Example: A retail organization successfully navigated this challenge during a customer experience AI implementation through several specific strategies:

Setting Realistic Expectations

Taking a proactive approach, the team explicitly prepared executives for the value trough before entering it. This preparation was done through explaining the typical implementation curve, identifying specific challenges they anticipated, and setting appropriate expectations for when different benefits would materialize.

Creating Visibility for Leading Indicators

Beyond lagging financial measures, they established leading indicators demonstrating progress even before full value materialization:

- System adoption rates by store and department
- Process adherence metrics showing implementation progress
- Customer feedback on specific experience elements
- Employee confidence measures from regular surveys

Implementing Staged Value Release

Rather than waiting for complete implementation before measuring results, they created a staged value-release approach. They identified specific benefits that would appear at different implementation points and highlighted each as it emerged.

Maintaining Leadership Engagement

They established a structured executive engagement programme during the value through period:

- Bi-weekly progress updates focused on milestone achievement
- Executive site visits to areas showing early adoption success
- Regular forums addressing implementation challenges
- Recognition events celebrating progress indicators

Reframing the Narrative

The team proactively shifted from a "results only" to a "journey and results" narrative during the trough period. This included emphasizing capability building, learning, and foundation development alongside emerging performance indicators.

These deliberate approaches maintained executive confidence through the challenging middle implementation period, preventing the premature redirection or abandonment that often occurs when initial enthusiasm is challenged by implementation reality.

Hugh Burgin, EY's Data and AI Leader for the Americas, reinforces why this approach is critical: "And each step along the way will not always be successful. It's about having that leadership commitment to the journey… And so, the C-suite sponsorship is so important, because number one, you have to be competitive. Number two, it's not always going to be easy. So, you need that sponsorship to keep the momentum."

QUICK WIN: Create a "value through preparedness plan" that identifies when your initiative might enter the implementation dip. Design specific strategies to maintain executive support during this period.

THE BOARD-LEVEL AI TRUST DASHBOARD

Beyond ongoing communication, sustainable executive and board support requires structured oversight tools that create confidence in both implementation progress and value realization.

Key Metrics for Building Executive Confidence

The most effective board-level dashboards balance multiple perspectives, creating comprehensive oversight without overwhelming detail.

Real-World Example: A financial services organization developed a highly effective board dashboard for an enterprise AI transformation that included five key dimensions:

Value Realization

- Business impact by strategic objective
- Financial returns against projections
- Customer and market response indicators
- Competitive positioning assessment

Implementation Progress

- Key milestone achievement against plan
- Resource utilization and budget status
- Adoption metrics across business units
- Technical deployment progress

Risk Management

- Compliance status across regulatory requirements
- Bias and fairness monitoring results
- Security and privacy assessment
- Model performance stability indicators

Capability Development

- Talent acquisition and retention metrics
- Skill building progress across organization
- Process maturity evolution
- Infrastructure capability advancement

Strategic Alignment

- Contribution to strategic objectives
- Market and competitive positioning

- Business model transformation indicators
- Future readiness assessment

This balanced dashboard provided comprehensive oversight without excessive detail, enabling board members to fulfil their governance responsibilities while maintaining appropriate strategic focus.

Microsoft's approach to AI governance exemplifies this balanced perspective. Alvaro Celis, HITEC Board Member and VP of Global ISV – Commercial Solutions at Microsoft, notes that "Microsoft's commitment to responsible AI dates back to 2017," and through "learnings and practices, we have implemented a robust governance process that guides the design, development, and deployment of AI in safe, secure transparent ways." The company has established itself as "an industry example through trainings, tools, programmes, and a strong governance structure" that extends beyond internal operations.

Visual Representation of Progress and Impact

Effective board dashboards use visual elements to create immediate clarity around complex implementation status.

Real-World Example: A healthcare organization implemented a particularly effective visual approach in their AI governance dashboard:

The Strategic Alignment Map

The team created a visual "heat map" showing how each AI initiative contributed to five strategic priorities, with colour-coding indicating both progress status and impact level. This single visualization allowed board members to quickly assess whether the AI portfolio was appropriately aligned with strategic objectives.

The Value Journey Visualization

Rather than relying on static metrics, they implemented a journey visualization showing each initiative's progression from initial implementation through capability building to value realization. This approach created appropriate awareness of when different types of returns would materialize.

The Risk Radar

They developed a radar visualization displaying risk levels across key dimensions (regulatory, ethical, technical and operational), with trend indicators showing whether each risk was increasing, stable, or decreasing. This approach enabled focused discussion on emerging concerns rather than static assessments.

These visualizations transformed board oversight from periodic status reviews to engaged strategic guidance by making complex implementation status immediately comprehensible.

KEY INSIGHT: The EU AI Act applies to all AI systems intended for use in EU markets, regardless of whether they are registered companies. The American approach to regulation varies significantly by state, with California exhibiting high concern for data privacy through the California Consumer Privacy Act and California Privacy Rights Act, while Texas prioritizes innovation with "much laxer privacy laws."

Balancing Technical and Business Indicators

Effective board dashboards must balance technical and business perspectives, creating comprehensive oversight without requiring specialized expertise.

Securing Executive Buy-In and Board Support

Real-World Example: A manufacturing organization achieved this balance through a tiered metric approach:

Level 1: Business Outcome Metrics

- Financial impact by initiative and portfolio
- Operational performance improvements
- Customer and market response indicators
- Competitive positioning measures

Level 2: Adoption and Usage Metrics

- Implementation coverage across facilities
- User engagement and activity levels
- Process change adherence
- Feedback and satisfaction indicators

Level 3: Technical Performance Metrics

- Model accuracy and reliability measures
- System availability and response times
- Data quality and coverage indicators
- Security and compliance status

This tiered approach provided comprehensive oversight while allowing board members to engage at appropriate depth based on their interests and expertise. The dashboard design emphasized business outcomes while making technical indicators available for those who wanted deeper understanding.

ACTION STEP: Design a tiered metric approach for your AI initiative, identifying the key business, adoption, and technical measures that will provide comprehensive oversight.

CREATING A SHARED VIEW OF SUCCESS

Perhaps most importantly, effective board dashboards create a shared definition of success that aligns executive and board expectations.

Real-World Example: A retail organization achieved this alignment through a "success criteria matrix" approach (see Table 7.4).

TABLE 7.4
Success Criteria Matrix

Dimension	Below Expectations	Meeting Expectations	Exceeding Expectations
Financial Returns	<8% revenue increase	8–12% revenue increase	>12% revenue increase
	<5% cost reduction	5–10% cost reduction	>10% cost reduction
Customer Impact	<5 point NPS increase	5–10 point NPS increase	>10 point NPS increase
	<10% engagement improvement	10–20% engagement improvement	>20% engagement improvement
Operational Efficiency	<15% productivity increase	15–25% productivity increase	>25% productivity increase
	<10% quality improvement	10–20% quality improvement	>20% quality improvement
Capability Building	Basic AI foundation	Solid AI foundation	Advanced AI foundation
	Limited talent development	Moderate talent development	Substantial talent development
Competitive Position	Keeping pace with industry	Leading within industry segment	Setting industry standards
	Limited differentiation	Notable differentiation	Substantial differentiation

This explicit success definition created shared understanding of performance targets and reduced the risk of misaligned expectations between the executive team and board.

QUICK WIN: Create a success criteria matrix for your AI initiative, defining clear standards for meeting or exceeding expectations across key dimensions.

CASE STUDIES: EXECUTIVE ENGAGEMENT SUCCESS

To illustrate these principles in action, let's examine how diverse organizations have successfully built and maintained executive and board support for their AI initiatives.

How a Financial Services Organization Secured $50M in Additional AI Funding

A global financial services organization provides our first instructive example. Initial AI investments showed promising results, but the team faced significant competition for capital as they sought additional funding for enterprise expansion.

Their approach demonstrates several critical success factors:

Strategic Framing Beyond Technology

Rather than positioning their request as "additional AI investment," they framed it as funding for "customer relationship transformation" – explicitly connecting the initiative to the organization's strategic priority of strengthening client relationships.

This framing shifted executive perception from technology spending to strategic investment, immediately changing the evaluation context.

Multi-Dimensional Value Demonstration

Instead of focusing solely on financial returns (which would compete directly with other capital requests), they demonstrated value across multiple dimensions:

- **Financial Impact:** 14% reduction in client acquisition costs, 22% increase in share of wallet
- **Competitive Necessity:** Detailed analysis of competitor capabilities and market positioning
- **Risk Mitigation:** Enhanced regulatory compliance and reduced operational risk
- **Capability Building:** Talent development and organizational transformation

This comprehensive approach transcended simple ROI comparison, creating a more compelling strategic case.

Evidence-Based Credibility

They built credibility through rigorous validation of initial results, including:

- Independent verification of financial impact by finance team
- Customer experience improvements confirmed through research
- Operational efficiency gains validated by process owners
- Competitive assessment verified through external analysis

This comprehensive validation addressed the skepticism that often surrounds AI results claims.

Phased Implementation with Clear Stage Gates

Rather than requesting the full $50M upfront, they structured a phased approach with explicit decision points:

- **Phase 1:** $12M for foundation building and initial implementation

- **Gate 1:** Evaluation against specific technical and business metrics
- **Phase 2:** $18M for broader deployment across core business lines
- **Gate 2:** Assessment of scaled results and implementation capabilities
- **Phase 3:** $20M for enterprise expansion and advanced capabilities

This approach reduced perceived risk while maintaining the comprehensive vision.

Executive-Specific Value Narratives

They created tailored narratives addressing each executive's specific priorities:

- **CEO:** Market differentiation and strategic positioning
- **CFO:** Financial returns and resource optimization
- **CRO:** Risk reduction and compliance enhancement
- **Business Unit Heads:** Specific operational and revenue impacts
- **CIO:** Technical architecture and capability building

These targeted communications ensured each influential stakeholder saw direct relevance to their priorities.

The result? Not only did they secure the full $50M, but their initiative was elevated to one of the organization's top five strategic priorities, with corresponding executive attention and protection during subsequent budget challenges.

PATTERN RECOGNITION: The most successful AI funding requests aren't positioned as technology investments but as strategic business initiatives that happen to leverage AI.

How a Public Sector Initiative Built Political Support

Political considerations, public scrutiny and complex stakeholder environments are among the unique executive challenges typically faced by government organizations. A state government agency overcame these barriers to secure sustained support for a citizen service AI initiative.

The approach for securing sustained support included several innovative elements:

Citizen-Centred Value Definition

Rather than focusing on internal efficiency or technology modernization, they defined ambitious but achievable goals explicitly from the citizen perspective.

- Dramatically reduce average case resolution time (ultimately targeting 60–70% reduction)
- Eliminate the majority of redundant documentation requirements (ultimately targeting a 70%+ decrease)
- Resolve most citizen inquiries on first contact (ultimately targeting 35–45% improvement)
- Achieve significantly higher citizen satisfaction (ultimately targeting 80%+ improvement in satisfaction scores)

This citizen-centred framing created support across political boundaries, transcending the partisan divisions that often derail government initiatives.

Cross-Agency Coalition Building

They recognized that sustained support required broader than single-agency backing. Their coalition approach included:

- Joint governance board with multiple agency representation
- Shared success metrics addressing diverse agency priorities

- Resource contribution model distributing investment
- Value attribution approach recognizing all participants

This coalition created resilience against leadership changes and budget fluctuations by distributing ownership across organizational boundaries.

Risk Mitigation through Phased Transparency

A deliberate transparency approach served to address the public scrutiny risk inherent in government AI through:

- Clear public communication about capabilities and limitations
- Phased implementation starting with lowest-risk applications
- Regular public reporting on performance and outcomes
- External ethics advisory involvement in governance

This approach prevented the negative publicity that often emerges when government AI initiatives are perceived as secretive or unaccountable.

Political Cycle Alignment

The initiative also benefited from a timeline that was explicitly designed to align with political cycles:

- Quick wins delivered within first budget cycle
- Major milestones aligned with key political checkpoints
- Value reporting synchronized with budget proposal timing
- Constituent impact stories collected for legislative briefings

This deliberate alignment ensured that tangible results could be demonstrated at politically significant moments.

Bipartisan Value Framing

Perhaps most importantly, they framed value in terms that resonated across political perspectives:

- Efficiency and cost savings appealing to fiscal conservatives
- Service quality and access improvements appealing to progressives
- Accountability and transparency measures addressing shared concerns
- Economic development through modernization appealing to business interests

This balanced value narrative created support across political divides, insulating the initiative from partisan budget battles.

Through these approaches, the initiative maintained executive and legislative support through two election cycles and multiple leadership changes, achieving an implementation continuity rare in government technology initiatives.

KEY INSIGHT: Research shows that between 2016 and 2023, the United States passed the highest number of AI-related bills (23), followed by Portugal (15) and Belgium (12). Political awareness of AI is growing rapidly across global jurisdictions.

How a Healthcare Provider Won Physician Leadership Buy-In

Healthcare organizations face particular challenges securing support from physician leaders who may view AI with scepticism or concern. A regional healthcare system overcame these barriers to build strong clinical leadership support for an AI diagnostic support initiative.

The approach demonstrates several healthcare-specific strategies:

Securing Executive Buy-In and Board Support

Evidence-Based Validation

Recognizing physicians' reliance on clinical evidence, the organization implemented a rigorous validation approach:

- Formal clinical trial of initial capabilities
- Peer-reviewed publication of methodology and results
- Side-by-side comparison with traditional diagnostic approaches
- Transparent sharing of both success cases and limitations

This evidence-based approach spoke directly to physician values and built credibility beyond what general business validation could achieve.

Clinical Leadership Co-Development

Rather than developing capabilities and then seeking physician support, respected clinical leaders were involved throughout the process:

- Physician design council guiding requirements and priorities
- Clinical specialists validating approach within each specialty
- Department chiefs participating in implementation planning
- Medical executive committee oversight of governance

This co-development created ownership rather than resistance among influential physician leaders.

Autonomy-Sensitive Implementation

Physicians' concerns about autonomy and judgement were addressed directly:

- Explicit positioning as "decision support" rather than replacement
- Transparent reasoning enabling physician evaluation
- Override capabilities with simple documentation
- Continuous learning from physician feedback and decisions

This sensitivity to autonomy concerns transformed what could have been seen as "threatening technology" into a valued support tool.

Patient Outcome Orientation

Value was framed primarily around patient outcomes rather than organizational efficiency:

- Substantially reduce diagnostic delays for critical conditions
- Improve treatment protocol selection accuracy
- Minimize adverse events from missed clinical interactions
- Reduce unnecessary testing and procedures

Supporting these goals were target ranges of 15–35% improvement that were based on comparable implementations.

Physician Experience Enhancement

Beyond patient outcomes, ambitious goals for improving the physician experience were set.

- Dramatically reduce time spent on documentation
- Automatically surface relevant research and literature
- Streamline access to complete patient information
- Minimize administrative interruptions during patient care

These goals were supported by target reductions of 25–40% in administrative time based on pilot studies.

This dual focus on patient outcomes and physician experience created a compelling value proposition for clinical leaders who might otherwise have resisted technology-driven change.

Through these approaches, potential physician resistance was transformed into active championship that proved essential for successful implementation across this complex clinical environment.

REFLECTION QUESTIONS: What are the unique professional priorities and concerns of key stakeholders in your organization? How can you address them directly in your AI value proposition?

How a Financial Services Institution Leveraged Board-Level AI Governance and Oversight

Our final case study examines how a global financial institution established effective board governance for enterprise AI initiatives, creating oversight that enabled rather than constrained innovation.

Several distinctive elements proved to be key to this approach:

Dedicated AI Committee Structure

A dedicated AI oversight committee within the organization's board governance structure was established:

- Focused charter addressing both innovation and risk dimensions
- Membership combining technology, business, and risk expertise
- Regular cadence with appropriate depth of engagement
- Clear relationship with full board reporting

This dedicated structure created space for appropriate oversight without forcing AI topics to compete with other governance priorities during general board meetings.

Balanced Capability Development

Board capabilities to provide effective oversight were systematically established:

- Educational sessions on core AI concepts and implications
- External expert perspectives on industry trends and standards
- Exposure to both technical and business dimensions
- Progressive depth building from concepts to specific applications

This capability development enabled informed oversight without requiring board members to become technical experts.

Risk-Calibrated Governance Framework

A tiered governance approach matching oversight intensity to risk profile was implemented:

- Streamlined review for lower-risk applications
- Enhanced scrutiny for consumer-facing or high-risk cases
- Comprehensive oversight for highest-risk applications
- Clear escalation criteria and processes

This calibrated approach provided appropriate control without creating innovation-dampening bureaucracy.

Outcome and Control Balanced Reporting

The board reporting balanced progress and risk dimensions:

- Value delivery against strategic objectives
- Implementation progress and adoption metrics
- Risk indicators and control effectiveness
- Competitive and market positioning

This balanced perspective prevented oversight from focusing exclusively on either opportunity or risk dimensions.

Forward-Looking Governance Evolution

Perhaps most distinctively, deliberate evolution was built into the governance approach:

- Regular reassessment of governance effectiveness
- Adjustment based on emerging applications and risks
- External calibration against evolving standards
- Capability building matching technology evolution

This evolutionary approach prevented governance from becoming fixed while technology and applications continued advancing.

Through these governance innovations, the institution created board oversight that provided genuine strategic guidance and appropriate risk management without becoming a bureaucratic obstacle to innovation, a balance many organizations struggle to achieve.

ACTION STEP: Assess your organization's AI governance approach using the five elements highlighted in this case study. Identify specific improvement opportunities based on gaps.

KEY TAKEAWAYS: BUILDING SUSTAINABLE EXECUTIVE SUPPORT

As we conclude our exploration of executive and board engagement, several key principles emerge that apply across industries and organization types:

SPEAK THE LANGUAGE OF LEADERSHIP

Successful AI initiatives translate technical capabilities into business outcomes that resonate with executive priorities. This translation isn't about simplification but about relevance, connecting AI to strategic objectives, competitive positioning, and organizational priorities.

BUILD MULTI-DIMENSIONAL VALUE CASES

The most compelling executive engagement transcends simple ROI calculations to address multiple value dimensions, such as strategic positioning, operational excellence, customer experience, talent implications, and risk management. This comprehensive approach creates broader and more durable support.

CREATE APPROPRIATE GOVERNANCE

Effective governance balances innovation enablement with risk management, providing appropriate oversight without creating paralysing bureaucracy. This balance requires governance mechanisms specifically designed for AI rather than simply applying traditional technology governance approaches.

Demonstrate Credible Results

Sustained executive support depends on rigorous validation of results claims. This requires measurement approaches that create credibility through independent verification, appropriate attribution, and realistic assessment of both successes and limitations.

Manage through Implementation Reality

Every significant AI initiative encounters challenges during implementation. Organizations that maintain executive support proactively manage expectations, create visibility into progress indicators, and maintain consistent communication through inevitable difficulties.

Align with Stakeholder-Specific Priorities

Different executives and board members bring distinct priorities and perspectives. Successful engagement addresses these differences through stakeholder-specific value narratives and communication approaches rather than one-size-fits-all messaging.

Build Ongoing Relationships, Not Just Approvals

The most sustainable executive support comes from continuous engagement rather than point-in-time approvals. Organizations that maintain regular communication, demonstrate ongoing progress, and involve executives appropriately throughout the journey are in a strong position for building the durable support essential for long-term success.

QUICK WIN: Schedule a 30-minute session with a key executive sponsor to review your implementation progress, focusing on business outcomes rather than technical details.

WHAT'S NEXT: BUILDING CROSS-FUNCTIONAL SUPPORT

With an understanding of how to secure executive and board support, we now turn to the equally critical challenge of building support across functions and departments.

Even the strongest executive sponsorship cannot ensure success without broad cross-functional support. In the next chapter, we'll provide practical guidance for building the organizational coalition necessary for sustainable AI implementation. We'll look at how to create alignment among diverse stakeholders whose cooperation is essential for successful implementation, including practical frameworks for transforming potential resistance into collaborative support throughout the organization.

KEY INSIGHT: As we shift from vertical support (executives and board) to horizontal support (cross-functional), the principles remain similar but the application changes dramatically. While executive support requires strategic alignment and governance, cross-functional support depends on operational integration and value translation at the daily work level.

ACTION STEP: Before moving to Chapter 8, download the executive support assessment from the book's online resource centre to evaluate your current executive engagement approach and identify specific improvement opportunities.

8 Building Cross-Functional Support for AI Initiatives

Chapter Roadmap

In this chapter, we will explore why AI demands unprecedented cross-functional collaboration – and how to transform traditional departmental silos into powerful engines of AI value creation. Among the strategies designed to build genuine alignment are practical frameworks for understanding diverse departmental perspectives, game theory techniques to transform conflict into cooperation, proven bridge-building approaches that create lasting collaboration, and methods for building sustainable mechanisms that maintain cross-functional momentum long after initial implementation. To illustrate these concepts in action, we look at case studies of organizations that successfully united marketing, operations, engineering, and compliance around AI initiatives.

THE UNCOMFORTABLE TRUTH ABOUT AI COLLABORATION

Have you ever sat in a room where everyone was speaking a different language while pretending to understand each other?

That's what happened to me recently at a leading healthcare organization's steering committee for a high-profile AI diagnostic project. The tech team was celebrating a milestone: their computer vision model had reached specialist-level accuracy for a specific diagnostic task. A genuine technical achievement.

Yet as I glanced around the room, I noticed something unsettling. The clinical director was scrolling through emails. The compliance officer was frowning at her notes. The patient experience lead looked openly confused.

When I asked them afterward about their muted reactions, their responses painted a picture that is more common than you might think.

The clinical director said, "Great, another tool my overworked doctors have to learn that doesn't fit our workflows."

The compliance officer commented, "[The technology guys] haven't addressed any of the privacy concerns I raised three months ago."

The patient experience lead added, "I still don't understand how this helps actual patients feel better cared for."

This debrief revealed that what was happening here wasn't just a communication problem – it was the result of a fundamental misalignment of what success even meant.

To the tech team, a more accurate model was the goal. To clinical leaders, workflow integration was the priority. To compliance, risk mitigation mattered most. To patient experience, improved care was the only metric that counted.

Same project. Same room. Completely different worlds.

REFLECTION POINT: Think about your recent AI initiatives. Have you noticed similar disconnects between technical achievements and business stakeholder reactions?

KEY INSIGHT: Most AI initiatives don't fail because of technical shortcomings – they fail because different departments define success in fundamentally different ways.

I see this everywhere. To the finance team, AI purely represents a cost-saving tool. Marketing sees it as a revenue generator. The operations group that wants stability; product leaders demand innovation. The compliance team is focused on risk, while the strategy office pushes for competitive advantage.

These diverse views are the expression of fundamentally different missions, operating under the false banner of "collaboration."

Despite all our talk about "cross-functional teams," most AI initiatives remain functionally fragmented. We create committees and hold meetings but continue to speak different languages and pursue different goals.

It's due to this failure to communicate – and the deeper misalignment of purpose, priorities, and even definitions of success – that many AI projects with strong technical foundations still fall short in delivering meaningful organizational value.

WHY AI DEMANDS UNPRECEDENTED COLLABORATION

Let's be clear about something: AI projects are not like your standard IT implementations.

Traditional IT systems typically operate within well-defined functional boundaries. Your finance system serves finance. Your CRM serves sales and marketing. Your inventory management system serves operations.

AI, in contrast, cuts across these boundaries by design. It learns from data generated across departments, impacts processes that span functions, and creates value that manifests throughout the organization.

Think about it. Even a "simple" AI application like automated contract analysis requires:

- Legal expertise to define what matters in contracts
- Finance input on cost implications
- Procurement knowledge of supplier relationships
- IT involvement for system integration
- Risk and compliance oversight for regulatory requirements
- Business unit context for operational impact

Unlike a departmental software implementation, AI systems rarely succeed when confined to a single function. Their very power comes from connecting and analysing data, processes, and insights across traditional boundaries.

The Dependency Paradox: The more powerful and transformative the AI, the more cross-functional dependency it creates.

Here's why: Simple AI tools might automate a single task within one department – like basic data entry or simple calculations. But transformative AI systems work by connecting data, processes, and insights across multiple areas of your business.

Consider a customer service AI agent that truly transforms the experience. It needs marketing data to understand customer segments, sales history to provide context, product information from engineering, billing data from finance, and operational workflows from customer service. The more sophisticated and valuable the AI becomes, the more it must draw from – and impact – different parts of your organization.

This creates an uncomfortable reality: the AI initiatives with the highest potential value are precisely the ones that require the most complex collaboration. A basic chatbot might be owned by customer service alone. But an AI system that can truly predict customer needs, personalize experiences, and optimize operations across channels? That demands finance, marketing, operations, IT, and customer service to work together in ways they've never done before.

Yet most organizations still approach AI through siloed structures, separated budgets, disconnected metrics, and fragmented leadership.

The result is predictable:

- Technical teams build impressive models that don't address real business problems
- Business teams demand solutions without understanding technical constraints
- Compliance flags risks that delay implementation indefinitely
- Finance questions investments when value remains unclear
- Users resist adoption when their needs weren't considered from the start

This isn't just inefficient. It's a recipe for failure.

WARNING SIGN: If your AI initiative has a single departmental "owner" with other functions in supporting roles, you're likely heading for trouble.

The hard truth is that successful AI requires a level of cross-functional collaboration that most organizations find uncomfortable, not because they don't want to collaborate, but because their structures, incentives, and cultures actively work against it.

The challenge isn't convincing people that collaboration matters. It's creating the conditions where real collaboration can actually happen.

CASE STUDY: BUILDING CROSS-FUNCTIONAL AI LITERACY AT ASANA

THE SITUATION: When Asana began exploring AI implementation, CEO Dustin Moskovitz faced a significant challenge: enthusiasm for AI varied widely across the organization, from excitement to open suspicion.

THE APPROACH: Rather than dismissing these varied perspectives as obstacles, Moskovitz recognized them as valuable inputs that needed alignment. His approach was both simple and profound: build a shared foundation of AI literacy.

> Internally, reactions to AI ranged from exhilaration to skepticism. We knew we needed to build literacy and hands-on experience to unite everyone around this transformation. So we launched an internal AI community and immersive workshops, encouraging all employees to tinker with the technology.
>
> – **Dustin Moskovitz,**
> *CEO and co-founder of Asana*

This approach went beyond education – to creating a shared reality where different perspectives could meaningfully connect. By giving everyone from HR to marketing to engineering the same foundational experience with AI, Asana began building bridges across traditional boundaries.

THE RESULTS: As Moskovitz noted, "Use cases for AI soon blossomed in every corner of the company," with teams across functions developing specialized AI solutions for their specific needs.

THE LESSON: Rather than erasing functional differences, cross-functional collaboration means creating the shared context where those differences become strengths rather than obstacles.

TRY THIS: Consider how you might create hands-on AI experiences for stakeholders across functions. What simple workshops or "tinkering sessions" could build shared understanding in your organization?

UNDERSTANDING DEPARTMENTAL PERSPECTIVES

Before we can build bridges, we need to understand the territory. Each department brings a distinct lens to AI initiatives: different priorities, concerns, metrics, and even language.

Finance has a value guardian perspective; operations look after reality checks; marketing and sales are tuned into customer voices; IT and technology bring a system architect view. HR and people have a cultural translator role; and legal, risk and compliance play on the defence team. What's more, different roles within each department often represent subsets of focus areas.

Let's examine these perspectives not just as "stakeholder requirements" but as fundamentally different worldviews that shape how teams engage with AI.

FINANCE: THE VALUE GUARDIANS

Primary Lens: Return on investment, cost management, risk mitigation

Key Questions They Ask

- What's the fully loaded cost of this initiative?
- When will we see financial returns?
- How does this compare to other investment opportunities?
- What financial risks does this create?

Language: NPV, ROI, payback period, cost avoidance, operating expense, capital allocation

Common Friction Points

- Demanding precise ROI forecasts for inherently uncertain innovations
- Focusing on short-term costs over long-term capabilities
- Requiring excessive validation before releasing resources
- Viewing AI purely as a cost-cutting tool

Requirements for Engaging Them: Clear financial models that balance short- and long-term impact, connection to strategic financial goals, rigorous but realistic measurement frameworks, and involvement in defining value metrics from the start.

OPERATIONS: THE EXECUTION REALITY CHECK

Primary Lens: Process efficiency, reliability, service delivery

Key Questions They Ask

- How will this affect our daily workflows?
- Can we maintain service levels during implementation?
- Does this create single points of failure?
- How do we support and maintain this?

Language: Uptime, throughput, cycle time, service levels, standard operating procedures

Common Friction Points

- Resistance to disrupting functioning processes
- Concern about technical failures impacting operations
- Scepticism about AI reliability in critical systems
- Focus on stability over innovation

Building Cross-Functional Support for AI Initiatives

Requirements for Engaging Them: Phased implementation plans, clear operational control mechanisms, involvement in design and testing, realistic transition approaches, and recognition of operational expertise.

Marketing and Sales: The Customer Voice

Primary Lens: Customer experience, competitive differentiation, revenue growth

Key Questions They Ask

- How does this improve the customer experience?
- Can we use this as a competitive differentiator?
- Will this help us acquire or retain customers?
- How do we message this to the market?

Language: Customer journey, engagement, conversion, market positioning, brand impact

Common Friction Points

- Pushing for speed to market over thoroughness
- Wanting customization that creates technical complexity
- Prioritizing flashy features over foundational capabilities
- Sometimes overpromising AI capabilities to the market

Requirements for Engaging Them: Clear connection to customer value, involvement in experience design, data on competitive positioning, and opportunities to shape the narrative.

IT and Technology: The System Architects

Primary Lens: Technical feasibility, integration, security, scalability

Key Questions They Ask

- How does this integrate with our existing systems?
- What are the security implications?
- Can our infrastructure support this?
- How do we maintain and govern this long-term?

Language: Architecture, infrastructure, APIs, technical debt, cybersecurity

Common Friction Points

- Focus on technical elegance over business practicality
- Concern about legacy system constraints
- Prioritizing security and stability over innovation
- Resistance to solutions not built to enterprise standards

Requirements for Engaging Them: Clear technical requirements, involvement in solution architecture, realistic integration paths, recognition of infrastructure constraints, and long-term maintenance considerations.

HR and People: The Cultural Translators

Primary Lens: Talent, skills, organizational impact, cultural acceptance

Key Questions They Ask

- How will this affect jobs and roles?
- What new skills do we need?
- How do we manage change and adoption?
- Does this align with our culture and values?

Language: Skills gap, change management, adoption, role design, organizational development

Common Friction Points

- Concern about workforce disruption
- Focus on organizational readiness over technical capability
- Resistance to solutions perceived as threatening jobs
- Prioritizing human judgement over automation

Requirements for Engaging Them: Transparent job impact assessment, skills development plans, change management strategies, and involvement in solution design to ensure human-AI balance.

Legal, Risk, and Compliance: The Reputation Custodians

Primary Lens: Regulatory compliance, risk management, legal protection

Key Questions They Ask

- Does this comply with relevant regulations?
- What new risks does this create?
- How do we ensure appropriate governance?
- Are we legally protected if something goes wrong?

Language: Regulatory requirements, liability, documentation, governance, controls

Common Friction Points

- Applying traditional risk frameworks to emerging technologies
- Focus on mitigation over enablement
- Detailed documentation requirements that slow progress
- Sometimes saying "no" without offering alternatives

Requirements for Engaging Them: Early involvement in risk assessment, participation in governance design, clarity on regulatory implications, and recognition of valid concerns without treating them as blockers.

SELF-ASSESSMENT: For your current AI initiative, rate (1 to 5) how well you understand each department's perspective and concerns. Where are your blind spots?

These perspectives aren't just separate preferences or priorities. They're fundamentally different realities shaped by professional expertise, departmental incentives, and measured outcomes.

It is not only important to acknowledge that these differences exist – it is also crucial to understand that each perspective has validity within its context. Finance isn't wrong to care about ROI. Legal isn't obstructing progress by focusing on compliance. Operations isn't being difficult when raising workflow concerns.

CRITICAL POINT: The challenge of cross-functional AI isn't getting people to abandon their perspectives. It's creating a shared reality where these perspectives can meaningfully connect.

THE TENSION TRIANGLE: BUSINESS, FINANCE, AND IT

David Sweenor identifies a fundamental conflict that complicates cross-functional collaboration. According to Sweenor, successful AI implementation requires balancing the conflicting priorities of three primary forces:

1. **Business units** seeking quick operational wins
2. **Finance departments** demanding clear ROI
3. **IT teams** concerned with infrastructure protection and security

It's the "Bermuda Triangle of AI," where initiatives get pulled in different directions, caught between conflicting forces, until they lose momentum entirely.

The tension point characteristic to this particular configuration reveals that even with dedicated leadership roles, the fundamental differences in departmental objectives can create significant challenges for cross-functional AI initiatives.

The conflict goes beyond differing priorities – to include different timelines, success metrics, and even value definitions. Business units want solutions that deliver immediate value. Finance requires clear financial returns that often take time to materialize. IT needs robust, secure implementations that don't compromise existing systems.

Counterintuitive Insight: Acknowledging these tensions explicitly tends to reduce rather than increase conflict. When departments recognize their differences as structural rather than personal, collaboration becomes easier.

AI initiatives get trapped in what I call the "tension triangle" when these three forces have different directions. Rather than moving forward and gaining speed, the push and pull causes efforts to lose momentum, clarity, and ultimately, value.

REFLECTION QUESTIONS: In your organization, which corner of the triangle typically has the most power in AI decisions? What impact does this have on outcomes?

BUILDING BRIDGES ACROSS FUNCTIONS

Understanding departmental perspectives is just the starting point. The real work is building bridges that facilitate genuine collaboration across these different worlds.

What I found particularly useful in my work are measures for creating steering committees that actually make decisions, techniques for developing truly shared success metrics and ways to create physical and virtual collaborative spaces. In addition, methods for building mutual understanding across departments as well as frameworks that facilitate structured collaboration help bolster chances of success.

Here are five bridge-building approaches that can transform cross-functional AI initiatives:

CREATING CROSS-FUNCTIONAL STEERING COMMITTEES THAT ACTUALLY WORK

Most organizations create steering committees for major AI initiatives – but many end up as information-sharing forums rather than true decision-making bodies. Here are some insights for building committees that drive real alignment:

- **Ensure Balanced Composition:** Include both horizontal functions (finance, IT, HR) and vertical business units directly impacted by the initiative.
- **Define Clear Decision Rights:** Explicitly state what decisions the committee makes that go beyond mere recommendations or reviews.

- **Establish Shared Success Metrics:** Create a balanced scorecard of outcomes that matter to all functions to prevent any single perspective from dominating.
- **Rotate Leadership:** Consider rotating the chair role among different functions to prevent one perspective from consistently framing the agenda.
- **Meet for Decisions, Not Updates:** Structure meetings around key decisions rather than status updates, keeping the focus on active collaboration rather than passive reporting.

A real-world example comes from a global pharmaceutical company. By restructuring AI governance, the organization created a three-tiered model: a strategic council for investment decisions, cross-functional working groups for implementation, and technical teams for execution. Each tier had balanced representation and clear decision authority, dramatically reducing the friction that had previously stalled AI initiatives.

Developing Shared Metrics and Success Definitions

When departments operate with different success metrics, collaboration becomes nearly impossible. To create genuinely shared definitions of success:

- **Start with Ultimate Outcomes:** Begin with the end-state impact the organization wants to achieve, not individual departmental goals.
- **Create Multi-Dimensional Scorecards:** Develop balanced metrics that include technical performance, business impact, user adoption, risk management, and financial returns.
- **Use Success Signals to Chart Progress:** Identify early signals of success that all functions can monitor, while maintaining focus on ultimate outcomes.
- **Link Departmental Metrics to Shared Goals:** Help each department see how their traditional metrics connect to and support the shared outcomes.
- **Make Trade-Offs Explicit:** Create forums where departments can openly discuss the inevitable trade-offs between speed, quality, cost, and risk.

PRACTICAL TOOL: Create a "value translation chart" that shows how departmental metrics (e.g., model accuracy, customer satisfaction, cost savings) contribute to shared organizational goals.

For a real-world example, consider a retail bank that implemented a "value balanced scorecard" for customer service AI. The scorecard tracked metrics across five dimensions: customer experience, operational efficiency, risk reduction, financial impact, and employee experience. Each department had primary responsibility for specific metrics, but all contributed to – and were evaluated on – the full scorecard.

Creating Cross-Functional Collaborative Spaces

Physical and virtual environments significantly impact how teams collaborate. To break down barriers:

- **Co-Locate Cross-Functional Teams:** When possible, physically locate team members from different functions together during critical project phases.
- **Create Visual Collaboration Spaces:** Use physical and digital visualization tools that make different perspectives visible and create shared understanding.
- **Implement Shared Knowledge Repositories:** Develop common information spaces that transcend departmental boundaries and create a single source of truth.
- **Use Regular Rhythm Events:** Establish cadences for cross-functional synchronization beyond formal steering committees.

- **Create Informal Connection Opportunities:** Don't underestimate the power of informal interactions to build relationships across functions.

A manufacturing company provides a real-world example. The firm created a physical AI Lab, where business, IT, operations, and data science team members worked together two days per week. The space was surrounded by visual analytics displaying shared metrics and progress. This arrangement cut implementation time by over 40% compared to previous initiatives.

BUILDING MUTUAL UNDERSTANDING AND RESPECT

Beyond structures, successful cross-functional collaboration requires a foundation of mutual respect and understanding:

- **Create Cross-Functional Shadowing Opportunities:** Have team members spend time embedded in other departments to gain firsthand experience of different perspectives.
- **Develop Shared Language Guides:** Create glossaries that translate between technical, business, legal, and financial terminology.
- **Implement Role Rotation in Meetings:** Periodically ask team members to present from another function's perspective.
- **Celebrate Diverse Expertise:** Explicitly recognize the value that different functional perspectives bring to the initiative.
- **Address Power Imbalances:** Be aware of informal power dynamics that may prioritize certain functions over others.

ACTION STEP: Identify one stakeholder whose perspective you don't fully understand. Schedule a "perspective exchange" session where you explore their priorities and concerns.

Innovative practices of a government agency implementing an AI-based citizen service platform provide a real-world example. The organization required all team members to complete a "cross-functional passport" programme where they spent time in each key department, earning "stamps" by demonstrating understanding of that function's priorities and constraints. This created a common foundation that dramatically improved collaboration.

USING STRUCTURED COLLABORATION FRAMEWORKS

Beyond general collaboration principles, specific frameworks can facilitate cross-functional alignment:

- **Design-Thinking Workshops:** Use human-centred design approaches to bring diverse perspectives together.
- **Value-Stream Mapping:** Collectively map processes across departmental boundaries to identify integration points and dependencies.
- **Scenario-Planning Exercises:** Develop shared future scenarios that help diverse functions align around common challenges and opportunities.
- **Decision-Modelling Frameworks:** Use structured approaches to make trade-offs and priorities explicit across functions.
- **Agile Ceremonies Adapted for Cross-Functional Contexts:** Modify agile practices like sprint reviews and retrospectives to include broader functional representation.

A real-world example comes from a financial services firm that struggled with cross-functional alignment. The company then implemented quarterly "value stream workshops" where representatives from each function collaboratively mapped how their AI initiatives created value from initial development through customer impact. This revealed critical handoff points between functions that had previously been invisible. The result? A 30% reduction in implementation friction.

CASE STUDY: KAREN STROUP'S ALIGNMENT APPROACH AT WEX

Karen Stroup, Chief Digital Officer at WEX, discovered something counterintuitive about cross-functional AI success: it needs to start at the top rather than with a single champion trying to drive adoption against resistance.

During a "Mind the Product" podcast episode, Stroup explained that the key to WEX's successful AI implementation was creating shared goals across the entire executive leadership team first:

> That alignment was actually incredibly important and continues to be so. So we take our goals from the CEO level, which is endorsed by the board, and we cascade them through the organization and ensure those cross-functional teams do have shared goals around AI. So, that was – I mean it sounds obvious but one of our really big secrets to success.

The breakthrough insight was making AI a collective priority with executive-level alignment. As Stroup notes, "We all had a shared goal on AI, and so the CTO and I are joined at the hip."

This approach ensured cross-functional alignment from the top down, with goals "endorsed by the board" and cascaded "through the organization and ensured those cross-functional teams do have shared goals around AI."

THE TAKEAWAY: Cross-functional alignment for AI often needs to start at the top, with executive leadership demonstrating the collaboration they want to see throughout the organization.

QUICK WIN: Schedule an executive alignment session focused solely on creating shared AI goals before diving into specific initiatives.

UNDERSTANDING DEPARTMENTAL PERSPECTIVES: THE GAME THEORY OF AI ALIGNMENT

One of the most powerful lenses for understanding cross-functional dynamics is game theory – the study of strategic interaction among rational decision-makers.

Traditional organizational structures create what game theorists call a "Prisoner's Dilemma," a situation where departments acting in their rational self-interest create worse outcomes for the organization as a whole:

- The compliance team has incentives to minimize risk rather than enable innovation
- The technology team is rewarded for technical excellence, not business adoption
- Business units are measured on short-term results, not long-term capability building
- Finance is evaluated on cost control, not value creation

Under these conditions, even well-intentioned leaders make decisions that optimize for their functional area at the expense of overall success.

A recognition of how organizational silos create a "Prisoner's Dilemma" can then inspire countermeasures. For example, there are five game theory strategies that can help foster collaboration. Creating incentive structures can promote cooperation – and gamification techniques can drive adoption.

Fundamental Challenge: How do we transform this competitive game into a cooperative one, where departments benefit more from collaboration than optimization of their individual metrics?

Using Game Theory to Navigate Conflicting Interests

As previously mentioned, Dr. Thuc Vu, entrepreneur and co-founder of OhmniLabs and Kambria, suggests turning to game theory to transform cross-departmental collaboration.

He proposes a shift in approach from a "one-shot game" to "repeat the game" in order to incentivize or even force players to collaborate.

Game theory offers several practical strategies for creating more collaborative dynamics:

1. **Change the Payoff Structure:** Modify incentives so that departments benefit more from collective success than individual optimization. This might include:
   - Shared bonuses tied to organizational AI outcomes
   - Recognition systems that celebrate cross-functional achievements
   - Career advancement linked to collaborative success
2. **Increase Interaction Frequency:** Game theory shows that repeated interactions foster cooperation. Create more frequent touchpoints between departments through:
   - Regular cross-functional working sessions
   - Shared project spaces (physical or virtual)
   - Rotation programmes that build relationship networks
3. **Extend Timeframes to Reach Further into the Future:** When departments see a longer-term relationship, cooperation becomes more valuable. Build this perspective by:
   - Creating multi-year AI roadmaps with clear interdependencies
   - Establishing long-lived cross-functional teams rather than temporary task forces
   - Developing capability platforms rather than one-off projects
4. **Create Information Transparency:** Cooperation improves when all players have access to the same information. Implement:
   - Shared dashboards visible to all departments
   - Open project documentation across functional boundaries
   - Joint risk/opportunity assessments
5. **Establish Credible Commitment Mechanisms:** Departments need assurance that others will uphold their end of collaborative agreements. Create this through:
   - Formal charters with explicit cross-functional commitments
   - Executive sponsorship of collaborative approaches
   - Clear resolution paths when collaboration breaks down

SUCCESS STORY: A telecommunications company restructured an AI investment approach from departmental budgets to a portfolio model with three tiers: enterprise platforms (funded centrally), cross-functional use cases (funded through shared budgets), and departmental applications (funded locally). This created natural incentives for collaboration on the highest-impact initiatives while maintaining departmental autonomy for specialized needs.

Creating Incentive Structures That Promote Collaboration

Beyond general game theory principles, specific incentive design can dramatically impact cross-functional dynamics:

- **Align Performance Metrics across Boundaries:** Ensure that departmental KPIs reinforce rather than contradict each other. For example, if IT is measured on system stability while the business is measured on innovation speed, this can lead to conflict.
- **Create Shared Ownership of Outcomes:** Design governance models where departments jointly own outcomes rather than individual components. This shifts the dynamic from "handoffs" to "co-creation."

- **Implement Joint Risk/Reward Structures:** Create mechanisms where departments share both the upside and downside of AI initiatives, aligning their risk tolerance and investment perspective.
- **Recognize Collaboration Explicitly:** Make cross-functional collaboration a distinct component of performance evaluation, not just an implied expectation.
- **Balance Functional Excellence with Enterprise Contribution:** Maintain clear functional expertise while creating pathways for that expertise to contribute to broader organizational goals.

PRACTICAL EXERCISE: Review your current metrics for an AI initiative. Do they create collaborative or competitive incentives? Where could you introduce shared metrics?

For a real-world example, let's highlight a retail company that implemented demand forecasting AI. It modified the incentive structure so that inventory managers, merchandising teams, and store operations all shared metrics around forecast accuracy, inventory levels, and customer availability. This replaced previous incentives that had led to conflicting behaviours across these functions.

GAMIFICATION TECHNIQUES TO DRIVE ADOPTION AND ENGAGEMENT

According to Jonathan Fields, CEO of Assembly, gamification techniques that incorporate rewards, storytelling, and healthy competition have been shown to improve teamwork metrics and break down departmental silos. His research indicates that gamified collaborative processes lead to "enhanced productivity, increased trust, and an overall supportive culture."

Beyond formal incentives, gamification elements can create engagement across functions:

- **Progress Visualization:** Create visual representations of shared journey maps that show advancement toward common goals.
- **Team-Based Challenges:** Design cross-functional challenges that require collaborative problem-solving around AI initiatives.
- **Recognition Systems:** Implement mechanisms to celebrate cross-functional wins and spotlight collaboration champions.
- **Skill-Building Pathways:** Create visual progression systems for developing both technical and collaborative capabilities.
- **Friendly Competition:** Where appropriate, create constructive competition between cross-functional teams addressing similar challenges.

A global consumer goods company provides a real-world example. The organization created an "AI adoption league" where cross-functional teams earned points for implementing AI use cases, with bonus multipliers for solutions that spanned multiple departments. The friendly competition dramatically increased engagement and led to teams actively seeking cross-boundary opportunities.

CASE STUDY: HOW GAMIFICATION HELPED ALIGN IT AND BUSINESS GOALS

THE CHALLENGE: When the marketing and sales departments at a major retailer found themselves in constant conflict over AI implementation priorities, their leadership turned to an unexpected solution: game theory.

THE APPROACH: Instead of continuing the cycle of competing requests and conflicting metrics, they designed what they called the "value alignment game."

Each department was asked to allocate fictional resources to various AI initiatives, but with a twist: they earned bonus points for initiatives that other departments also prioritized. The initial

Building Cross-Functional Support for AI Initiatives

rounds revealed stark differences in priorities, but as the game progressed, departments began to strategically align their choices, seeking opportunities for mutual benefit.

THE RESULT: What began as an exercise became the foundation for an actual AI roadmap – one built on identified areas of overlapping value rather than departmental competition.

This approach transformed what had been a contentious planning process into a collaborative one, revealing opportunities for mutual benefit that had been obscured by departmental competition.

GAMIFICATION IDEA: Create a simple "alignment points" system where teams earn recognition when they identify and address another department's priority concerns within their AI plans.

THE CROSS-FUNCTIONAL VALUE CANVAS

TOOL OVERVIEW:

- What it is: A structured framework for aligning departments around shared AI value
- When to use it: Early in planning and at key decision points
- Who should participate: Representatives from all key stakeholder departments
- Expected outcome: A unified understanding of value, concerns, and collaboration points

To operationalize cross-functional alignment, organizations need practical tools that make different perspectives visible and actionable. One of the most effective tools I've developed and implemented is the cross-functional value canvas.

This canvas serves as both a diagnostic tool to identify misalignment and a planning tool to create shared value understanding across departments.

CANVAS COMPONENTS

The cross-functional value canvas consists of five key sections:

1. **Shared Purpose Statement**
   A clear articulation of the ultimate outcome the AI initiative aims to achieve, transcending departmental boundaries.
2. **Departmental Value Mapping**
   For each key function (typically five to seven departments), capture:
   - Primary value expected from the initiative
   - Key concerns or risks from their perspective
   - Success metrics that matter to this function
   - Essential contributions needed from this function
3. **Value Conflicts and Synergies**
   Explicitly identify:
   - Potential conflicts between departmental values
   - Natural synergies that create mutual benefit
   - Required trade-offs that need negotiation
4. **Integration Points**
   Map the critical touchpoints where functions must collaborate:
   - Shared decisions that require joint input
   - Handoff points between departments
   - Data and insight flows across boundaries
   - Resource dependencies between functions

5. **Shared Value Narrative**
   Develop a unified story that integrates departmental perspectives:
   - The collective case for change
   - Integrated vision of success
   - Common language that bridges functional terminology
   - Shared risk/reward framework

FULL RESOURCE: You will find a complete version of this canvas as part of the suite of templates and tools online.

How to Use the Canvas: A Step-by-Step Guide

Step 1 of 6: Individual Department Mapping
Begin by working with each key department individually to capture their perspective on value, concerns, metrics, and contributions.

Step 2 of 6: Cross-Functional Workshop
Bring representatives from all functions together to share their individual perspectives and identify conflicts, synergies, and integration points.

Step 3 of 6: Negotiation and Alignment
Facilitate structured discussion of priority-setting and trade-offs across functions, creating explicit agreements on how conflicts will be resolved.

Step 4 of 6: Integration Planning
Develop specific plans for collaboration at each integration point, including decision rights, communication protocols, and joint accountability.

Step 5 of 6: Narrative Development
Collaboratively create a shared value story that all functions can authentically support and communicate within their areas.

Step 6 of 6: Regular Refresh
Revisit the canvas at key milestones to assess alignment and make adjustments as the initiative evolves.

Facilitator Tip: When conducting the cross-functional workshop, give each department equal time to present their perspective before moving to discussion. This prevents dominant voices from framing the conversation.

Real-world example comes from a global bank that used the cross-functional value canvas to align seven departments around a customer service AI initiative. The process revealed that while IT was focused on a technical architecture that could scale globally, regional business units were primarily concerned with local customization. This fundamental misalignment had been invisible in previous planning. The canvas process led to a hybrid architecture approach that balanced standardization with flexibility, resolving what had been months of unproductive conflict.

FROM CONFLICT TO COLLABORATION: PRACTICAL STRATEGIES

Understanding departmental perspectives and using tools like the cross-functional value canvas provides the foundation for alignment. But translating this understanding into sustained collaboration requires addressing specific organizational friction points, including territorial concerns and resource competition.

On this foundation, organizations can demonstrate mutual benefits and impact, create genuine shared ownership of outcomes and build lasting collaborative mechanisms.

Addressing Territorial Concerns and Resource Competition

Territorial behaviour in organizations isn't just politics – it is a rational response to resource scarcity and performance pressures. Here are measures designed to overcome these natural barriers:

- **Create Dedicated Cross-Functional Resources:** Establish pools of funding, talent, and technology resources specifically for cross-functional AI initiatives, reducing the zero-sum competition.
- **Implement Transparent Resource Allocation:** Make the process for distributing AI resources explicit and merit-based, reducing perceived favouritism.
- **Recognize Departmental Expertise Contributions:** Create formal mechanisms to acknowledge when departments contribute specialized knowledge to enterprise initiatives.
- **Address the "Attribution Problem":** Develop ways to fairly recognize departmental contributions to shared outcomes without diminishing the collective achievement.
- **Create Balanced Governance Structures:** Ensure resource decisions involve balanced input from affected functions, not top-down allocation.

Consider the experience of a manufacturing company as a real-world example. The firm established an "AI value fund" with resources contributed by all major departments. Allocation decisions were made by a cross-functional committee using a transparent scoring model that balanced departmental and enterprise priorities. This dramatically reduced the resource conflicts that had previously stalled initiatives.

Creating Shared Ownership of Outcomes

Beyond resource allocation, successful AI initiatives require genuine shared ownership across functions:

- **Implement Joint Accountability Models:** Create explicit structures where multiple departments share responsibility for outcomes, not just activities.
- **Define Ownership at the Outcome Level:** Move from functional ownership of components to shared ownership of ultimate outcomes.
- **Create Cross-Functional Delivery Teams:** Form persistent teams with members from different functions who develop shared identity and purpose.
- **Establish Integrated Governance:** Implement decision structures where key choices require multi-functional agreement rather than sequential approvals.
- **Recognize Collective Achievements:** Celebrate successes as team wins rather than departmental accomplishments.

LEADERSHIP INSIGHT: True collaboration happens when departments see successful outcomes as reflecting positively on themselves – rather than competing for credit.

A real-world example relates to a financial services firm, which reorganized AI governance from a sequential approval process (business case → IT assessment → risk review → financial approval) to an integrated model where cross-functional teams collaboratively developed proposals addressing all dimensions simultaneously. This reduced approval time from months to weeks and created solutions with stronger cross-functional support.

Demonstrating Mutual Benefit and Impact

Perhaps the most powerful driver of collaboration is a clear understanding of how working together creates more value than working separately:

- **Quantify the Collaboration Premium:** Measure and communicate the additional value created through cross-functional approaches compared to siloed alternatives.
- **Develop Function-Specific Value Narratives:** Create clear articulations of how each department benefits from the collaborative approach.
- **Implement Early Win Strategies:** Identify and deliver quick cross-functional successes that demonstrate mutual benefit.
- **Create Visibility into Interdependence:** Help departments see how their success depends on integration with other functions.
- **Share Success Stories:** Actively communicate examples where cross-functional collaboration led to superior outcomes.

For a real-world example, consider the outcomes of a telecommunications company that analysed the results of AI initiatives over a two-year period and found that projects with strong cross-functional governance delivered 3.2 times more value than those managed within single departments. This analysis became a powerful tool for overcoming resistance to more collaborative approaches.

CASE STUDY: TELSTRA'S PARTNERSHIP APPROACH

The Situation: While many organizations struggle with internal cross-functional collaboration, Telstra took a different approach by creating a strategic joint venture with Accenture.

The Approach: This partnership enabled the company to streamline AI and data providers while accelerating the AI roadmap across multiple business areas. The company pursues an ambitious goal that relates to an agentic AI future, where key tasks are optimised and teams work with intelligent AI agents.

The Insight: The partnership demonstrates how external collaboration can sometimes facilitate internal cross-functional implementation. By bringing in outside expertise with experience in enterprise-wide transformation, Telstra was able to create a more integrated approach than it might have achieved working solely with internal resources.

The Lesson: Sometimes the path to internal cross-functional alignment runs through external partnerships that bring fresh perspective and established integration methodologies. The key is ensuring these partnerships enhance rather than replace internal collaboration capabilities.

CONSIDER THIS: Could an external partner help facilitate cross-functional alignment in your organization by providing neutral expertise and integration experience?

BUILDING LASTING COLLABORATIVE MECHANISMS

Beyond resolving specific conflicts, organizations need sustainable structures that foster ongoing collaboration. There are four key mechanisms that help drive success: implementing regular alignment rituals; developing cross-functional career paths; and building collaboration into organizational design.

- **Create Formal Integration Roles:** Establish positions specifically responsible for cross-functional coordination and alignment. As Colin Reeves explains, chief AI officers are "typically mandated to modernise processes with AI, ensuring that AI is used with ethics

Building Cross-Functional Support for AI Initiatives

and governance in mind, and in building an AI-first culture." This responsibility necessarily spans departmental boundaries.
- **Implement Regular Alignment Rituals:** Design recurring processes that bring functions together around shared priorities and challenges.
- **Develop Cross-Functional Career Paths:** Create advancement opportunities that reward integration skills and diverse functional experience.
- **Build Collaboration into Organization Design:** Structure teams and reporting relationships to facilitate rather than impede cross-functional work.
- **Create Shared Digital Environments:** Implement technology platforms that enable visibility and collaboration across departmental boundaries.

QUICK WIN: Establish a monthly "cross-functional AI review" where representatives from each key department share updates, challenges, and opportunities for collaboration.

A real-world example comes from a healthcare system that established a team of "AI value integrators." These professionals with both technical understanding and business acumen facilitated collaboration between clinical, operational, and technical functions. These roles became highly sought-after career steps, creating a pipeline of leaders with cross-functional perspective.

THE HIDDEN COSTS OF FAILED COLLABORATION: LESSONS FROM THE EUROPEAN SPACE AGENCY

Sometimes, a dramatic failure can serve to reveal the true importance of cross-functional collaboration. The European Space Agency experienced a catastrophic failure when "the explosion of the $600 million Ariane 5 within 96 seconds of its maiden flight" was attributed to "poor inter-team collaboration" despite each component being independently sound.

This case illustrates a critical insight: technical excellence within functions isn't enough when these functions don't collaborate effectively. Each team had built components that functioned perfectly when tested individually but failed when integrated, because teams hadn't adequately communicated about how those components would interact.

In contrast, "NASA, which had historically struggled with disconnected silos and endured failures of its own, moved to avoid such a painful recurrence by bringing its third-party suppliers in-house." NASA recognized that physical and organizational proximity could foster the kind of ongoing communication needed for complex systems integration.

WARNING SIGN: If your teams are saying, "It sometimes feels much easier to share information with my network across industry than to exchange data among colleagues within my own firm," you have a serious collaboration problem.

The lesson is stark: the cost of failed cross-functional collaboration isn't just inefficiency – it can be complete system failure, even when individual components work perfectly.

CASE STUDIES IN CROSS-FUNCTIONAL SUCCESS

SECTION OVERVIEW: This section presents four detailed case studies showing successful cross-functional AI initiatives:

Theory and frameworks provide the foundation, but real-world examples bring cross-functional collaboration to life. Here are four detailed case studies that illustrate different aspects of successful cross-boundary AI initiatives. With a lens on cross-functional integration, we revisit the examples of the marketing and operations collaboration of a retail company; the engineering and supply chain integration of a manufacturing firm; the clinical and administrative alignment in healthcare; and the risk, compliance, and innovation partnership in financial services.

RETAIL: HOW MARKETING AND OPERATIONS UNITED THROUGH AI

The Challenge
A major retail chain was struggling to implement an AI-driven demand forecasting system. The marketing department wanted highly granular customer segmentation to drive personalized campaigns, while operations needed stable aggregate forecasts to manage inventory and staffing. Previous attempts had satisfied neither group.

The Breakthrough Approach
Rather than treating these as competing requirements, the company implemented a "dual-layer" forecasting architecture:

- A shared data foundation that both teams contributed to and maintained
- A granular prediction layer optimized for marketing personalization
- An aggregate stability layer optimized for operational planning
- A cross-functional governance team that prioritized model improvements
- Shared metrics that balanced personalization impact and operational stability

Key Success Factors

- Recognition that both perspectives had validity rather than forcing a single approach
- Technical architecture that accommodated different functional needs
- Shared data infrastructure that created mutual dependency
- Governance model with balanced influence
- Leadership that valued integration over functional optimization

The Results

- 23% improvement in forecast accuracy
- 18% increase in campaign effectiveness
- 12% reduction in inventory costs
- Culture shift from competition to collaboration between previously adversarial departments

MANUFACTURING: BRIDGING ENGINEERING AND SUPPLY CHAIN

The Challenge
A global manufacturing company was implementing an AI-based predictive maintenance system. The engineering team was focused on technical performance and equipment reliability, while the supply chain team was primarily concerned with parts availability and service logistics. These different priorities had created parallel initiatives with limited integration.

The Breakthrough Approach
The company established a "maintenance value stream" approach that:

- Created integrated teams spanning engineering, operations, and supply chain
- Developed a shared digital twin environment where both groups could see impact
- Implemented unified maintenance planning that optimized across both domains
- Created co-located "reliability centres" with representatives from all functions
- Established shared metrics around total cost of ownership rather than functional KPIs

Building Cross-Functional Support for AI Initiatives

Key Success Factors

- Physical co-location that fostered ongoing interaction
- Unified technical environment that created shared visibility
- Metrics that transcended functional boundaries
- Leadership emphasis on end-to-end optimization
- Regular cross-training between engineering and supply chain roles

The Results

- 35% reduction in unplanned downtime
- 28% decrease in maintenance parts inventory
- 15% improvement in technician utilization
- Breakdown of longstanding cultural barriers between "technical" and "operational" functions

HEALTHCARE: CLINICAL AND ADMINISTRATIVE ALIGNMENT

The Challenge

A healthcare system was implementing AI for patient flow optimization. Clinical leaders were primarily concerned with quality of care and clinical outcomes. Administrative teams were focused on efficiency and cost management. These different priorities had stalled previous improvement efforts.

The Breakthrough Approach

The organization implemented a "patient journey" framework that:

- Formed cross-functional teams organized around patient pathways rather than departments
- Created a shared visualization of how clinical and operational factors affected patient experience
- Developed integrated metrics that linked quality, experience, and efficiency
- Implemented joint rounds where clinical and administrative leaders reviewed cases together
- Established a balanced governance structure with equal clinical and operational influence

Key Success Factors

- Focus on patient experience as a unifying priority
- Visual management that made interdependencies clear
- Leadership modelling of collaborative behaviour
- Balanced influence in decision-making
- Regular forums for joint problem-solving

The Results

- 31% reduction in patient wait times
- 22% improvement in staff satisfaction
- 16% increase in patient experience scores
- 12% decrease in operating costs
- Transformational change in clinical-administrative relationships

PATTERN RECOGNITION: Notice how each successful case started with a unifying goal that transcended departmental boundaries (dual-layer forecasting architecture, maintenance value stream, patient journey framework).

FINANCIAL SERVICES: RISK, COMPLIANCE, AND INNOVATION PARTNERSHIP

The Challenge
A global bank was struggling to implement AI-based fraud detection. Innovation teams wanted to rapidly deploy advanced algorithms, while risk and compliance functions were concerned about explainability, bias, and regulatory requirements. The result was a standoff that left the bank vulnerable to emerging fraud patterns.

The Breakthrough Approach
The bank developed a "responsible innovation framework" that:

- Created integrated teams including risk, compliance, and technical experts
- Implemented a staged deployment model with appropriate controls at each level
- Developed explainability tools that satisfied both technical and regulatory needs
- Established clear thresholds for different risk levels with corresponding governance
- Created a dedicated AI ethics committee with diverse representation

Key Success Factors

- Early involvement of risk and compliance in design rather than after-the-fact review
- Technical approaches that addressed regulatory concerns by design
- Shared language that bridged technical and regulatory terminology
- Governance scaled to risk level rather than one-size-fits-all
- Leadership emphasis on both protection and innovation

The Results

- 42% improvement in fraud detection
- 65% faster deployment of new models
- Zero regulatory findings in subsequent audits
- Cultural transformation from "risk as blocker" to "risk as driver of transformation"
- Framework became an industry benchmark for responsible AI innovation

VALUES AS FOUNDATION FOR CROSS-FUNCTIONAL COLLABORATION

KEY CONCEPT: Shared values create the foundation for successful cross-functional collaboration.

At the heart of effective cross-functional collaboration lies a deeper element that's easy to overlook: shared values. As John Maxwell, author, leadership development expert and host of The Maxwell Leadership Podcast, explains, "Values are like a magnet, they attract like-minded people. They're an identity, they define and identify who the team is."

Yet values serve different functions in organizational collaboration. Maxwell describes values as determining "the foundation of the team" and defining "personal value… as something that influences and guides my behaviour" while organizational value is "something that influences and guides the team's behaviour."

Sam Silverstein, writer, consultant and accountability expert, takes a more structural view, noting that values function as "the glue" that connects people in an organization, distinguishing between organizations where "people get along really great" versus those with "sideload communication."

Deeper Insight: When departments struggle to collaborate on AI initiatives, this often happens because they're operating from different foundational values, not just different priorities or incentives.

This tension between values as attraction versus connection points to different approaches for fostering cross-functional collaboration. Creating effective cross-functional collaboration for AI requires not just aligning metrics and incentives, but also finding or creating shared values that can serve as a foundation for joint work.

This doesn't mean departments must share all values – finance can still prioritize fiscal responsibility while marketing values customer experience – but they need enough common ground to build meaningful collaboration.

REFLECTION QUESTIONS: What values could serve as common ground across departments in your organization's AI initiatives? Customer focus? Evidence-based decision making? Innovation with integrity?

BREAKING THE SILOS: LIGHTFUL'S AI SQUAD APPROACH

Lightful, a technology company, demonstrates how cross-functional collaboration can be built into the fabric of AI development. Rather than approaching AI as a primarily technical effort with occasional input from business stakeholders, a dedicated cross-functional "AI squad" – comprising members from technology, design, and product departments – was created.

Their collaborative problem-solving approach enabled them to create AI tools that genuinely addressed user needs rather than just technical possibilities. By bringing together diverse expertise from the start, they avoided the common pattern where technical teams build impressive capabilities that don't solve real business problems.

THE KEY INSIGHT: Cross-functional collaboration isn't something that happens after technical work is completed; it needs to be integrated into the development process itself. When teams with different perspectives collaborate from inception to implementation, they create solutions that naturally integrate different functional requirements rather than trying to reconcile competing priorities after the fact.

CONCLUSION: THE COLLABORATION IMPERATIVE

As we've explored throughout this chapter, building cross-functional support isn't just a nice-to-have for AI initiatives – it is a fundamental requirement for success.

The most powerful AI technologies will fail to deliver value if they remain trapped in functional silos. Conversely, even relatively simple AI applications can create transformative impact when they successfully bridge organizational boundaries.

Let me leave you with five key principles for cross-functional AI success:

1. **Integration Creates More Value Than Optimization:** The greatest AI opportunities don't come from optimizing individual functions but from connecting processes, insights, and capabilities across traditional boundaries.
2. **Collaboration Must Be Designed, Not Assumed:** Cross-functional alignment doesn't happen naturally in most organizations – it requires intentional structures, processes, and incentives for overcoming institutional barriers.
3. **Perspective Diversity Is a Feature, Not a Bug:** Different functional viewpoints aren't obstacles to be overcome but essential inputs that create more robust and valuable solutions when properly integrated.
4. **Shared Ownership Drives Sustained Success:** AI initiatives thrive when multiple functions genuinely own outcomes together, not when one function drives while others comply.
5. **Leaders Must Model Integration:** Cross-functional collaboration mirrors leadership behaviour. When executives operate in silos, the organization will follow suit regardless of formal structures.

TAKING ACTION: Choose one technique from this chapter to implement in your next AI planning meeting. Start small but start now.

As AI becomes increasingly central to organizational success, the ability to build genuine cross-functional alignment will separate leaders from laggards. The frameworks and approaches in this chapter provide a roadmap for creating the collaborative foundation that AI value requires.

In the next chapter, we'll explore how to extend this alignment beyond formal departments to build broader organizational adoption and trust – turning cross-functional support into genuine organizational momentum.

9 Fostering Employee Trust and Adoption

Chapter Roadmap

In this chapter, we look at why people are the critical foundation of AI value. From there, we come to understand legitimate concerns driving resistance, look at the adoption flywheel framework, provide actionable trust-building strategies – and show real-world success stories that bring these concepts to life.

THE HUMAN SIDE OF AI VALUE

Here's a truth that gets lost amid all the technical spectacle: AI doesn't create value until people actually use it.

Not just tolerate it or grudgingly comply with it, but genuinely embrace, trust and incorporate it into their daily work.

Consider this scene I witnessed at a global logistics company. The data team had spent nine months building an impressive route optimization system using cutting-edge algorithms. The technical metrics were outstanding: 23% more efficient routes, potential fuel savings in the millions, significantly reduced delivery times.

Yet six months after deployment, nothing had changed. The numbers showed minimal improvement. Why? Because the drivers weren't using it.

When I spoke with the drivers, their reasons were illuminating:

"The system doesn't know about the construction on 7th Street that's been going on for months."

"It keeps sending me to delivery locations during their lunch breaks when no one's available to receive packages."

"The last time I followed its suggestion, I ended up on a road too narrow for my truck. Had to back up half a mile."

"It doesn't account for which customers are flexible and which ones need precision timing."

The AI system had been designed with remarkable technical sophistication but not enough empathy for the people expected to use it. The drivers weren't being stubborn or technophobic – they were making rational choices based on their expertise and experience.

KEY INSIGHT: The most sophisticated AI in the world creates zero value if people don't trust it enough to use it.

This pattern repeats across organizations of all types. The healthcare AI that physicians quietly work around. The customer service system that agents minimize on their screens while handling calls their own way. The inventory optimization tool that store managers override with manual orders.

Such projects could be considered technical successes that failed at adoption.

They represent perhaps the most overlooked aspect of AI value: the critical gap between capability and adoption. Between what AI can do and what people will actually use it to do.

PATTERN RECOGNITION: In organizations with failed AI implementations, look for signs of what I call "passive rejection" – systems that are technically live but practically ignored.

UNDERSTANDING EMPLOYEE PERSPECTIVES ON AI

Before we can build trust and adoption, we need to understand the legitimate concerns that drive resistance. These aren't irrational fears to be dismissed but valid perspectives that must be addressed.

REAL CONCERNS BEHIND AI RESISTANCE

Employee concerns about AI typically fall into four categories, each requiring a different response:

1. **Capability Concerns: "Will It Work?"**
   - Does the AI actually perform as promised?
   - Can it handle the exceptions and edge cases that make up real work?
   - Is it reliable in high-pressure situations?
   - Does it integrate with existing systems and workflows?
2. **Control Concerns: "Who's in Charge?"**
   - Does the AI make recommendations or decisions?
   - Can employees override the system when needed?
   - Who's accountable when things go wrong?
   - How transparent is the reasoning behind AI suggestions?
3. **Career Concerns: "What Happens to Me?"**
   - Will this replace parts of my job that I value?
   - Do I have the skills to work with this new technology?
   - How will my performance be measured with this new system?
   - Does this change my career trajectory and opportunities?
4. **Culture Concerns: "How Does This Change Who We Are?"**
   - Does this technology align with our values and identity?
   - Will this change what's valued in our organization?
   - How will this affect our team dynamics and relationships?
   - Does this undermine the human elements that matter in our work?

REFLECTION QUESTIONS: Which of these four concern categories is most prevalent in your organization, and how might that impact your AI implementation approach?

Beyond managing change resistance, understanding these concerns is important for recognizing legitimate issues that, if addressed, lead to better solutions.

THE CREATIVE PROFESSIONAL'S DILEMMA: IDENTITY AT STAKE

At KEO International Consultants, architects and designers weren't merely wary of new technology – they saw AI as a fundamental threat to their professional identity. As Damir Jaksic from KEO explains, "In our field of architecture and design, creativity is highly valued. Architects, designers, and interior designers feel that AI-powered tools would stifle their creative expression or undermine the uniqueness of their design."

This wasn't just about job security – it was about the very essence of what makes someone a creative professional. KEO's leadership recognized this deeper concern and completely reframed the approach. Rather than positioning AI as an efficiency tool, they demonstrated how it could enhance creativity by handling routine tasks and providing inspiration, particularly valuable in an industry facing talent shortages.

Jaksic observed, "AI is coming up at the right time in my view to help increase productivity." This statement illustrates how the messaging evolved to address the real underlying concern – not that AI would replace jobs, but that it would transform the creative nature of those jobs.

WARNING SIGN: When employees express concerns about AI quality or reliability, these topics often mask deeper fears about professional identity and value. Listen for what's not being said.

Anxiety about Job Displacement and Skills Relevance

Let's address the elephant in the room. Many employees do worry about AI's impact on their jobs, and ignoring this concern doesn't make it go away.

This anxiety typically manifests in three distinct forms, each requiring a different approach:

Replacement Anxiety: The fear that AI will eliminate one's role.
- Most acute in roles with highly structured, repetitive tasks
- Often expressed indirectly through questioning of AI reliability
- May lead to unconscious sabotage of implementation

Deskilling Anxiety: Concern that AI will transform roles to require less expertise.
- Common among knowledge workers and professionals
- Focused on loss of professional identity and status
- May manifest as insistence on maintaining current processes

Relevance Anxiety: Worry about whether one's skills will remain valuable.
- Present across all levels, including leadership
- Centres on career trajectory and future opportunities
- Often expressed as concern about "keeping up"

QiFang Sun, CIO at Collectius, captured the stark duality in employee reactions to AI: "One end is, they think [that AI is] so powerful that it's going to replace the humans – they are afraid that they might lose a job," while at "the other end is, people view it as a black box. They don't see the explainability and they don't trust it."

This perfect storm of both fear and scepticism creates a complex challenge that cannot be solved by simply highlighting benefits.

SURPRISING STATISTIC: Research by Hamilton (2025) identified that 63% of employee resistance to AI stems from concerns about job displacement and skills relevance rather than the technology itself. Organizations that addressed these concerns directly saw 78% higher adoption rates than those focusing solely on technical training.

Addressing these anxieties requires more than reassurance – it demands concrete plans for how roles will evolve rather than disappear, how expertise will be elevated rather than eliminated, and how new skills will be developed.

ACTION STEP: Create a "skills evolution map" for each affected role, showing:

1. Which current tasks AI will support
2. Which new responsibilities will emerge
3. What skills development will be provided
4. How performance evaluation will evolve

Trust in AI Systems and Decisions

Beyond job concerns, employees often question whether they can – or should – trust AI systems, particularly in consequential domains.

This trust equation has several components:

Performance Trust: Confidence in the AI's basic capabilities.
- Does it consistently do what it claims?
- How does it handle edge cases and exceptions?
- Does its performance degrade gracefully or catastrophically (in other words, when the system encounters unfamiliar situations, does it gradually become less accurate while simultaneously signalling its uncertainty – or does it suddenly produce completely unreliable results without warning)?

Process Trust: Belief in how the AI works.
- Is the reasoning transparent?
- Can users understand why specific recommendations are made?
- Are there signs to tell when the system is uncertain or guessing?

Purpose Trust: Alignment with intended objectives.
- Is the AI optimizing for the right outcomes?
- Does it balance competing priorities appropriately?
- Are its objectives aligned with user and organizational values?

People Trust: Confidence in the humans behind the AI.
- Who built this and do they understand our context?
- Who maintains it and responds when issues arise?
- Who makes decisions about how it evolves?

Daniel Larsson, Senior Architect at Cristie Nordic, discovered that many misconceptions about AI stem from a fundamental misunderstanding of its purpose. He explains, "The purpose of LLMs [large language models] is to keep the conversation going. It's supposed to lead you on, ask questions and interact [with you]. Once you understand what it's built for, it takes away the issue with hallucinations."

While this insight may seem controversial, it highlights the fact that building trust often begins with simply clarifying what the system is actually designed to do, rather than letting fears and misconceptions fill the knowledge gap.

PATTERN RECOGNITION: The most successful AI implementations build trust by addressing all four dimensions – performance, process, purpose, and people – rather than focusing solely on accuracy metrics.

QUICK WIN: Create a one-page "trust sheet" for your AI system that clearly explains:

- What it does and doesn't do
- How it makes recommendations
- When users should override it
- Who to contact with concerns

The Human Side of AI Transformation

Finally, we must recognize that AI adoption isn't just a technical or even rational process – it's profoundly emotional and social.

The foundation for successful AI transformation often lies in values alignment – ensuring that new technology supports rather than undermines what people care most about in their work. This principle applies whether you're hiring new people or implementing new systems.

Jeffery Crabb, CEO of Diamond C, learned this lesson the hard way when growing his business: "When we scaled our business, we hired for function but not for values. I learned that hiring without a focus on values alignment creates misalignment and weakens the core of the culture, making it hard to keep everyone connected."

This insight applies even more so to AI implementations, where technology that doesn't align with core values can create profound disconnection and resistance.

Employees don't experience AI implementations as abstract technological changes. They experience them as:

Identity Shifts: Changes in "who I am" in the organization.
- From expert to advisor
- From decision-maker to decision-supporter
- From independent contributor to system partner

Relationship Recalibrations: Transformations in work connections.
- Between colleagues as responsibilities shift
- Between managers and teams as oversight changes
- Between employees and customers as interaction patterns evolve

Value Realignments: Changes in what matters and why.
- What constitutes "good work" in an AI-enabled environment
- Which skills and contributions receive recognition
- How success is defined and measured

Case Study: The Identity Crisis in Retail Banking

A retail bank implemented an AI-powered customer service system that changed how branch employees engaged with customers. Despite strong technical performance, adoption stalled because the system shifted employee identity from "financial expert" to "technology facilitator" – a change that undermined their sense of professional value.

The breakthrough came from reframing the system as enhancing rather than replacing expertise. The team created new recognition programmes for "augmented advisory excellence," celebrating how employees added human value beyond the AI. Within three months, the same employees who had been avoiding the system became its champions, and customer satisfaction scores reached an all-time high.

KEY INSIGHT: Human dimensions are often invisible in implementation plans but can make or break adoption. The most technically successful AI can fail if it requires identity shifts, relationship changes, or value realignments that people aren't prepared for or willing to make.

THE ADOPTION FLYWHEEL

Understanding employee perspectives is the foundation – but driving adoption requires a structured approach that builds momentum over time. I've found the "adoption flywheel" model to be particularly effective in moving organizations from initial awareness to full integration of AI capabilities.

Unlike linear adoption models, the flywheel recognizes that adoption is cyclical and self-reinforcing when properly designed. Each successful turn of the wheel generates energy for the next, creating accelerating momentum rather than constant pushing.

The adoption flywheel: Awareness → Understanding → Use → Value → Belief → Advocacy

Let's explore each stage of this flywheel and how to optimize it for different organizational contexts.

STAGE 1: AWARENESS – GETTING ON THE RADAR

The adoption journey begins with basic awareness, with ensuring people know the AI solution exists, understand its basic purpose, and recognize it as relevant to their work.

Key Challenges at This Stage

- Information overload and competing priorities
- Preconceptions based on prior technology experiences
- Unclear relevance to daily work realities
- Anxiety triggered by initial messaging

Effective Approaches

- Connect to felt pain points rather than technical capabilities
- Use storytelling to make benefits tangible and relatable
- Involve respected peers in initial communications
- Address concerns directly rather than ignoring them
- Create low-pressure opportunities for initial exposure

WARNING SIGN: If your awareness campaign focuses primarily on the technical sophistication of your AI rather than its relevance to daily work, you're likely creating more resistance than interest.

ACTION STEP: Map the top three pain points for each department that will use your AI system, and ensure all awareness communications directly address how the system will help with those specific challenges.

STAGE 2: UNDERSTANDING – CREATING COMPREHENSION

Once aware, employees need to develop a functional understanding of what the AI does, how it works (at an appropriate level), and how it fits into their work context.

Key Challenges at This Stage

- Varying technical literacy across the organization
- Misconceptions about AI capabilities and limitations
- Difficulty connecting abstract functions to practical applications
- Minimal motivation to invest in learning

Effective Approaches

- Create layered explanations for different knowledge levels
- Use visual models that simplify complex functionality
- Provide concrete examples relevant to specific roles
- Offer hands-on demonstration opportunities
- Develop simple mental models that build on existing knowledge

KEO International Consultants developed a brilliantly simple approach called "Summer of Innovation" – a series of accessible lunchtime sessions taught by senior leaders that broke down complex AI concepts into digestible pieces. Rather than overwhelming employees with technical details, the team provided just enough context for people to experiment with AI on their own, fostering understanding through practical exploration.

This approach recognized that understanding doesn't come from comprehensive training but from supported experimentation that builds confidence and competence.

Fostering Employee Trust and Adoption

QUICK WIN: Create role-specific "day-in-the-life" scenarios showing exactly how the AI system fits into normal workflows. Include screenshots, simple decision trees, and before/after comparisons.

Stage 3: Use – First Engagement with the System

The critical jump from understanding to actual use represents the most significant adoption hurdle. This initial usage experience often determines whether adoption will continue or stall.

Key Challenges at This Stage

- Friction in initial interaction with the system
- Disruption to established work patterns
- Performance expectations versus reality
- Uncertainty about appropriate reliance level

Effective Approaches

- Design low-risk entry points with high success probability
- Provide in-context guidance during initial usage
- Create "safety nets" for initial trial periods
- Celebrate and recognize early usage attempts
- Ensure immediate support for questions and issues

QiFang Sun at Collectius found an ingenious way to create positive first experiences with AI. Rather than implementing a complex system with high stakes, they started with a multilingual communication tool that helped Thai colleagues write business correspondence in English more confidently. This approach provided immediate, visible benefits with minimal risk, building positive associations with AI that encouraged further exploration.

PATTERN RECOGNITION: The most successful first AI experiences involve:

1. High probability of success
2. Low consequence of failure
3. Immediate visible benefit
4. Minimal disruption to existing workflows

WARNING SIGN: If your AI implementation requires users to dramatically change their workflows from day one, you're likely creating an adoption barrier that's difficult to cross.

Stage 4: Value – Experiencing Tangible Benefits

For adoption to continue and grow, users must experience personal value – benefits that matter to them specifically, not just to the organization.

Key Challenges at This Stage

- Value often emerges over time, not immediately
- Different roles experience different benefits
- Organizational benefits may not translate to personal ones
- Value may be offset by initial learning costs

Effective Approaches

- Identify and highlight role-specific value drivers

- Create visibility into emerging benefits, even small ones
- Collect and share authentic peer value stories
- Reduce friction costs that offset perceived benefits
- Build in early wins that create immediate small benefits

Michael Kors' implementation of Mastercard's "Shopping Muse" AI assistant shows how value must be visible at multiple levels. While the organization measured value through a 15 to 20% increased conversion rate, it ensured that frontline employees could see how the system improved their personal metrics and customer interactions. This dual value perspective – organizational and personal – was critical to sustained adoption.

KEY INSIGHT: Value doesn't just happen – it must be intentionally designed, measured, and made visible to the people whose adoption drives that value.

REFLECTION QUESTIONS: What specific value metrics matter most to different roles in your organization, and how will you make the impact of AI visible as implementation progresses?

STAGE 5: BELIEF – DEVELOPING TRUST IN THE SYSTEM

As value experiences accumulate, users develop deeper belief in the AI system's capabilities and reliability, transitioning from cautious usage to meaningful reliance.

Key Challenges at This Stage

- Building trust requires consistent positive experiences
- Single negative experiences can disproportionately affect trust
- Trust builds at different rates for different individuals
- Different aspects of trust may develop unevenly

Effective Approaches

- Make system reliability and limitations transparent
- Create visibility into ongoing improvements
- Establish clear resolution paths when issues arise
- Share success patterns and evidence across peer groups
- Recognize and address trust variations across teams

QUICK WIN: Create a simple feedback channel where users can report both successes and challenges with the AI system, and ensure they receive a response within 24 hours showing how their input is being used.

STAGE 6: ADVOCACY – BECOMING PROMOTERS WITHIN THE ORGANIZATION

The flywheel reaches its highest momentum when users become active advocates, promoting the AI system to peers and driving organic adoption growth.

Key Challenges at This Stage

- Converting passive acceptance into active promotion
- Giving advocates appropriate tools and information
- Sustaining enthusiasm as the system becomes normalized
- Balancing authentic advocacy with formal messaging

Fostering Employee Trust and Adoption

Effective Approaches

- Create formal and informal advocacy opportunities
- Provide advocates with supporting materials and data
- Recognize and reward advocacy contributions
- Connect advocates across the organization
- Involve advocates in future development decisions

Damir Jaksic from KEO International Consultants captured this perfectly: "One of the most important parts is recognizing and rewarding innovation [by] acknowledging employees who embrace AI. [This is to] incentivize others to follow suit."

This insight shows that advocacy doesn't happen automatically but must be actively cultivated through appropriate recognition systems that celebrate those who champion new approaches.

ACTION STEP: Identify potential AI champions within each department, not based on seniority but on peer influence and openness to change. Provide them with early access, additional support, and opportunities to shape the implementation.

BARRIERS AT EACH STAGE AND HOW TO OVERCOME THEM

For the adoption flywheel to build momentum, we must identify and address the specific barriers that can stall progress at each stage.

Awareness Barriers

- Information noise and competition for attention
- Negative preconceptions about AI
- Fear triggered by initial messaging
- Lack of perceived relevance

Understanding Barriers

- Technical complexity beyond user background
- Insufficient explanation for specific roles
- Lack of concrete examples relevant to daily work
- Cognitive overload from too much information

Use Barriers

- High friction in initial interactions
- Workflow disruption without clear benefit
- Fear of mistakes or looking incompetent
- Inadequate support during learning curve

Value Barriers

- Benefits too delayed or abstract
- Personal costs that offset organizational benefits
- Misalignment with what users actually value
- Inconsistent or unreliable performance

Belief Barriers

- Trust damaged by early negative experiences
- Lack of transparency about limitations

- Inconsistent performance across use cases
- Unclear responsibility when issues arise

Advocacy Barriers

- No clear channels for sharing experiences
- Lack of recognition for influence efforts
- Insufficient information to support peer questions
- Organizational culture that discourages peer influence

SHOCKING STATISTIC: Writer's 2025 Survey revealed that 31% of employees admit to actively undermining their company's generative AI strategies, with higher rates among Gen Z and millennial workers. One in ten workers reported tampering with performance metrics to make AI appear less effective.

This resistance is largely driven by fears that AI will diminish their value and creativity (33%), take over their jobs (28%), or increase their workload (24%). These statistics highlight that barrier identification and addressing root concerns isn't just good practice – it is essential for preventing active resistance.

ACTION STEP: Barrier Removal Framework

1. **Targeted Barrier Assessment:** Identify specific barriers at each flywheel stage through user research, not assumption.
2. **Stage-Appropriate Interventions:** Address barriers with interventions designed for that specific stage rather than general adoption support.
3. **User-Segment Customization:** Recognize that different user groups face different barriers and require tailored approaches.
4. **Continuous Barrier Monitoring:** Track changes in adoption barriers as implementation progresses and users gain experience.
5. **Barrier Removal Resources:** Allocate specific resources to addressing the highest-impact barriers rather than spreading change management efforts evenly.

ROLE-SPECIFIC ADOPTION TECHNIQUES

The adoption flywheel spins differently for different roles within the organization. Let's examine how to optimize adoption approaches for key organizational segments:

Executive Adoption

Executives often have limited direct interaction with AI systems but significant influence over organizational adoption through their signalling and resource allocation.

Key Adoption Drivers

- Strategic alignment with personal priorities
- Competitive positioning and market relevance
- Measurable impact on key performance indicators
- Reputational benefits within and beyond the organization

Effective Approaches

- Connect AI to strategic narratives they already support
- Create short, high-impact exposure experiences

- Provide competitive intelligence on peer adoption
- Design "impact dashboards" showing value in their terms
- Create opportunities for visible leadership association

Bogdan Nita from World Vision International could not be more clear about this necessity: "Everything starts with the top. If you don't have the right buy-in at the top-level organization – starting even with the board of directors, but then moving to the executive leadership team – below that level, nothing will ever happen."

PERCEPTION GAP ALERT: Research reveals a stunning 30-point difference in how executives and employees view AI adoption success. While 70% of leaders believe their AI approach is strategic and successful, only 40% of frontline workers agree – a fundamental disconnect that threatens AI initiatives.

QUICK WIN: Create a one-page "executive AI impact dashboard" that shows direct connections between AI initiatives and the strategic priorities each executive personally champions.

Manager Adoption

Middle managers often form a critical adoption layer as they translate executive vision into team implementation and can either amplify or dampen adoption signals.

Key Adoption Drivers

- Impact on team performance metrics
- Resource efficiency and allocation benefits
- Team satisfaction and retention
- Personal productivity and decision support
- Visibility and recognition opportunities

Effective Approaches

- Emphasize how AI supports their management challenges
- Provide team-level impact visibility
- Create peer forums for sharing management approaches
- Develop manager-specific usage guidance
- Address their role in supporting team adoption

KPMG's success in integrating AI tools like GPT and Copilot came from embedding these technologies within their "Hearts Not Heads" model. By breaking down silos between accounting, tax, and advisory services, they enabled seamless data sharing and unified workflows. This approach centred on positioning AI as an augmentation to professional expertise rather than a replacement, giving managers clear ways to demonstrate this benefit to their teams.

ACTION STEP: Create manager-specific AI implementation toolkits that include:

1. Team adoption roadmaps
2. FAQ documents for common team concerns
3. Success metrics at both team and individual levels
4. Recognition frameworks for early adopters

Frontline Adoption

Frontline employees often have the most direct interaction with AI systems – and their adoption directly determines operational value realization.

Key Adoption Drivers

- Daily work simplification and friction reduction
- Performance improvement without additional effort
- Enhanced capabilities and decision support
- Protection from errors and mistakes
- Recognition for expertise and adaptation

Effective Approaches

- Focus intensely on workflow integration and usability
- Create clear visibility into personal benefit
- Provide extensive practice opportunities before full deployment
- Implement robust support systems for questions and issues
- Recognize and reward adaptation efforts

PATTERN RECOGNITION: When a telecommunications company implementing an AI customer service system conducted "barrier mapping" workshops with different employee segments, they discovered that the technical support team worried about diagnostic accuracy while the billing team feared customer reactions to AI-generated explanations. This difference in concerns required completely different adoption strategies for each group.

By developing barrier-specific interventions for each team, the organization achieved 82% adoption across all teams within three months, far exceeding their previous best technology adoption rate of 60% over six months.

KEY INSIGHT: For AI implementations, it's important to recognize that there is no "one-size-fits-all" path to adoption. Different roles have different concerns and different value drivers – and they require different support systems.

THE AI TRUST-BUILDING PLAYBOOK

While the adoption flywheel provides the overall framework, organizations need specific tactics to build trust as the foundation of sustainable adoption. The following playbook provides practical approaches proven to build AI trust across different contexts.

Transparent Communication Strategies

Beyond being an ethical principle, transparency is a practical trust-building approach. Effective transparency strategies include:

1. **Capability Transparency:** Be honest about what the AI can and cannot do.
   - Clearly communicate both capabilities and limitations
   - Avoid overpromising and setting unrealistic expectations
   - Update capability descriptions as the system evolves
   - Differentiate between current and planned functionality
2. **Confidence Transparency:** Make uncertainty visible.
   - Implement confidence scores for AI-generated recommendations (such as loan approvals, medical diagnoses, product suggestions, or maintenance alerts)
   - Clearly indicate when the system is operating outside normal parameters
   - Distinguish between high- and low-confidence predictions
   - Provide guidance on appropriate reliance based on confidence

Fostering Employee Trust and Adoption

3. **Purpose Transparency:** Be clear about why the AI exists.
   - Communicate the specific problems the AI aims to solve
   - Explain how success is measured and evaluated
   - Clarify whose needs the system is designed to meet
   - Update purpose as organizational priorities evolve
4. **Process Transparency:** Explain how decisions are made.
   - Provide appropriate explanations for key recommendations
   - Create visibility into major data sources
   - Offer routes to additional information for those who want it
   - Balance detail with usability in explanations

Case Study: The "Trust Layer" in Financial Services

A financial services firm implementing an AI-based loan recommendation system created what they called a "trust layer" in their user interface. This included:

- Confidence indicators for each AI-generated recommendation
- Plain-language explanations of major factors influencing decisions
- Clear indications when applications fell outside normal parameters
- One-click access to detailed reasoning when needed

This transparency approach increased both adoption rates and appropriate reliance levels, with loan officers showing better judgement about when to follow or override recommendations. Six months after implementation, loan processing time decreased by 37% while exception rates dropped by 28%.

QUICK WIN: Create a simple "AI transparency card" for your system that honestly addresses:

- What it does well
- Where it may struggle
- How confident it is in different situations
- When human judgement should override it

INVOLVEMENT AND CO-CREATION APPROACHES

Nothing builds trust more effectively than meaningful involvement in the creation process. Effective involvement strategies include:

1. **User Research Participation:** Involve users in defining needs and requirements.
   - Conduct thorough contextual inquiry before design
   - Engage users in problem definition, not just solution testing
   - Create ongoing feedback mechanisms throughout development
   - Close the loop by showing how input influenced decisions
2. **Design Collaboration:** Give users influence over how the AI works.
   - Implement participatory design processes with diverse users
   - Create prototyping sessions to test concepts before building
   - Develop multiple design options for user evaluation
   - Balance user preferences with technical constraints transparently
3. **Testing Partnership:** Make users active partners in evaluation.
   - Create structured testing roles for representative users
   - Implement graduated exposure from test to production
   - Provide influence over readiness decisions
   - Recognize and reward testing contributions

4. **Continuous Improvement Voice:** Sustain involvement after launch.
   - Establish formal channels for ongoing improvement ideas
   - Create visibility into how feedback shapes the roadmap
   - Involve users in prioritizing enhancement opportunities
   - Recognize contributions to system evolution

University Health transformed healthcare AI adoption through comprehensive stakeholder involvement. By engaging clinical staff from the earliest planning stages and maintaining their input throughout implementation, the team ensured the AI was perceived as augmenting clinical expertise rather than replacing it. This approach broke down traditional barriers between clinical and administrative departments, improving both patient outcomes and operational efficiency.

REFLECTION QUESTIONS: How early and meaningfully are end users involved in your AI initiatives? Are they merely informed of decisions, consulted for opinions, or genuinely empowered to influence the solution?

ACTION STEP: The Co-Creation Framework

1. Identify representative users across different roles and experience levels
2. Involve them in problem definition before discussing solutions
3. Create structured forums for ongoing input throughout development
4. Ensure feedback visibly shapes the implementation
5. Recognize contributions to building better solutions

Skills Development and Career Path Clarity

Trust in AI is deeply connected to trust in one's own ability to work effectively with the technology. Successful organizations address this through:

1. **Role Evolution Mapping:** Show how roles will change, not disappear.
   - Create clear "before and after" role descriptions
   - Highlight how AI elevates rather than eliminates expertise
   - Illustrate new value-adding responsibilities
   - Provide examples of how similar roles have evolved
2. **Skills Gap Identification:** Help people understand what they need to learn.
   - Conduct role-specific skills gap assessments
   - Create personalized learning roadmaps
   - Differentiate essential from optional new capabilities
   - Highlight transferable skills from current roles
3. **Learning Path Development:** Provide clear routes to building necessary capabilities.
   - Create structured learning journeys appropriate to roles
   - Offer multiple learning modalities (formal, peer, experiential)
   - Recognize and celebrate skills development milestones
   - Allocate protected time for learning new capabilities
4. **Career Opportunity Creation:** Show growth paths in an AI-enhanced organization.
   - Develop and communicate new career pathways
   - Create transition roles that bridge current and future states
   - Highlight success stories of role evolution
   - Connect skills development to advancement opportunities

Fostering Employee Trust and Adoption

COUNTERINTUITIVE INSIGHT: A telecommunications company implementing AI across customer operations discovered that allocating 10% of work time to skills development didn't reduce productivity – it dramatically increased it. Creating "AI apprentice" positions, where employees could develop new capabilities, not only accelerated adoption but improved retention as employees saw clear growth opportunities rather than threats.

QUICK WIN: Create "AI career pathways" documents for key roles showing how responsibilities will evolve, what new skills will be valued, and what new opportunities will emerge as AI capabilities mature.

CREATING PSYCHOLOGICAL SAFETY

Fear is the enemy of adoption. Organizations that successfully build AI trust establish psychological safety through:

1. **Leadership Modelling:** Have leaders demonstrate safe experimentation.
   - Executives publicly engage with AI systems
   - Leaders acknowledge their own learning curves
   - Management celebrates learning, not just success
   - Hierarchy relaxed during implementation phases
2. **Learning-Focused Metrics:** Measure growth, not just performance.
   - Track and celebrate adoption progress, not just outcomes
   - Implement graduated performance expectations
   - Create safe periods for experimentation
   - Initially separate learning metrics from performance evaluation
3. **Mistake Normalization:** Make it safe to struggle and learn.
   - Explicitly expect and accept initial mistakes
   - Create forums for sharing challenges without judgement
   - Develop processes for learning from errors
   - Celebrate vulnerability and growth mindsets
4. **Support Accessibility:** Ensure help is readily available.
   - Implement multiple support channels (peer, expert, technical)
   - Create "no question too small" support environments
   - Provide just-in-time assistance during critical moments
   - Recognize and reward support-seeking behaviour

Case Study: From Blame to Learning

An multinational insurance company transformed from a blame culture to one of learning by shifting from "Root Cause Analysis (RCA)" to "Lessons Learned as a Practice (LLaaP)." According to those involved, this approach was "not compulsory and was unsupervised" and shifted "authority away from senior management figures," which was "especially useful given the current level of distrust felt across teams."

The transformation was so effective that a team member who had "previously agonised over a lack of empowerment" admitted to sometimes feeling "over-empowered." By creating psychological safety around AI implementation, employees felt free to experiment, share challenges, and collaboratively build better approaches without fear of judgement.

WARNING SIGN: If employees are reluctant to provide feedback about AI systems, rarely admit confusion, or always claim the system is working perfectly, you likely have a psychological safety problem that will undermine honest adoption.

ACTION STEP: Create a "learning, not perfection" campaign during AI rollout that:

1. Explicitly celebrates early feedback from users
2. Shares leadership's own learning challenges
3. Creates informal forums for discussing AI struggles
4. Recognizes those who help improve the system

DEVELOPING ORGANIZATION-WIDE AI LITERACY

For sustainable adoption, organizations need to build a foundation of AI literacy. This doesn't mean turning everyone into technical experts – but creating sufficient understanding for appropriate trust and usage.

CORE CONCEPTS EVERYONE SHOULD UNDERSTAND

Baseline AI literacy that all employees should develop includes:

1. **AI Capabilities and Limitations**
   - The difference between narrow and general AI
   - What current AI can and cannot do
   - How AI predictions differ from human judgements
   - Common failure modes and how to detect them
2. **Data Foundations**
   - How AI learns from data
   - The relationship between data quality and AI performance
   - Basic understanding of potential biases in data
   - How their actions affect data quality
3. **Appropriate Reliance**
   - When to rely on AI recommendations
   - When to question or override suggestions
   - How to provide effective feedback
   - What to do when something seems wrong
4. **Organizational Context**
   - How AI fits into organizational strategy
   - What problems the organization is trying to solve with AI
   - Their role in successful AI implementation
   - How value is measured and recognized

KEY INSIGHT: AI literacy isn't about technical knowledge; it is about practical understanding. Focus on the concepts that enable appropriate trust and effective usage rather than technical details.

GOVERNANCE GAP: McKinsey's 2024 Report highlighted that only 18% of organizations have an enterprise-wide council authorized to make decisions on responsible AI governance. This governance gap signals the need for comprehensive educational approaches that ensure responsible decision-making at all levels.

ROLE-SPECIFIC AI EDUCATIONAL NEEDS

Beyond baseline literacy, different roles require different knowledge depths:

Executive Education

- Strategic implications of AI for industry and competition
- Investment and resource allocation frameworks
- Risk governance and ethical considerations
- Organizational change management requirements

Management Education

- Integration of AI into operational workflows
- Performance monitoring and intervention approaches
- Team capability development strategies
- Balancing human and AI contributions

Professional/Knowledge Worker Education

- Domain-specific AI applications and limitations
- Complementary skill development
- Effective human-AI collaboration approaches
- Quality assessment of AI outputs

Technical Team Education

- AI development and implementation methodologies
- Human-centred design for AI
- Explainability and transparency techniques
- Ethical dimensions of AI development

Case Study: World Vision International's Tiered Education Approach

Bogdan Nita, CIO at Singapore-based World Vision International, launched an "AI Academy" with content tailored to specific departments. As he explained, "For example, when we try to upskill our finance people, we always use models from the finance space. It's always very important to know your target audience and make sure you adjust the context to respond to that specific target audience."

Instead of generic AI training, the organization created department-specific courses that showed how AI would affect daily work in each area. For finance personnel, for example, finance-specific AI models were used to make learning relevant, demonstrating how practical, targeted education could overcome resistance to new technologies.

ACTION STEP: Create a tiered AI literacy programme with:

1. Organization-wide baseline modules
2. Department-specific application modules
3. Role-specific skill development
4. Leadership-focused governance modules

EXPERIENTIAL LEARNING APPROACHES

Abstract concepts rarely drive adoption. Effective AI literacy programmes use experience-based approaches:

1. **Scenario-Based Learning**
   - Develop realistic situations employees will encounter
   - Create safe practice environments reflecting actual work

- Implement graduated difficulty as confidence grows
- Provide immediate feedback on decisions and actions
2. **Peer Demonstration**
   - Have respected colleagues demonstrate practical usage
   - Create opportunities to observe real work using AI
   - Develop "day-in-the-life" scenarios with the technology
   - Encourage authentic sharing of both benefits and challenges
3. **Guided Experimentation**
   - Create structured opportunities to try capabilities
   - Provide scaffolding that prevents damaging mistakes
   - Design progressive challenges that build confidence
   - Celebrate experimentation regardless of outcome
4. **Problem-Based Learning**
   - Start with relevant problems, not technical features
   - Have teams solve real challenges using AI tools
   - Focus on outcomes rather than perfect usage
   - Connect learning directly to performance improvement

QUICK WIN: Create an "AI-sandbox" environment where employees can experiment with the system using historical data without impacting live operations or feeling performance pressure.

Building a Common AI Language

Miscommunication frequently undermines adoption. Organizations need to develop:

1. **Shared Terminology**
   - Create accessible definitions of key AI concepts
   - Develop organization-specific terminology where needed
   - Address common misconceptions explicitly
   - Update language as technology and usage evolve
2. **Communication Bridges**
   - Build translation mechanisms between technical and business language
   - Create role-based communication frameworks
   - Develop metaphors and analogies that build understanding
   - Implement visual communication tools for complex concepts
3. **Feedback Vocabularies**
   - Develop specific language for different types of system issues
   - Create structured ways to communicate confidence levels
   - Implement clear terms for different types of overrides
   - Build common expressions for improvement suggestions

A government agency implementing AI citizen services created an "AI translation guide" that mapped technical terms to operational language. For instance, "false positives" became "unnecessary flags," and "model confidence" became "recommendation strength." They also created visual indicators for common concepts like system confidence and unusual cases. This common language dramatically reduced miscommunication between technical and operational teams and improved appropriate usage.

PATTERN RECOGNITION: Organizations with successful AI implementations invest in creating shared language that bridges technical and operational perspectives, reducing the friction that comes from miscommunication.

Fostering Employee Trust and Adoption

MANAGING RESISTANCE, BUILDING SUPPORT

Even with strong trust-building and literacy approaches, organizations will encounter resistance. Effective adoption strategies directly address resistance rather than trying to overcome it.

IDENTIFYING AND ADDRESSING RESISTANCE PATTERNS

Resistance to AI adoption typically follows predictable patterns, each requiring a different response:

Experience-Based Resistance

- **Signs:** References to previous technology disappointments
- **Root Cause:** Past negative experiences with similar initiatives
- **Effective Response:** Acknowledge past issues, explain specific differences, provide credible evidence of improvement

Knowledge-Based Resistance

- **Signs:** Misconceptions about how AI works or what it does
- **Root Cause:** Insufficient or incorrect information
- **Effective Response:** Provide accessible, relevant information, create safe spaces for questions, offer hands-on experience

Personal Impact Resistance

- **Signs:** Concerns about negative personal impacts
- **Root Cause:** Perceived threat to role, status, or values
- **Effective Response:** Address legitimate concerns directly, create clear benefits for the resistant individuals, involve them in shaping implementation

Identity-Based Resistance

- **Signs:** Fundamental rejection based on professional identity
- **Root Cause:** Perceived conflict with core values or self-concept
- **Effective Response:** Connect to deeper professional values, show how AI enhances rather than replaces identity, create identity-affirming roles in the new approach

Dan Sherratt, VP of Creative and Innovation at Poppins, captures this identity-based resistance perfectly: "There will always be people, agencies, creatives that use these tools to devalue work. I would hope that there would always be a value for quality and craftsmanship, and I don't think that will always come from purely AI-generated work."

This perspective reveals that creative professionals' resistance often stems from deeply held values about craftsmanship and quality rather than just fear of change. Addressing such concerns requires showing how AI can enhance rather than replace those values.

REFLECTION QUESTIONS: What type of resistance is most common in your organization, and how might your current approach be failing to address the root cause?

QUICK WIN: Create a "resistance pattern guide" for your implementation team that helps them identify the four resistance types and provides specific responses for each.

WORKING WITH INFORMAL INFLUENCERS AND NETWORKS

Formal authority drives compliance but informal influence has the power to drive true adoption. Successful organizations identify and engage these informal networks through:

1. **Influence Mapping**
   - Identify respected voices across different teams and levels
   - Recognize both positive influencers and potential blockers
   - Map informal communication networks
   - Understand who people turn to for advice and validation
2. **Early Influencer Engagement**
   - Involve key influencers in early planning and testing
   - Give them an authentic voice in shaping implementation
   - Provide them with firsthand experience before others
   - Listen genuinely to their concerns and suggestions
3. **Peer Advocacy Development**
   - Create structured roles for informal leaders
   - Provide them with additional information and support
   - Create forums where they can share experiences
   - Recognize and reward their influence contributions
4. **Network Activation**
   - Leverage existing communities of practice
   - Use informal channels alongside formal communications
   - Create opportunities for peer-to-peer learning
   - Design adoption approaches that spread through networks

COUNTERINTUITIVE INSIGHT: A global consulting firm discovered that implementation success had almost no correlation with formal positional authority but strongly correlated with informal influence. By asking a simple question – "Who do you turn to for advice about new tools?" – they identified mid-level consultants with outsized influence on adoption patterns.

These informal influencers became the core of an "AI Pioneers" programme with early access, direct input to developers, and recognition for supporting peers. Adoption spread through the organization at nearly twice the rate of previous technology implementations that relied primarily on formal channels.

ACTION STEP: Informal Influence Mapping

1. Identify key opinion leaders through a simple survey: "Who do you turn to for advice about new tools?"
2. Create an "AI champions" group with these informal influencers
3. Provide them with early access and additional support
4. Create channels for them to share their experiences with peers
5. Recognize and reward their contributions to broader adoption

Creating Positive AI Experiences

Perception is heavily influenced by experiences, particularly those shaped by early interactions. Organizations can shape these experiences through:

1. **First Impression Design**
   - Create intentionally positive initial interactions
   - Start with high-success-probability use cases
   - Ensure robust support during first experiences
   - Design memorable moments that challenge negative expectations
2. **Quick Win Identification**
   - Find opportunities for immediate small benefits

Fostering Employee Trust and Adoption

- Ensure first tasks have high success rates
- Create visibility into early positive outcomes
- Build momentum through cumulative small wins

3. **Friction Reduction**
   - Eliminate unnecessary steps and complications
   - Provide streamlined interfaces for common tasks
   - Create shortcuts for experienced users
   - Continuously improve based on usage patterns
4. **Positive Feedback Loops**
   - Make improvements visible and attributable to user input
   - Celebrate and share success stories
   - Create momentum through demonstrated progress
   - Build cycles of improvement visible to users

Case Study: The "First Five" Programme in Healthcare

A healthcare system implementing an AI clinical documentation assistant faced significant physician resistance until completely redesigning the initial experience. Through a "First Five" programme, the organization carefully curated clinicians' first interactions to ensure success, focusing on the five most common and straightforward documentation scenarios.

They provided extra support during these interactions and created immediate visibility into time saved. This approach resulted in 92% of clinicians choosing to continue using the system after their initial experiences, compared to 45% adoption with the previous approach of comprehensive but overwhelming training.

The key insight: First impressions matter more than feature completeness.

WARNING SIGN: If your AI implementation starts with comprehensive training on all features rather than carefully designed first experiences, you're likely creating an overwhelming rather than encouraging first impression.

QUICK WIN: Identify the simplest, highest-value use case for your AI system and design a "quick start" guide that gets users to that specific value with minimal steps.

BUILDING CHAMPIONS AT ALL LEVELS

Distributed advocacy creates more sustainable adoption than centralized promotion. Effective champion development includes:

1. **Champion Identification**
   - Look beyond formal leadership for potential advocates
   - Identify early adopters with peer credibility
   - Recognize those with complementary perspectives
   - Find individuals who naturally connect groups
2. **Champion Development**
   - Provide deeper knowledge and behind-the-scenes access
   - Create opportunities to shape the implementation approach
   - Offer recognition that matters in their professional context
   - Build their confidence as knowledgeable resources
3. **Champion Support**
   - Equip them with information to answer common questions
   - Create backchannels for addressing concerns and issues
   - Provide tools to make advocacy easier
   - Protect them from potential negative consequences

4. **Champion Network Building**
   - Connect advocates across departments and levels
   - Create forums for sharing approaches and experiences
   - Develop community identity and purpose
   - Evolve their role as adoption matures

ACTION STEP: Create a multi-level champion strategy with:

1. Executive sponsors who remove barriers
2. Manager champions who align team priorities
3. Frontline champions who provide peer support
4. Regular forums connecting champions across levels

CREATING A SUPPORTIVE ENVIRONMENT FOR AI ADOPTION

Beyond specific adoption techniques, the broader organizational environment significantly impacts AI acceptance and use. Organizations need to shape this environment intentionally rather than assuming existing conditions will support new technology adoption.

Cultural Elements That Enable AI Success

Organizational culture – including the shared assumptions, values, and norms that guide behaviour – can either accelerate or obstruct AI adoption. Key cultural elements that enable success include:

1. **Learning Orientation**
   - Valuing growth and development over perfection
   - Treating mistakes as learning opportunities
   - Encouraging experimentation and intelligent risk-taking
   - Celebrating knowledge acquisition alongside performance
2. **Collaborative Mindset**
   - Valuing cross-functional integration
   - Recognizing collective achievement alongside individual contribution
   - Sharing information openly rather than hoarding
   - Encouraging mutual support across boundaries
3. **Evidence-Based Decision-Making**
   - Valuing data alongside experience and intuition
   - Willingness to challenge assumptions with evidence
   - Comfort with probabilistic rather than certain answers
   - Balancing quantitative and qualitative insights
4. **Adaptability**
   - Embracing rather than resisting change
   - Valuing flexibility in processes and approaches
   - Seeing evolution as improvement rather than criticism
   - Focusing on outcomes rather than established methods

Denise Wiseman, a healthcare expert, identifies how existing cultures can undermine AI adoption: "[The] siloed nature of our healthcare, and how those silos are really preventing us from making the change that so many people are speaking up about."

She further observes that competitive metrics often overshadow collaborative goals: "Our primary focus too often is on competition, whether it is among individuals, among teams and departments [or] one organisation to another, etcetera. We're focused on the red-green scorecards rather than [on how we are] looking holistically… and how's my contribution helping our community or society at large?"

SURPRISING PATTERN: A retail organization found that stores with what they called a "learning store" culture – where trying new approaches was explicitly valued over perfect execution – achieved AI adoption rates 40% higher than locations without this cultural foundation. The difference wasn't in the technology or training, but in the cultural environment that shaped how people responded to the technology.

REFLECTION QUESTIONS: How might your organization's current culture be enabling or inhibiting AI adoption, and what specific cultural elements could you strengthen to create a more supportive environment?

LEADERSHIP BEHAVIOURS THAT BUILD TRUST

Leaders at all levels shape adoption through their behaviours, not just their words. Trust-building leadership behaviours include:

1. **Authentic Engagement**
   - Demonstrating personal use of AI tools
   - Sharing honest reactions, including challenges
   - Showing genuine interest in user experiences
   - Being present during implementation, not just announcements
2. **Consistent Messaging**
   - Aligning words and actions around AI priorities
   - Maintaining focus rather than chasing the next thing
   - Connecting AI to existing priorities rather than adding more
   - Providing a stable narrative through implementation challenges
3. **Transparent Decision-Making**
   - Clearly explaining the rationale for AI investments
   - Being honest about challenges and constraints
   - Sharing what is known, unknown, and in progress
   - Including diverse perspectives in visible decision processes
4. **Supportive Accountability**
   - Setting clear but reasonable expectations
   - Providing resources that match expectations
   - Addressing obstacles rather than just demanding results
   - Recognizing effort and progress, not just outcomes

Case Study: The CEO Who Walked the Talk

When a financial services firm implemented an AI advisory system, their CEO took an unusual approach. Instead of simply announcing the initiative, he spent a full day using the system alongside frontline advisors, openly sharing his learning curve and asking questions.

He continued to allocate time each month to experience system updates firsthand, creating a powerful signal that learning was expected and valued. This authentic engagement created a ripple effect through management layers, with leaders at all levels demonstrating similar behaviours.

Employee surveys showed that this visible leadership engagement was the single strongest factor in building trust in the new approach, far outweighing formal communications or incentives.

WARNING SIGN: If leaders are highly visible when announcing AI initiatives but invisible during implementation struggles, they are sending a clear signal that success matters more than learning.

ACTION STEP: Create specific "leadership visibility" moments throughout your AI implementation where senior leaders:

1. Use the technology alongside frontline staff
2. Share their own learning challenges
3. Ask genuine questions about user experiences
4. Demonstrate how feedback shapes improvements

Reward Systems That Encourage Adoption

Formal and informal rewards powerfully shape behaviour. To encourage AI adoption, organizations should align rewards through:

1. **Adoption-Specific Recognition**
   - Creating visibility for early adopters and champions
   - Recognizing learning and adaptation efforts
   - Celebrating milestone achievements in adoption
   - Highlighting how AI enables higher-value contributions
2. **Performance-Metric Alignment**
   - Adjusting metrics to reflect transition periods
   - Creating balanced scorecards that include adoption measures
   - Recognizing appropriate AI use, not just outcomes
   - Evolving metrics as adoption matures
3. **Career-Path Integration**
   - Creating advancement opportunities linked to AI capabilities
   - Recognizing AI expertise in promotion considerations
   - Developing specialized roles that bridge technical and domain knowledge
   - Highlighting success stories of AI-related career enhancement
4. **Team-Based Incentives**
   - Rewarding collective adoption milestones
   - Creating shared goals across functional boundaries
   - Recognizing support and knowledge-sharing
   - Celebrating team-level innovation in AI application

KEY INSIGHT: A professional services organization discovered that adding "technology leverage" as a specific dimension in their performance review process dramatically accelerated AI adoption. By featuring AI proficiency in promotion criteria and showcase assignments, they sent a clear signal that embracing new technologies was valued.

QUICK WIN: Create quarterly "AI impact awards" that recognize innovative applications, outstanding support, and successful adoption across different departments.

Knowledge-Sharing and Collaboration Structures

The flow of information and expertise significantly impacts adoption. Organizations can optimize these flows through:

1. **Communities of Practice**
   - Creating forums for practitioners to share approaches
   - Facilitating cross-functional learning groups
   - Developing specialized interest communities around specific applications
   - Building connection between technical and domain experts

Fostering Employee Trust and Adoption

2. **Knowledge Repositories**
   - Implementing systems to capture and share learnings
   - Creating accessible libraries of use cases and examples
   - Developing searchable solution databases
   - Building cumulative organizational knowledge
3. **Collaboration Platforms**
   - Providing tools for asynchronous knowledge sharing
   - Creating visibility into questions and solutions
   - Facilitating connection across organizational boundaries
   - Building cumulative reference resources
4. **Cross-Functional Rhythms**
   - Establishing regular touchpoints across teams
   - Creating forums for sharing challenges and solutions
   - Implementing structured retrospective processes
   - Building continuous improvement mechanisms

Case Study: The AI Application Exchange

A global manufacturing company created what they called an "AI Application Exchange," a digital platform where teams could share how they were using their predictive maintenance system in different facilities around the world.

The platform included success stories, solution templates, implementation guides, and forums for questions. Instead of each site reinventing approaches or struggling with similar challenges, teams could learn from each other's experiences and build on proven successes.

By making adaptation visible across sites, they accelerated adoption of advanced use cases by an estimated 18 months compared to location-by-location discovery.

ACTION STEP: Create a simple "AI knowledge hub" where teams can:

1. Share success stories and use cases
2. Document solutions to common challenges
3. Access implementation guides and templates
4. Connect with others using similar applications

CASE STUDIES IN ORGANIZATIONAL ADOPTION

Let's examine several detailed case studies that illustrate successful approaches to building AI trust and adoption across different contexts:

HOW DIGITAL TRANSFORMATION OVERCAME RESISTANCE AT A GLOBAL INSURANCE COMPANY

The Organization: A global insurance company implementing an AI claims processing system

Initial Situation: After investing millions in an AI claims assessment platform, initial rollout met widespread resistance. Claims adjusters viewed the system as threatening their expertise and judgement. Six months after launch, usage remained below 20%, and management was considering abandoning the project.

The Breakthrough Approach: Rather than pushing harder, the implementation team fundamentally shifted their approach:

1. **Reframing the Purpose:** They repositioned the AI from "automated claims processing" to "enhanced adjuster capabilities" – shifting from replacement to augmentation messaging.

2. **Deep User Research:** They conducted extensive interviews and observation sessions with adjusters to understand their actual concerns, discovering that loss of professional identity was a bigger issue than job security.
3. **Expertise Recognition:** They created a formal process for adjusters to provide feedback on AI recommendations, explicitly valuing their judgement and using it to improve the system.
4. **Graduated Control:** They implemented a tiered approach where adjusters started with full review of all AI recommendations, gradually reducing oversight as both the system and their confidence improved.
5. **Career Evolution:** They created new roles for "AI claims specialists" who combined technical understanding with claims expertise, creating advancement paths.

Key Success Factors

- Recognizing and addressing identity concerns, not just practical issues
- Giving adjusters meaningful influence over system evolution
- Creating visible career opportunities in the new approach
- Building champions within the adjuster community
- Measuring and celebrating adjuster contributions to system improvement

The Results

- Adoption reached 87% within four months of the approach change
- Claims processing time decreased by 34%
- Adjuster job satisfaction increased from pre-implementation levels
- The company now has a waiting list for their AI claims specialist roles
- Several adjusters who were initial resistors became the most effective champions

PATTERN RECOGNITION: The turning point came when the implementation team stopped seeing resistance as something to overcome and started seeing it as valuable feedback about what needed to change in their approach.

BUILDING CITIZEN SERVICE AGENT SUPPORT

The Organization: A government tax authority implementing an AI citizen support system

Initial Situation: The tax authority launched an AI assistant to help citizen service agents answer taxpayer questions more efficiently. Despite strong technical performance, agents avoided using the system, continuing to rely on manual knowledge base searches. Management initially attributed this to general resistance to change.

The Breakthrough Approach: Through careful analysis, the implementation team discovered more specific adoption barriers and addressed them through:

1. **Trust-Building through Transparency:** They redesigned the interface to show confidence levels for recommendations and links to source documentation, addressing agents' concerns about accuracy.
2. **Contextual Integration:** They embedded the AI directly into the existing case management workflow rather than requiring agents to use a separate system.
3. **Agent-Centred Metrics:** They shifted from measuring cost reduction to tracking "agent empowerment metrics" like time saved, citizen satisfaction, and complex-case resolution.
4. **Tiered Implementation:** They identified initial use cases with high success probability, creating positive first experiences before expanding to more complex scenarios.
5. **Peer Support Network:** They established a network of experienced agents who provided support and shared success stories, creating adoption through trusted colleagues rather than management mandate.

Fostering Employee Trust and Adoption

Key Success Factors

- Addressing the specific trust concerns (accuracy and verifiability)
- Integrating the tool rather than disrupt existing workflows
- Creating metrics that mattered to agents, not just management
- Building adoption through peer influence
- Designing for initial success experiences

The Results

- Adoption reached 92% of agents within seven months
- Average case resolution time decreased by 26%
- First-contact resolution rates improved by 18%
- Agent satisfaction scores increased significantly
- The system became agent-requested rather than management-pushed

KEY INSIGHT: The successful turning point came when the team stopped measuring solely organizational benefits (cost reduction) and started measuring value that mattered to agents.

SHOP FLOOR AI ADOPTION IN MANUFACTURING

The Organization: A global automotive manufacturer implementing AI-based quality control

Initial Situation: The company invested in an advanced computer vision system to identify defects in their production line. Despite proven technical accuracy, production teams ignored or overrode system recommendations, causing implementation to stall with minimal value delivery.

The Breakthrough Approach: The company reimagined the implementation through:

1. **Operator Co-Development:** Involving experienced quality inspectors in training and refining the AI system explicitly signalled that their expertise was valued.
2. **Explainability Focus:** A redesign of the interface showed specifically what the system was detecting and why, addressing the "black box" concern.
3. **Complementary Capabilities:** By positioning the AI to handle consistent, visible defects while elevating human inspection for subtle or novel issues, role clarity was enhanced.
4. **Progressive Trust-Building:** A "trust-building mode" was implemented, where the system made recommendations but humans maintained decision authority, gradually shifting as confidence grew.
5. **Shared Success Metrics:** Balanced metrics were created that showed how the combined human-AI approach outperformed either alone.

Key Success Factors

- Valuing and incorporating operator expertise rather than replacing it
- Creating explainability appropriate to the shop floor context
- Defining clear and complementary roles for humans and AI
- Building trust through progressive experience rather than mandate
- Measuring combined performance rather than AI versus human

The Results

- Defect detection improved by 32% over previous approaches
- False positive rates decreased by 28% compared to initial AI implementation
- Production line efficiency increased by 15%
- Quality inspector job satisfaction improved
- The company now has operators actively suggesting new applications for computer vision

QUICK WIN: Create clear "human-AI partnership" definitions for each role affected by your AI implementation, showing specifically how the technology will complement rather than replace human expertise.

UNIVERSITY HEALTH: CLINICAL AI INTEGRATION SUCCESS

The Organization: A healthcare system implementing AI technologies for radiologists
Initial Situation: University Health aimed to implement AI technologies to assist radiologists in cancer detection and treatment outcome prediction but faced potential resistance from clinical staff concerned about AI replacing clinical judgement.
The Breakthrough Approach: The implementation team created a comprehensive strategy:
1. **Extensive Stakeholder Involvement:** Clinical staff was involved from the earliest planning stages, ensuring their input shaped system requirements and implementation approach.
2. **Supportive Positioning:** AI was consistently framed as an augmentation to clinical expertise rather than a replacement, emphasizing how it would handle routine cases while allowing specialists to focus on complex diagnoses.
3. **Silo-Breaking Integration:** Workflows were deliberately designed to bridge traditional barriers between clinical and administrative departments.
4. **Dual-Value Demonstration:** Metrics were designed to show improvements in both patient outcomes and operational efficiency, connecting AI adoption to both clinical and administrative priorities.
5. **Gradual Authority Transition:** A phased approach was implemented where radiologists maintained complete oversight initially, with AI autonomy gradually increasing as trust developed.

Key Success Factors

- Positioning AI as enhancing rather than replacing clinical expertise
- Breaking down departmental silos through integrated workflows
- Demonstrating value across both clinical and administrative metrics
- Building input channels for ongoing clinical feedback and improvement
- Creating clear visibility into patient outcome improvements

The Results

- Successful implementation across both clinical and administrative processes
- Improved cancer detection rates and treatment outcome prediction
- Increased operational efficiency and reduced administrative burden
- Breakdown of longstanding silos between departments
- Creation of a model for future healthcare AI implementations

PATTERN RECOGNITION: Across all these success stories, the common thread isn't technical excellence alone but human-centred implementation that addresses the legitimate concerns, professional identities, and value needs of the people expected to use the technology.

CONCLUSION: THE HUMAN FOUNDATION OF AI VALUE

As we've explored throughout this chapter, even the most technically sophisticated AI creates zero value if people don't trust and use it. This means that the human side of AI has to be more than a secondary consideration – it has to be considered as a foundation upon which technical value rests.

Fostering Employee Trust and Adoption

Let's conclude with five key principles for building the human foundation of AI success:

1. **Trust Precedes Adoption:** People will not use what they don't trust, and trust is built through experience, transparency, and involvement rather than technical performance alone.
2. **Value Must Be Personal, Not Just Organizational:** For sustainable adoption, individuals must experience benefits that matter to them, not just abstract organizational outcomes.
3. **Identity Shapes Acceptance:** How AI affects people's sense of professional identity and value dramatically impacts their willingness to adopt, often more than practical considerations.
4. **Influence Flows through Networks, Not Charts:** Adoption tends to spread through informal influence networks that may bear little resemblance to formal reporting structures.
5. **Environment Shapes Behaviour More Than Intention:** The cultural context, leadership behaviours, reward systems, and collaboration structures surrounding AI significantly influence adoption.

FINAL ACTION STEP: The Trust and Adoption Assessment – Take 15 minutes to honestly evaluate your current AI implementation approach:

1. Are you focusing as much on human adoption as technical performance?
2. Have you mapped the specific concerns of different roles and stakeholders?
3. Is your adoption strategy aligned with your organizational culture?
4. Have you identified and engaged informal influencers?
5. Are your metrics measuring both technical and human success?

By addressing human dimensions with the same rigour typically reserved for technical implementation, organizations can dramatically accelerate the journey from AI capability to genuine value.

DOWNLOADABLE RESOURCES

To support your AI value efforts, visit the online resource centre at www.valuesofai.com to access:

- The AI trust dashboard template
- The adoption flywheel implementation guide
- The role-evolution-mapping tool
- The barrier-identification worksheet
- The champion-development framework

These practical tools will help you apply the concepts from this chapter to your specific organizational context.

10 Communicating AI's Societal and Ethical Values

Chapter Roadmap

In this chapter, we'll explore why external trust is as crucial as internal adoption for AI success – and how to map and address diverse stakeholder expectations. Among the measures designed to advance trust-building and buy-in are practical frameworks for responsible AI communication; strategies for managing evolving regulatory requirements; case studies of organizations that turned ethical AI into competitive advantage, and approaches for balancing data privacy with innovation.

WHEN EXTERNAL TRUST BECOMES AN INTERNAL CRISIS

The news alert lit up phones across the executive team at 6:47 a.m. A major outlet was running a front-page story about how the hospital's AI system was making life-and-death decisions about patient care priorities.

By 7:15 a.m., the phones wouldn't stop ringing. Patient advocacy groups were demanding emergency meetings. Community leaders were calling for investigations. Social media was erupting with stories from families who now questioned every medical decision that had affected their loved ones.

By 8:30 a.m., three regulatory bodies had launched formal inquiries. The state health department was sending investigators. The hospital's legal team was fielding calls from lawyers representing affected families. Local politicians were issuing statements distancing themselves from the "algorithmic healthcare scandal."

By 10 a.m., staff were afraid to go to work. Protesters had gathered outside the hospital's main entrance. Clinical staff who had enthusiastically adopted the AI system just weeks before were now questioning every recommendation it made. The nurses' union was demanding an emergency meeting. Two board members had called for the CEO's resignation.

The technical team was in shock. The AI system had reduced wait times for critical cases by 37%. Patient outcomes had improved. Clinical staff satisfaction was at an all-time high. Building the system had included testing for bias, involving clinicians at every step, and following every technical best practice.

But none of that mattered now: the narrative had gone off the rails – and was completely out of their control.

"We built this to save lives," the lead data scientist said, staring at her computer screen filled with resignation emails from key team members. "How did helping patients become a crisis that's destroying everything we've worked for?"

The answer reveals a brutal truth about AI implementation: technical excellence and internal success mean nothing if external stakeholders don't trust your intentions, understand your methods, or believe in your commitment to their values. Once that trust is broken, it spreads like wildfire, consuming years of work in a matter of hours.

If you think this sounds like a worst-case scenario that would never happen to a well-run organization, think again. The uncomfortable truth is that there are pervasive commonalities between this hospital and most organizations. Similar crises have destroyed real initiatives at companies you'd recognize, led by smart, capable and well-intentioned people.

Strip away specific details and you'll recognize a pattern, where organizations across industries have been blindsided as their AI initiatives touch stakeholder lives.

There are countless public accounts of similar scenarios:

- **Apple and Goldman Sachs faced accusations** that their Apple Card algorithm gave women lower credit limits, though regulators ultimately found no evidence of discrimination.
- **Amazon's facial recognition system Rekognition** drew intense criticism when American Civil Liberties Union (ACLU) testing showed that it incorrectly matched 28 members of Congress with mugshots, with nearly 40% of false matches being people of colour.
- **Target and Walmart's AI-powered surveillance systems** created customer concerns about privacy and monitoring, particularly around facial recognition and behavioural tracking in stores.
- **Manufacturing companies implementing AI automation** faced community resistance when MIT and Boston University research projected that AI could replace as many as two million manufacturing workers by 2025.

PATTERN RECOGNITION: In each of these cases – Apple Card, Amazon Rekognition, retail surveillance, and manufacturing automation – the backlash wasn't because the AI systems failed technically; it was because the organizations hadn't prepared for how they would be perceived by external stakeholders.

WARNING SIGN: Notice how quickly these situations escalated? ACLU's Rekognition testing made national headlines. Apple Card faced regulatory scrutiny. Manufacturing companies encountered organized resistance. The first indication of an external trust problem is often not gradual feedback but a sudden public crisis.

Every organization that's faced this kind of meltdown had smart people, good intentions, and technical systems that actually worked. What they didn't have was a plan for when external trust collapsed.

REFLECTION QUESTIONS: Looking at these real cases, what external stakeholders could derail your AI initiative through mistrust or misunderstanding, and what steps have you taken to proactively engage them before a crisis hits?

THE EXTERNAL STAKEHOLDER LANDSCAPE

Unlike traditional technology implementations that primarily affected internal users, AI creates ripple effects that extend far beyond organizational boundaries. That's why it is so important to pay attention to the complex ecosystem of external stakeholders with legitimate interests in AI initiatives, understand their distinct concerns, and identify the counterintuitive patterns that determine how trust is built or broken.

MAPPING THE AI STAKEHOLDER ECOSYSTEM

"Ethics is so important and it's becoming even more important," says Cindy Hoots, Chief Digital Officer at AstraZeneca.

Recognizing the significance of public perception, AstraZeneca became "one of the first pharmaceutical companies to put out an ethical use statement around the way in which we would use AI, and use people's personal data," Hoots explains. "We take it so seriously in terms of making sure that we don't ever breach that trust that we have with the people that share their data with us."

Key external stakeholders include:

1. **Direct Users and Customers**
   - Recipients of AI-influenced products and services
   - People whose data feeds AI systems
   - Individuals directly impacted by AI decisions or recommendations
2. **Regulatory and Oversight Bodies**
   - Industry-specific regulators with varying AI focuses
   - Data protection and privacy authorities
   - Consumer protection agencies
   - Anti-discrimination and rights enforcement bodies
   - Emerging AI-specific regulatory authorities
3. **Partners and Suppliers**
   - Technology and service providers in your AI stack
   - Business partners affected by your AI systems
   - Supply chain participants impacted by AI decisions
   - Companies whose data integrates with yours
4. **Industry Ecosystem**
   - Competitors watching your AI approach
   - Industry associations developing standards
   - New entrants disrupting traditional boundaries
   - Professional bodies concerned with practice standards
5. **Community and Civil Society**
   - Local communities where you operate
   - Advocacy groups for potentially affected populations
   - Research institutions studying AI impacts
   - Media organizations reporting on AI developments
   - General public forming opinions about AI use

ACTION STEP: Create a comprehensive stakeholder map specific to your AI initiative, identifying all external groups who might be affected by or have influence over your implementation.

THE STAKEHOLDER PARADOX: WHO HAS THE MOST IMPACT?

KEY INSIGHT: The stakeholders with the greatest potential impact on your AI success often have the least direct relationship with your organization.

Consider this stakeholder influence pattern:

- **High Familiarity/Low Impact:** Those who understand your specific AI implementation in detail (technical vendors, internal teams) have limited power to affect public perception.
- **Medium Familiarity/Medium Impact:** Direct customers and users have moderate understanding and moderate influence.
- **High Familiarity/High Impact:** Current and former insiders (employees, contractors, advisors) who understand your actual practices and can credibly expose gaps between public communications and reality.
- **Low Familiarity/High Impact:** Distant stakeholders like regulators, media, and public opinion often have very limited understanding of your specific approach but enormous influence over your freedom to operate.

This creates a fundamental communication challenge: those who can most affect your AI's societal acceptance may have the least context about your specific implementation.

Communicating AI's Societal and Ethical Values

UNDERSTANDING STAKEHOLDER CONCERNS AND PRIORITIES

Different stakeholder groups bring distinct concerns to AI implementations. Effective communication requires understanding these varying perspectives:

Customer and User Concerns

- Will my data be used responsibly and securely?
- Is this system fair and unbiased in how it treats me?
- Do I understand how decisions affecting me are made?
- Can I control or override automated processes?
- Am I getting genuine value from this technology?

Regulatory Concerns

- Does this implementation comply with existing regulations?
- Are data protection principles being upheld?
- Could this create or amplify discriminatory outcomes?
- Is there appropriate human oversight and accountability?
- Can the organization explain how decisions are made?

Community Concerns

- Will this technology displace jobs in our community?
- Does it affect vulnerable populations disproportionately?
- Is it being deployed with meaningful consent?
- Who benefits and who bears the risks?
- Does it reflect our social values and priorities?

Partner and Supplier Concerns

- How does this affect our relationship and interactions?
- Are responsibilities and liabilities clearly defined?
- Do we have compatible approaches to data and algorithms?
- How are benefits and risks shared across the ecosystem?
- Can we align our ethical frameworks and standards?
- Could this change our value proposition or our market positioning?
- Will this AI implementation put us at a competitive disadvantage?
- Are we being (intentionally or unintentionally) excluded from AI-enabled opportunities?

QUICK WIN: Develop a simple "concerns matrix" that lists each stakeholder group and their top 3 to 5 questions about your AI system. Use this to guide your communication priorities.

THE EXPECTATIONS GAP

KEY INSIGHT: External stakeholders don't evaluate your AI against a standard of perfection; they evaluate it against their own expectations, which can, to a certain extent, be shaped by your communication.

I've observed that organizations facing the fiercest backlash aren't necessarily those with the most problematic AI implementations. They're often the ones who haven't paid enough attention to the gap between stakeholder expectations and reality.

Organizations that overpromise AI capabilities, downplay limitations, or gloss over ethical considerations create expectation gaps that can eventually become trust crises. Conversely, organizations that communicate honestly about both capabilities and constraints often receive greater latitude when challenges inevitably arise.

This creates a seemingly paradoxical best practice: transparent communication about limitations and risks doesn't increase your vulnerability – it actually creates resilience by aligning expectations with reality.

Real-world example: A financial institution implementing an AI credit decision system took the unusual step of proactively publishing their fairness metrics, acknowledged limitations, and oversight processes before regulatory requirements mandated disclosure. When a later industry-wide controversy erupted about algorithmic bias, they faced significantly less scrutiny and maintained stronger customer trust than competitors who had remained opaque until forced to disclose.

REFLECTION QUESTIONS: Where might your current communications be creating unrealistic expectations about your AI system's capabilities or understating its limitations?

The Evolving Regulatory Environment

The regulatory landscape for AI is evolving rapidly, creating compliance challenges that directly impact communication requirements.

As extensive research on AI Governance and Ethics reveals, we are increasingly seeing distinct regional approaches emerging:

Global Regulatory Divergence

- The EU's comprehensive AI Act creating risk-based obligations
- China's approach emphasizing national security and social harmony
- Emerging sector-specific and state-level regulations in the US
- Regional approaches emphasizing different values and priorities

This regulatory variation creates challenges for multinational organizations. While Middle Eastern leaders increasingly frame AI as a boardroom-level strategic imperative, European firms must simultaneously navigate the complex obligations of the EU AI Act. This includes a shift from reactive compliance to proactive integration – where legal, ethical, and strategic dimensions converge within the AI design process.

Cross-Cutting Regulatory Themes

- Transparency and explainability requirements
- Fairness and non-discrimination standards
- Data protection and privacy frameworks
- Risk assessment and management obligations
- Human oversight and intervention requirements

Industry-Specific Regulatory Focus

- Financial services emphasizing fairness and explainability
- Healthcare focusing on safety and effectiveness
- Transportation prioritizing reliability and safety
- Public services centring on fairness and non-discrimination
- Critical infrastructure emphasizing security and resilience

My research on AI Governance and Ethics highlights a fundamental tension in AI governance between trust-centred and innovation-centred approaches. As an example, while Hugh Burgin, EY's Data and AI Leader for the Americas, emphasizes that "trust is super important" and organizations are "trying to put in the guardrails and frameworks that allow their AI solutions to be trustworthy," there's pushback from innovation advocates who argue that "strict rules could hinder innovation."

WARNING SIGN: Organizations that view regulatory compliance as merely a checkbox exercise often miss the opportunity to build it into their value proposition and competitive advantage.

Public Perception Challenges and Opportunities

Beyond specific stakeholder concerns, organizations face broader challenges in how AI is perceived:

AI Perception Asymmetries

- Failures receive disproportionate attention compared to successes
- Abstract risks often generate more concern than concrete benefits
- Specialized applications often get conflated with general AI narratives
- Nuanced implementations get reduced to simplistic framings
- Technical complexity creates barriers to accurate understanding

Media Coverage Patterns

- Emphasis on controversy over routine implementation
- Tendency to personify AI with agency it doesn't possess
- Focus on extreme scenarios rather than more common use cases
- Challenges presenting nuanced technical concepts accurately
- Attraction to narratives that align with existing anxieties

Opportunity Areas in Public Perception

- Education about actual capabilities and limitations
- Transparency that builds understanding and trust
- Demonstrable benefits that connect to human values
- Meaningful engagement rather than one-way communication
- Leadership in responsible implementation and governance

The Specificity Shield

KEY INSIGHT: Organizations often make a critical mistake in external AI communication: they communicate about AI in general rather than their specific implementation.

This creates several problems:

- It connects your specific application to broad AI anxieties.
- It fails to differentiate your approach from problematic ones.
- It misses opportunities to highlight your specific safeguards.
- It makes your communication generic rather than relatable.

There is a counterintuitive approach that works better: be specific about your particular implementation, use case, and governance. This specificity can shield you from generic AI concerns rather than connecting you to them.

Real-world example: A retail company implementing computer vision for inventory management initially faced employee and customer concerns about surveillance and privacy. Their early communications about "AI implementation" inadvertently connected to broader societal concerns about facial recognition and tracking. When they shifted to extremely specific communication about "shelf inventory scanning" with explicit statements about what data was and wasn't collected, concerns dropped dramatically. Specificity created clarity that generality couldn't achieve.

ACTION STEP: Audit your AI communications to replace generic AI terms with specific descriptions of your actual implementation, focusing on precise functionality and boundaries that stakeholders can easily understand.

RESPONSIBLE AI COMMUNICATION FRAMEWORK

Understanding the stakeholder landscape – including groups with distinct concerns and priorities as well as stakeholders with potentially disproportionate public sway – can enable organizations, and teams to come up with a structured framework for responsible AI communication that builds trust while supporting innovation.

CORE PRINCIPLES FOR ETHICAL AI MESSAGING

Effective external communication builds from these foundational principles:

1. **Proportionality**
    - Match communication depth to impact level
    - Provide greater transparency for higher-stakes systems
    - Scale explanation to the significance of effects
    - Adjust detail based on potential consequences
2. **Accessibility**
    - Adapt communication to different knowledge levels
    - Layer information from summary to detail
    - Use plain language for core concepts
    - Provide multiple formats for diverse needs
3. **Accuracy**
    - Make claims supported by evidence
    - Acknowledge limitations and constraints
    - Avoid overstatement of capabilities
    - Present balanced assessment of benefits and limitations
4. **Accountability**
    - Clearly identify responsible parties
    - Explain oversight and governance mechanisms
    - Provide recourse for concerns or issues
    - Demonstrate ongoing monitoring and improvement
5. **Context-Sensitivity**
    - Adapt messaging to audience concerns and knowledge
    - Consider cultural and social context of reception
    - Acknowledge relevant historical and ethical issues
    - Respect audience autonomy and values

REFLECTION QUESTIONS: Which of these principles might be most challenging for your organization to implement, and why?

THE FACTS FRAMEWORK FOR RESPONSIBLE AI COMMUNICATION

The foundational principles can be operationalized through the FACTS framework: a structured approach to designing external communications about AI.

Communicating AI's Societal and Ethical Values

F – Functionality: What does the system actually do?
- Explain core capabilities in plain language
- Describe the specific use case and purpose
- Clarify what's automated versus human-supported
- Outline key limitations and constraints

A – Alignment: How does this connect to values and priorities?
- Link to organizational mission and values
- Connect to stakeholder priorities and concerns
- Demonstrate relevant ethical considerations
- Show how it advances shared goals

C – Controls: How is the system governed and overseen?
- Explain risk assessment and management approaches
- Describe human oversight and intervention points
- Outline testing and validation processes
- Clarify recourse mechanisms for issues

T – Transparency: What visibility is provided into operation?
- Explain what can be disclosed about functioning
- Describe available explanations for outcomes
- Outline what information users/subjects receive
- Clarify audit and verification approaches

S – Security: How are data and systems protected?
- Describe data protection approaches
- Explain access controls and safeguards
- Outline security testing and validation
- Clarify incident response procedures

Google's What-If Tool provides a practical example of how these principles can be implemented. This interactive visual tool "promotes ethical AI development by enhancing transparency and fairness by identifying potential biases in machine learning models, enabling developers to make informed decisions and mitigate unintended biases, resulting in more equitable and trustworthy AI systems."

ACTION STEP: Apply the FACTS framework to your AI initiative, documenting your approach to each element. Identify any gaps where your current communication plan falls short.

Real-world example: A healthcare organization used the FACTS framework to create a patient-facing explanation of their clinical decision support AI. The layered communication began with a one-page overview addressing each FACTS element, with links to progressively deeper information for those who wanted it. Patient trust scores increased by 28% after implementation, and clinicians reported greater comfort discussing the system with patients.

Balancing Innovation and Responsibility

Organizations face a perceived tension between innovation and responsible communication. This tension becomes a genuine dilemma when it is framed as a binary choice. A more effective approach recognizes that responsibility enables sustainable innovation.

The False Dichotomy Problem

Many organizations approach AI communication through limiting binary frames:

- Open communication creates risk vs. minimal disclosure enables innovation
- Transparency slows progress vs. opacity allows faster development and the ability to pivot
- Addressing concerns constrains possibilities vs. ignoring concerns permits exploration

This framing creates unnecessary trade-offs and can ultimately undermine both innovation and trust.

The Innovation-Trust Cycle
KEY INSIGHT: Responsible communication and innovation reinforce each other in a positive cycle rather than competing.

1. **Responsible Communication** → builds trust and understanding
2. **Trust** → creates social license for greater innovation
3. **Innovation within Trust Boundaries** → delivers value and credibility
4. **Credibility** → enables more ambitious responsible communication
5. **Cycle continues** → expanding the scope of trusted innovation

Strategies for Balancing Innovation and Responsibility
1. **Tiered Transparency**
   - Create proportional disclosure based on risk level
   - Provide more detailed communication for higher-impact systems
   - Scale explanation to the significance of effects
   - Focus detailed disclosure on areas of greatest concern
2. **Participatory Approaches**
   - Involve key stakeholders in defining disclosure standards
   - Create feedback mechanisms to refine communication
   - Establish ongoing dialogue rather than one-way disclosure
   - Use stakeholder input to prioritize transparency efforts
3. **Progressive Revelation**
   - Begin with core information needed for informed consent
   - Add detail as stakeholder interest and system impact grow
   - Create information layers from summary to technical detail
   - Allow stakeholders to determine their desired depth
4. **Living Documentation**
   - Treat disclosure as evolving rather than static
   - Update as systems, understanding, and standards change
   - Create versioned transparency that shows evolution
   - Build communication systems that improve over time

QUICK WIN: Identify one aspect of your AI system where providing greater transparency would serve to build competitive advantage rather than undermining it.

Real-world example: A financial services company implementing AI for fraud detection created a "transparency roadmap" that evolved with their system. They began with a clear explanation of purpose, general approach, and oversight – then progressively added more detailed documentation on performance, fairness metrics, and model governance as the system proved its value and stakeholders requested more information. This progressive approach allowed them to focus initial communication on building basic trust while evolving toward more sophisticated transparency as the relationship with stakeholders matured.

Addressing Concerns Proactively

Effective AI communication anticipates and addresses concerns before they become crises where possible. This proactive strategy requires systematic approaches for identifying and responding to potential issues, while balancing resource constraints and strategic considerations.

The Concern Horizon Framework

This structured approach helps organizations map potential concerns across timeframes:

Immediate Concerns (0–3 months)

- Direct questions from current users and customers
- Issues raised by employees and partners
- Media inquiries about your implementation
- Comparison with competitor approaches

Emerging Concerns (3–12 months)

- Evolving regulatory requirements
- Industry standard developments
- New research on similar techniques
- Advocacy group focus in your domain

Horizon Concerns (1–3 years)

- Academic research directions
- Early regulatory signals
- Evolving public discourse
- Leading indicator controversies

WARNING SIGN: Organizations that only respond to immediate concerns often find themselves constantly in reactive mode, missing the opportunity to shape the narrative around emerging issues.

The Anticipation-Alarm Balance

Organizations face a delicate balance in proactive communication:

- Address concerns too early and you may raise issues not yet on stakeholder radars
- Wait too long and you appear reactive rather than responsible
- Focus on remote possibilities and you create unnecessary anxiety
- Limit scope too much and you miss emerging concerns

The solution isn't picking a single point on this spectrum, but developing a graduated approach that maps concern response to concern maturity and likelihood.

The Concern Response Matrix

This decision (see Table 10.1) tool helps determine appropriate response levels to potential concerns.

This graduated approach allows organizations to stay ahead of concerns without overreacting to speculative issues.

ACTION STEP: Establish a quarterly "concern horizon review" process to systematically identify and assess potential issues before they affect your AI implementation.

Real-world example: A retail company using AI for inventory optimization created a quarterly "concern horizon review" that systematically examined emerging issues in their domain. When early academic research raised questions about unintended environmental impacts from similar systems, they were able to assess their specific approach, develop appropriate safeguards, and proactively communicate with environmental stakeholders before the issue became broadly controversial. This proactive stance positioned them as responsible leaders rather than reactive followers when the concern eventually gained wider attention.

TABLE 10.1
The Concern Response Matrix

Concern Level	Monitoring Response	Communication Response	Action Response
Speculative (theoretical but unproven)	Track research and discourse	Develop internal position	Research mitigation options
Emerging (evidence building but uncertain)	Establish formal tracking	Prepare contingent messaging	Develop preliminary safeguards
Probable (likely to affect your systems)	Implement specific monitoring	Proactively address with key stakeholders	Implement preventative measures
Active (currently impacting similar systems)	Continuous monitoring	Transparent communication with all stakeholders	Deploy full mitigation approach

BUILDING CREDIBILITY THROUGH TRANSPARENCY

Transparency is often treated as a binary state – you either have it or you don't. In reality, effective transparency is multidimensional and strategic.

The Transparency Dimensions Framework

This model identifies five key dimensions of AI transparency, each requiring distinct approaches:

1. **Purpose Transparency:**
   - Clear articulation of problems being addressed
   - Explanation of intended outcomes and benefits
   - Disclosure of limitations and constraints
   - Connection to organizational mission and values
2. **Process Transparency:** How the system works
   - Appropriate explanation of technical approach
   - Description of data sources and usage
   - Clarification of human-AI interaction points
   - Explanation of testing and validation approaches
3. **Performance Transparency:** How well it works
   - Disclosure of relevant performance metrics
   - Explanation of how performance is measured
   - Clarification of variation across contexts
   - Communication of improvement processes
4. **People Transparency:** Who is responsible
   - Identification of accountable parties
   - Explanation of oversight structures
   - Clarification of development approach
   - Description of expert involvement
5. **Protection Transparency:** How risks are managed
   - Explanation of risk assessment approaches
   - Description of safeguards and controls
   - Clarification of monitoring and audit processes
   - Information on recourse mechanisms

PATTERN RECOGNITION: Organizations that excel in just one dimension of transparency while neglecting others often face scepticism rather than trust. All dimensions must work together to create credibility.

The Strategic Transparency Advantage

KEY INSIGHT: Organizations that proactively define what transparency means in their context gain several advantages over those that treat it as a defensive necessity.

Many organizations approach transparency defensively, disclosing only what's required or demanded. This treats transparency as a cost to be minimized.

A counterintuitive alternative approach treats transparency as a strategic asset. Organizations that proactively define what transparency means in their context gain several advantages:

- They set the terms of the transparency conversation rather than reacting to external definitions.
- They can design disclosure approaches that align with their values and strengths.
- They build trust capital before controversies arise.
- They demonstrate leadership that differentiates them from competitors.
- They help shape emerging standards rather than simply complying with them.

QUICK WIN: Identify one aspect of your AI system where you could provide visibility beyond what's required, particularly in an area aligned with your organizational values.

Real-world example: A technology company implementing AI recruiting assistance tools developed a "transparency by design" approach. Rather than waiting for regulatory requirements, they created a public explanation of their five-dimensional transparency approach, including specific commitments in each area. When a competitor faced controversy over lack of disclosure, they were able to point to their established framework rather than scrambling to develop a reactive position. This proactive transparency became a competitive advantage in selling to enterprise clients concerned about ethical AI use.

MANAGING EXTERNAL PERCEPTIONS

Beyond developing responsible communication frameworks that build on core principles of proportionality, accessibility, accuracy, accountability, and context-sensitivity, organizations need specific strategies for engaging with key external stakeholders.

A foundational understanding – that trust and innovation reinforce rather than compete with each other; that proactive concern identification prevents reactive crisis management; and that transparency is multidimensional and can be a strategic advantage – provides a strong incentive to act on these insights. Now let's examine approaches for four critical stakeholder groups.

Media Engagement Strategies

Media coverage significantly shapes broader perception of AI initiatives. Effective media engagement requires specific approaches for this unique stakeholder group:

1. **Narrative Development**
   - Create clear, accurate, and accessible explanations
   - Develop authentic stories that illustrate purpose and impact
   - Prepare concrete examples that demonstrate benefits
   - Build contextual framing that situates your approach appropriately
2. **Spokesperson Preparation**
   - Train technical experts in accessible communication
   - Prepare for predictable questions and concerns

- Develop analogies and examples that clarify complex concepts
- Ensure consistent messaging across different representatives
3. **Relationship Building**
   - Develop connections with relevant journalists before crises
   - Provide background education on your approach
   - Share context that helps accurate reporting
   - Build credibility through consistent accessibility
4. **Issue Management**
   - Monitor coverage for inaccuracies and misconceptions
   - Respond proportionally to problematic framing
   - Provide corrective information respectfully
   - Use issues as opportunities for education

The Media Preparedness Toolkit

Organizations should develop these specific resources for effective media engagement:

- **Core Narrative Document:** Concise explanation of what you're doing and why
- **Technical Translation Guide:** Explanations of key concepts in accessible language
- **FAQ Compilation:** Answers to common and difficult questions
- **Visual Assets:** Graphics that explain complex concepts simply
- **Example Library:** Concrete illustrations of system operation and benefits
- **Expert Directory:** Knowledgeable spokespersons for different aspects

PATTERN RECOGNITION: Organizations that develop media relationships only during crises often face more critical coverage than those who invest in ongoing education and relationship-building with journalists.

Real-world example: A government agency implementing AI for benefit eligibility determination created a comprehensive media kit that included plain-language explanations, visual decision flow diagrams, anonymized before-and-after case examples, and background on their testing and oversight processes. They proactively briefed key journalists, resulting in significantly more accurate coverage than similar initiatives at other agencies that relied on reactive media approaches.

Customer and Partner Communication

Direct stakeholders require communication approaches that balance transparency with practical usability:

1. **Layered Information Design**
   - Create tiered disclosure from summary to detail
   - Allow stakeholders to determine their information depth
   - Design for different knowledge levels and interests
   - Integrate information into natural interaction points
2. **Value-Centred Framing**
   - Connect AI capabilities to stakeholder priorities
   - Explain benefits in terms that matter to them
   - Address concerns from their perspective
   - Balance capability explanation with value demonstration
3. **Interaction-Based Understanding**
   - Use actual interactions to build understanding
   - Create natural opportunities to learn through experience

- Make explanations easily accessible without disrupting workflows
- Design for learning over time, not just initial disclosure
4. **Relationship-Based Trust**
   - Recognize that trust transfers from people to systems
   - Maintain human connections alongside AI implementation
   - Create clear accountability and contact points
   - Demonstrate understanding of stakeholder contexts

WARNING SIGN: If your AI communication relies primarily on lengthy legal documents or technical specifications, you're likely failing to meet most stakeholders' information needs effectively.

The Disclosure-Usability Balance

Organizations face a genuine tension in customer communication:

- Comprehensive disclosure can create information overload.
- Simplified communication may omit important details.
- Technical accuracy can reduce understandability.
- Accessible language may sacrifice precision.

Resolving this tension requires moving beyond a single communication artefact to a layered approach that serves diverse needs.

The Progressive Disclosure Framework

This approach structures information in tiers that balance comprehensiveness with usability:

Tier 1: Essential Understanding

- What the system does and why
- How it affects the user directly
- Key benefits and potential concerns
- How to get more information

Tier 2: Functional Understanding

- How the system works in general terms
- What data is used and how
- Key limitations and constraints
- How to provide feedback

Tier 3: Detailed Understanding

- Technical approach and methodology
- Performance metrics and testing approach
- Governance and oversight mechanisms
- Ongoing improvement processes

ACTION STEP: Audit your current AI communication materials to ensure you have appropriate content for each tier of understanding, from essential overview to detailed technical information.

Real-world example: A financial services company implementing an AI financial advisor created a progressive disclosure system. Their customer interface included simple explanations of recommendations with "learn more" options, a comprehensive support centre with more detailed explanations, and a full technical paper for those wanting deep understanding. This layered approach satisfied both regulatory requirements and diverse customer information needs without overwhelming most users.

COMMUNITY AND PUBLIC OUTREACH

Broader communities affected by AI implementation require engagement approaches that build understanding and trust beyond direct usage relationships.

1. **Community Education**
   - Create accessible learning opportunities about AI capabilities
   - Develop materials appropriate for different knowledge levels
   - Address common misconceptions proactively
   - Build basic AI literacy that supports informed discussion
2. **Participatory Engagement**
   - Involve community representatives in appropriate governance
   - Create feedback mechanisms for community input
   - Hold dialogue forums for two-way communication
   - Demonstrate how input influences decisions
3. **Impact Transparency**
   - Clearly communicate how AI affects the community
   - Share both benefits and potential concerns
   - Provide ongoing updates as systems evolve
   - Create visibility into outcomes and performance
4. **Relationship Investment**
   - Build connections before controversies arise
   - Engage with community leaders and organizations
   - Demonstrate commitment to community well-being
   - Invest in long-term trust-building

The Community Engagement Spectrum

Different situations require different engagement approaches. This spectrum helps organizations select appropriate methods:

Inform: One-way communication providing balanced information
- Community briefings and presentations
- Accessible educational materials
- Transparent impact reporting
- Clear explanations of decisions

Consult: Two-way communication seeking feedback
- Surveys and input mechanisms
- Public comment opportunities
- Feedback forums and channels
- Consultation with community representatives

Involve: Interactive participation in development
- Advisory groups with community members
- Co-design workshops for aspects of implementation
- Participatory testing and evaluation
- Ongoing dialogue forums

Collaborate: Joint decision-making in defined areas
- Shared governance mechanisms

- Collaborative oversight structures
- Joint problem-solving processes
- Mutual accountability frameworks

KEY INSIGHT: The level of community engagement should increase with the potential impact of the AI system on the community. Higher-impact implementations warrant more collaborative approaches.

Real-world example: The AskNivi platform in Kenya demonstrates how understanding community needs drives successful AI implementation. Rather than deploying cutting-edge generative AI, they chose "a retrieval-based approach leveraging large language models' interpretative capabilities while avoiding exposing users to generated content." This conscious choice reflected a deep understanding of healthcare stakeholders' concerns about AI-generated medical information. The result – enabling three million Kenyan women to access healthcare information via SMS – shows how stakeholder-centred design creates both social impact and sustainable success.

Regulatory and Policy Engagement

Engaging effectively with regulatory and policy stakeholders requires specialized approaches that demonstrate compliance while shaping evolving standards.

1. **Compliance Demonstration**
   - Document adherence to existing requirements
   - Develop clear evidence of responsible practice
   - Create comprehensive but navigable documentation
   - Demonstrate ongoing monitoring and improvement
2. **Standards Participation**
   - Engage in developing industry and technical standards
   - Contribute expertise to standard-setting bodies
   - Pilot evolving best practices
   - Share lessons learned from implementation
3. **Policy Education**
   - Provide technical context to policymakers
   - Explain practical implementation considerations
   - Demonstrate responsible approaches to emerging issues
   - Create accessible explanations of complex systems
4. **Constructive Advocacy**
   - Advocate for effective rather than minimal regulation
   - Propose practical approaches to legitimate concerns
   - Engage with broad stakeholder perspectives
   - Focus on governance that enables responsible innovation

The Regulatory Engagement Advantage

KEY INSIGHT: Organizations that treat regulators as partners rather than adversaries gain significant competitive advantages in AI implementation.

Many organizations view regulatory engagement defensively, seeing regulators primarily as constraints on innovation. This creates an adversarial dynamic that ultimately harms both innovation and regulatory effectiveness.

A counterintuitive alternative approach treats regulators as crucial stakeholders in building sustainable AI ecosystems. Organizations that engage constructively with regulatory development gain several advantages:

- They help shape regulations that work better in practice.
- They build relationships that facilitate communication during issues.
- They gain early insight into regulatory direction.
- They demonstrate leadership that builds broader trust.
- They develop compliance approaches aligned with their values.

REFLECTION QUESTIONS: Does your organization approach regulatory engagement as a defensive necessity or as a strategic opportunity to help shape more effective standards?

Real-world example: A healthcare organization implementing AI clinical decision support took a proactive approach to regulatory engagement. Rather than waiting for formal requirements, they initiated regular briefings with relevant regulatory bodies, sharing their governance approach and seeking input on evolving standards. When regulations were eventually formalized, they had already implemented most requirements and had relationships that facilitated efficient certification. This proactive stance gave them a significant time-to-market advantage over competitors who had taken a minimal-compliance approach.

MANAGING THE EVOLVING REGULATORY ENVIRONMENT

Societal conditions are evolving alongside increasing AI adoption. As discussed, a proactive approach for managing external perceptions – for example, with the media, customers, partners and the larger community – can help build a strategic advantage. This also applies to an active engagement in the regulatory environment for AI, which is developing rapidly yet unevenly across regions and industries. Organizations need strategic approaches to navigate this complexity effectively while maintaining competitive advantage.

NAVIGATING REGIONAL REGULATORY DIFFERENCES

The global regulatory landscape is fragmented, with different regions emphasizing distinct concerns and approaches. This creates complexity but also opportunity for organizations that develop coherent cross-regional strategies.

Key Regional Regulatory Patterns

European Union

- Comprehensive, risk-based regulation through the AI Act
- Strong emphasis on fundamental rights and human oversight
- Significant transparency and documentation requirements
- Substantial penalties for non-compliance

In the European Union, the AI Act introduces risk-based categorization, transparency mandates, and serious penalties for non-compliance. Leaders are expected to go beyond minimum legal requirements and adopt what the Holistic AI Team describes as "trustworthy-by-design" principles. This means not just adhering to rules, but actively aligning AI with societal values, human oversight, and organizational accountability from the outset.

United States

- Sector-specific regulation through existing agencies

- Emerging state-level requirements creating compliance complexity
- Emphasis on voluntary standards and self-regulation
- Focus on particular high-risk domains and applications

According to the London School of Economics industry insights report on AI in Law & the Legal Profession, "the American strategy aims to promote flexible adaptation to emerging technology and market demands while avoiding unduly restrictive laws," creating a significantly different approach compared to the EU's comprehensive framework.

China

- Security and social harmony seen as primary concerns
- Strong emphasis on alignment with national priorities
- Sector-specific rules with particular focus on certain domains
- Significant data localization and sovereignty requirements

Middle East: Research suggests that Middle Eastern countries, notably the United Arab Emirates, leverage their centralized governance models to accelerate AI development. Thomas Kuruvilla, Managing Partner at Arthur D Little Middle East, also notes that "AI is more than a buzzword in the Middle East; it's a strategic imperative that's receiving boardroom attention" and leaders are "not just adopting AI but are strategically deploying it to unlock a new frontier of possibilities."

The research further notes that Saudi Arabia, under Crown Prince Mohammed bin Salman, has engaged in significant collaborations, such as the partnership between Groq and Saudi Aramco to build large AI infrastructure, demonstrating a state-driven approach to AI adoption.

PATTERN RECOGNITION: Regions with the most restrictive regulatory approaches often emphasize protecting existing values, while those with more permissive approaches prioritize creating new competitive advantages.

The Regulatory Alignment Matrix

This framework helps organizations identify core requirements across jurisdictions and develop efficient compliance approaches.

1. **Identify Universal Requirements**
   - Map regulations across all relevant jurisdictions
   - Identify common requirements that apply everywhere
   - Develop baseline compliance approaches for universal elements
   - Create core documentation that serves multiple regulators
2. **Address Regional Variations**
   - Identify regionally specific requirements
   - Develop modular approaches for different jurisdictions
   - Create region-specific documentation supplements
   - Establish processes for tracking regulatory divergence
3. **Engage with Regulatory Development**
   - Monitor evolving requirements across regions
   - Participate in consultation processes where appropriate
   - Contribute to standard development in key areas
   - Provide practical implementation perspective
4. **Build Flexible Compliance Infrastructure**
   - Design governance systems adaptable to changing requirements
   - Create documentation that can evolve efficiently
   - Develop compliance verification processes that scale
   - Build communication approaches usable across jurisdictions

ACTION STEP: Create a "regulatory consilience map" that identifies core requirements common across all jurisdictions where you operate, forming the foundation of your compliance approach.

Real-world example: A global technology company developed a "regulatory consilience" approach for their AI products. They mapped requirements across 20 jurisdictions and identified eight core principles that addressed fundamental requirements across all regions. They then created modular compliance packages that began with these universal elements and added jurisdiction-specific components as needed. This approach reduced duplicate compliance work by 60% while ensuring comprehensive coverage across their markets.

STRATEGIC APPROACHES TO ADDRESSING JOB DISPLACEMENT CONCERNS

Few issues generate more intense stakeholder concern than AI's potential impact on employment. Organizations need thoughtful strategies to address these legitimate concerns responsibly.

The Job Impact Communication Framework

The following structured approach helps organizations communicate honestly about employment impacts.

1. **Impact Assessment**
   - Conduct thorough analysis of potential effects
   - Distinguish between displacement, transformation, and creation
   - Identify particularly vulnerable roles and teams
   - Develop realistic timelines for potential changes
2. **Transition Planning**
   - Create concrete plans for affected employees
   - Develop reskilling and redeployment approaches
   - Establish clear support mechanisms and resources
   - Build timeline and process transparency
3. **Opportunity Development**
   - Identify new roles created by AI implementation
   - Develop pathways from affected roles to new opportunities
   - Create training and development resources
   - Establish transition support mechanisms
4. **Stakeholder Engagement**
   - Engage transparently with employees about plans
   - Communicate honestly with communities and governments
   - Involve workforce representatives in planning
   - Create ongoing dialogue rather than one-time announcement

WARNING SIGN: Organizations that frame job impacts as "not our problem" or rely on vague statements like "the market will adjust" typically face stronger resistance than those who acknowledge their responsibility in the transition.

The Honesty-Anxiety Balance

Organizations face a genuine tension in job impact communication:

- Complete transparency may create unnecessary anxiety
- Minimizing impacts damages credibility when changes occur
- Different stakeholders require different levels of detail
- Projections contain inherent uncertainty

Communicating AI's Societal and Ethical Values

This tension cannot be resolved through a single communication approach, but instead requires a graduated strategy that balances honesty and timing.

The Staged Disclosure Approach

This model provides a balanced framework for job impact communication.

Stage 1: Strategic Intent

- Communicate overall direction and rationale
- Provide general timeframes for implementation
- Establish principles guiding workforce decisions
- Create initial engagement mechanisms

Stage 2: Impact Framework

- Share assessment methodology and approach
- Provide categorized impact projections
- Outline support and transition principles
- Deepen engagement with affected groups

Stage 3: Specific Plans

- Communicate concrete transition approaches
- Provide individual-level clarity where possible
- Detail support resources and access methods
- Implement comprehensive engagement

Stage 4: Implementation Transparency

- Report regularly on actual versus projected impacts
- Share success stories and learnings
- Provide ongoing support visibility
- Maintain continuous dialogue

REFLECTION QUESTIONS: If your AI implementation will impact existing jobs, have you created a clear, staged communication plan that balances honesty with appropriate timing?

Real-world example: A telecommunications company implementing AI customer service systems developed a four-year workforce transformation plan. Rather than making vague statements or concealing potential impacts, a staged communication approach was implemented. It began with strategic intent, then provided increasingly specific information as plans developed. The organization established a "Career Transition Centre" two years before significant role changes began, giving employees time to develop new skills or make informed decisions. This transparent approach increased retention among high-performing team members who appreciated the honesty and planning horizon while significantly reducing community concerns compared to competitors who were less forthcoming.

BUILDING PUBLIC TRUST FOR AI THAT REPLACES ENTIRE FUNCTIONS

Some AI implementations fundamentally transform or replace entire functions rather than simply augmenting them. These situations require specialized communication approaches to build and maintain trust.

Function Transformation Trust Framework

This model provides guidance for high-impact transformation communication.

1. **Purpose Clarity**
   - Explain the genuine need for transformation
   - Connect to legitimate organizational and societal benefits
   - Address why incremental approaches are insufficient
   - Demonstrate alignment with broader values and goals
2. **Impact Honesty**
   - Provide transparent assessment of changes
   - Acknowledge real costs alongside benefits
   - Explain distribution of impacts across stakeholders
   - Demonstrate comprehensive impact consideration
3. **Responsible Transition**
   - Detail concrete approaches for affected stakeholders
   - Show meaningful support for those negatively impacted
   - Create visible commitments with accountability
   - Demonstrate ongoing responsibility beyond implementation
4. **Governance Credibility**
   - Establish robust oversight appropriate to impact level
   - Include diverse perspectives in governance mechanisms
   - Create transparency into decision processes
   - Implement meaningful constraints and principles

QUICK WIN: Identify one specific function your AI will transform (not just augment) and create a communication plan that specifically addresses purpose clarity, impact honesty, responsible transition, and governance credibility.

The Societal Value Exchange

For function-replacing AI, organizations must articulate a clear societal value exchange that demonstrates benefits beyond narrow organizational interests. This requires:

- Explaining how efficiency gains translate to broader benefits
- Demonstrating how transformation creates new opportunities
- Showing how affected stakeholders share in positive outcomes
- Connecting to legitimate societal needs and priorities

THE SPECIFICITY SHIELD

KEY INSIGHT: Being specific about what is changing (and what isn't) creates less resistance than using vague, general language intended to minimize the scope of change.

Organizations often make a critical mistake when communicating about function-replacing AI: they try to minimize or generalize the scope of change to reduce controversy. This approach typically backfires, creating greater suspicion and resistance.

The counterintuitive alternative is to be as specific as possible since specificity:

- Prevents conflation with broader AI narratives
- Focuses discussion on actual rather than imagined impacts
- Demonstrates thorough understanding and planning
- Creates clarity that builds credibility

Real-world example: When a logistics company implemented autonomous vehicles for specific warehouse operations, vague language about "automation enhancements" was initially used. This created widespread anxiety about job losses across all operations. When they shifted to specific communication about "automated pallet transport in three specific warehouse zones," with clear explanations of which functions would and wouldn't be affected, employee and community concern decreased dramatically. This specificity created a "shield" that prevented the limited implementation from being conflated with broader automation narratives.

CREATING TRANSPARENCY AROUND DATA USAGE AND MODEL DECISION-MAKING

Stakeholders increasingly expect transparency about how AI systems use data and make decisions. Effective communication in this area requires structured approaches that balance technical accuracy with accessibility.

In *Making Data Work*, I've explained why this is crucial: "Our data offers insights into who we are, where we live, work and travel, how much money we have and what we own, and about the people and issues we care about." Given such sensitive information, stakeholders naturally want to understand how it's being used.

This concern is heightened by incidents like the high-profile Cambridge Analytica scandal, where data inappropriately harvested from up to 87 million Facebook profiles was used to provide analytical assistance to the 2016 presidential campaign. Research suggests that firms typically lose over 5% of their market value within the first five days of a single trust-damaging data leak or hack.

The Layered Transparency System

This framework provides a graduated approach to data and decision transparency.

Layer 1: Essential Understanding (For All Stakeholders)

- What data is collected and why
- How it's used in general terms
- What decisions or recommendations the system makes
- How these affect different stakeholders
- Who has access to different data elements

Layer 2: Functional Understanding (For Interested Stakeholders)

- More specific data elements and sources
- How data flows through the system
- General explanation of decision approach
- Key factors influencing outcomes
- Limitations and constraints

Layer 3: Technical Understanding (For Highly Engaged Stakeholders)

- Detailed data specifications
- Model architecture and approach
- Performance metrics and validation methods
- Explanation methodology and limitations
- Governance and oversight mechanisms

ACTION STEP: Audit your AI system's data usage explanations to ensure they address all three layers of transparency, from essential understanding to technical details.

The Transparency-Comprehensibility Trade-Off

Organizations face a fundamental tension in transparency efforts.

- Technical accuracy often reduces understandability
- Simplified explanations may sacrifice important nuance

- Different stakeholders need different explanation types
- Some aspects may be proprietary or sensitive

Rather than choosing a single point on this spectrum, effective transparency systems embrace this tension by creating multiple explanation types for different needs.

The Multi-Modal Explanation Approach

This framework provides diverse explanation types to serve different purposes.

1. **Purpose-Based Explanations**
   - Focus on why the system exists
   - Explain what problems it is designed to address
   - Connect to meaningful goals and outcomes
   - Simple, accessible terminology, for all stakeholders
2. **Process-Based Explanations**
   - Describe how the system operates in general
   - Explain data flows and decision points
   - Outline human involvement and oversight
   - Moderately technical terminology, for interested stakeholders
3. **Feature-Based Explanations**
   - Identify key factors influencing specific decisions
   - Show relative importance of different inputs
   - Explain how features interact and combine
   - Somewhat technical terminology, for impacted stakeholders
4. **Counterfactual Explanations**
   - Show how different inputs would change outcomes
   - Explain decision thresholds and what pushes outcomes in different directions
   - Demonstrate system constraints and limitations
   - Modestly technical, practical terminology, for affected users
5. **Technical Explanations**
   - Provide detailed methodological documentation
   - Explain model architecture and approach
   - Document performance metrics and limitations
   - Highly technical terminology, for auditors and specialists

KEY INSIGHT: Different stakeholders need different types of explanations, not just different levels of detail. The most effective transparency systems provide multiple explanation modalities rather than just simplifying or using a single approach.

Real-world example: A financial institution implemented a multi-modal explanation system for their credit decision AI. What was provided were purpose explanations in marketing materials, process explanations in customer agreements, feature-based explanations for specific decisions, counterfactual explanations for declined applications, and technical documentation for regulators. This layered approach satisfied diverse stakeholder needs without forcing a one-size-fits-all explanation that would inevitably fail some audiences.

CASE STUDIES OF ORGANIZATIONS THAT TURNED ETHICAL AI INTO COMPETITIVE ADVANTAGE

From different regional regulatory approaches and job displacement concerns to data and decision transparency requirements, organizations have to navigate a complex and ever-evolving external ecosystem to mitigate potential stumbling blocks to their AI journey.

Communicating AI's Societal and Ethical Values

All these considerations fall under the umbrella of ethical AI requirements, which organizations can either view as constraints or transform into genuine competitive advantages.

Here are three detailed case studies that illustrate different approaches to this opportunity.

FINANCIAL SERVICES: TRANSPARENCY AS TRUST DIFFERENTIATOR

The Organization: A mid-sized wealth management firm implementing AI-based investment recommendations

Initial Situation: The industry was facing growing scrutiny about "black box" algorithms making financial recommendations without adequate explanation. Regulators were beginning to signal interest in transparency requirements, and customer research showed growing unease about algorithmic decisions affecting their finances.

The Breakthrough Approach: Rather than waiting for regulatory requirements or following industry norms of minimal disclosure, the firm developed a "Transparent AI" initiative with several pioneering elements:

1. **Recommendation Explanations:** Plain-language explanations – showing the 3 to 5 key factors influencing each investment recommendation – were created.
2. **Investment Philosophy Documentation:** A detailed explanation of the principles guiding their AI, including both optimizations and constraints, was published.
3. **Performance Reporting:** Clear metrics comparing AI-recommended portfolios with traditional approaches, including both successes and limitations, were provided.
4. **Client Control Tools:** Interfaces allowing clients to set their own preferences and constraints that would override algorithmic defaults were developed.
5. **Independent Verification:** A respected third party was engaged to audit the explanation system and verify its accuracy.

Key Success Factors

- Moving beyond compliance to genuine transparency
- Creating explanations that balanced accuracy and accessibility
- Giving clients meaningful control and preference expression
- Building external verification of the approach
- Treating transparency as a positive differentiator rather than a regulatory burden

The Results

- Client acquisition rates increased 28% compared to industry average
- Client retention improved by 23%, particularly among high-net-worth segments
- Regulatory relationships strengthened, creating smoother approval for subsequent innovations
- The approach became a case study highlighted by industry associations
- Competitors were forced to play catch-up with similar transparency initiatives

PATTERN RECOGNITION: The organization turned what most competitors viewed as a regulatory burden into a market differentiation opportunity by embracing transparency as a feature rather than a constraint.

HEALTHCARE: COMMUNITY ENGAGEMENT AS SOCIAL LICENSE

The Organization: A regional healthcare system implementing AI for clinical decision support and resource allocation

Initial Situation: Healthcare AI was facing increasing scrutiny about bias, privacy, and the appropriate balance between algorithmic and human judgement. Previous implementations at other systems had faced community backlash, regulatory scrutiny, and adoption resistance from clinicians concerned about patient perception.

The challenge was similar to what IBM Watson for Oncology faced. Despite Watson's 2011 Jeopardy! success, its healthcare application stumbled because "Watson's AI recommendations were inconsistent with local clinical practices, making it difficult to implement in regions with different treatment standards or drug availability." As Ron Schmelzer, Principal Analyst at Cognilytica, noted: "We've talked to people in the healthcare industry and they say, 'Well, we're actually some of the most conservative adopters of technology. I don't know why people thought we would have been the first to adopt.'"

The IBM Watson for Oncology case represents a $4 billion failure because it neglected contextual implementation requirements and stakeholder needs despite its technical capabilities, highlighting the critical importance of stakeholder engagement.

1. **The Breakthrough Approach:** Rather than treating AI as a purely technical implementation, the healthcare system developed a "Community Voice in AI" programme:
2. **Diverse Advisory Council:** A council that included patients, community advocates, clinical staff, ethicists, and technical experts to provide oversight was established.
3. **Tiered Engagement Process:** Engagement opportunities – ranging from broad community forums and focused patient group discussions to clinical staff workshops – were created.
4. **Values-Explicit Framework:** An explicit framework was developed and published, showing how the AI approach aligned with core values of equity, quality, autonomy, and transparency.
5. **Progress Reporting:** Regular public reporting on implementation, performance, challenges, and lessons learned was established.
6. **Continuous Feedback Loops:** Ongoing mechanisms were in created to accommodate concerns and suggestions to influence system evolution.

Key Success Factors

- Treating community engagement as strategic rather than ceremonial
- Creating meaningful influence for diverse stakeholders
- Explicitly connecting AI to core values and principles
- Establishing transparency as an ongoing commitment
- Building governance with credible independence

The Results

- Implementation proceeded with strong community support despite controversies affecting competitors
- Clinical adoption reached 87%, substantially above industry average
- Regulatory reviews completed with minimal friction
- The system became a reference model highlighted by healthcare quality organizations
- Media coverage shifted from potential concerns to the innovative engagement approach

QUICK WIN: Identify one key community stakeholder group affected by your AI implementation and create a specific engagement plan that gives them meaningful input into the system's governance.

RETAIL: ETHICAL AI AS BRAND ALIGNMENT

The Organization: A consumer retail company implementing AI for personalization, inventory management, and pricing

Initial Situation: Retail AI applications were raising growing concerns about manipulation, privacy, and fairness. Consumer research showed increasing scepticism about how companies were using data and algorithms, with particular concerns about differential pricing and invasive tracking.

The Breakthrough Approach: The company recognized alignment between existing brand values of transparency and fair treatment and the opportunity to extend these values to their AI approach:

1. **Brand-Aligned AI Principles:** Clear principles for AI development and use were created to highlight explicit connections to established brand values.
2. **Customer Data Dashboard:** Interfaces were developed that gave customers visibility and control over their data and personalization preferences.
3. **Fair Pricing Commitment:** Clear boundaries on algorithmic pricing, including explicit limitations on personalized pricing were established and published.
4. **Transparent Personalization:** Simple indicators were created that showed when and how content was being personalized for individual customers.
5. **Values-Based Marketing:** A responsible AI approach was integrated into the broader brand narrative and marketing materials.

Key Success Factors

- Authentic connection to existing brand positioning
- Creating visible manifestations of values in customer experience
- Establishing meaningful self-regulation in controversial areas
- Making responsible AI part of their market identity
- Treating ethical principles as competitive assets rather than constraints

The Results

- Customer trust scores increased by 31% compared to industry average
- Digital channel engagement improved by 27% due to greater confidence
- Employee pride and retention strengthened, particularly among technical teams
- The approach generated positive media coverage and industry recognition
- The company established thought leadership in an area of growing consumer interest

REFLECTION QUESTIONS: How could your organization's existing brand values and mission be authentically expressed through your AI implementation approach?

These case studies illustrate a crucial insight: organizations that move beyond compliance to make responsible AI communication a strategic priority can create significant competitive advantages. By authentic alignment with values, meaningful stakeholder engagement, and transparent practices, they build trust capital that creates business value while advancing ethical implementation.

BALANCING DATA PRIVACY AND INNOVATION

As demonstrated, a proactive approach to ethical implementation can help organizations move beyond compliance to strategic value creation, unlocking measurable business benefits. Another area in AI communication that can be fraught with tension is data privacy.

Organizations must navigate this complex topic through approaches that respect privacy while enabling innovation.

COMMUNICATING DATA USE CLEARLY AND EFFECTIVELY

Clear communication about data practices builds trust and reduces misconceptions. Effective data communication includes:

1. **Purpose Specificity**
   - Explain why specific data is collected
   - Connect data use to stakeholder benefits
   - Avoid vague or overly broad purpose statements
   - Update explanations as uses evolve
2. **Accessible Language**
   - Use plain language free of technical jargon
   - Provide concrete examples of data usage
   - Create visual explanations where helpful
   - Layer information from summary to detail
3. **Granular Disclosure**
   - Differentiate between data types and sensitivity
   - Explain different processing and protection approaches
   - Clarify data flows and access controls
   - Provide appropriate detail on retention and deletion
4. **Context-Sensitive Communication**
   - Adapt information to usage context
   - Provide relevant details at decision points
   - Create ongoing visibility beyond initial disclosure
   - Layer information to avoid overwhelming users

WARNING SIGN: If your data usage explanations don't clearly connect specific data elements to specific benefits for users, you're likely creating unnecessary privacy concerns.

The Data Transparency Card System
This framework creates standardized, comparable data use explanations.

Basic Information Card

- Who is collecting the data and why
- Essential categories of data collected
- Primary uses and benefits
- How to learn more or control usage

Detailed Practice Card

- Specific data elements collected
- Third parties with access
- Retention and deletion practices
- Security and protection approaches
- Rights and control mechanisms

Technical Specification Card

- Complete data dictionaries
- Processing methodologies
- Detailed protection mechanisms
- Compliance documentation
- Governance and oversight structures

ACTION STEP: Create a standardized set of data transparency cards for your AI system that provide consistent, layered explanation of data practices across all products and services.

Real-world example: A technology company created a standardized "Data Use Cards" system for all their AI products. Each product had a consistent set of cards from basic to technical that explained data practices in standardized formats. This consistency allowed users to more easily understand and compare practices across services, building trust through predictability and accessibility. The system became so successful that several industry peers adopted compatible formats, creating a de facto standard that simplified understanding for consumers.

BUILDING TRUST THROUGH TRANSPARENT PRACTICES

Beyond clear communication, organizations build privacy trust through practices that demonstrate respect for data. Key approaches include:

1. **Control Mechanisms**
   - Provide meaningful choices about data use
   - Create granular rather than all-or-nothing options
   - Make controls accessible and understandable
   - Respect choices consistently across systems
2. **Visibility Systems**
   - Create ongoing transparency into data usage
   - Provide access to personal data when appropriate
   - Show how data influences experiences and decisions
   - Create audit trails of access and processing
3. **Minimization Practices**
   - Collect only what's necessary for stated purposes
   - Limit retention to appropriate timeframes
   - Anonymize or aggregate where possible
   - Demonstrate commitment to values-based practices beyond compliance
4. **Protection Verification**
   - Implement appropriate security measures
   - Verify effectiveness through testing and audit
   - Provide appropriate transparency into approaches
   - Demonstrate ongoing attention to emerging risks

KEY INSIGHT: Trust is built not just through what you say about data practices, but through what users actually experience when they interact with your system's privacy controls and transparency features.

THE INNOVATION-PROTECTION BALANCE

Organizations face a fundamental tension between data utilization for innovation and data restriction for protection. This creates difficult decisions about what data to collect, how long to retain it for, and what uses to enable.

In *Making Data Work*, I've explored this tension between data privacy and value creation in greater detail. While privacy concerns have increased, organizations need data to create value.

This tension can be resolved in mutually beneficial ways, as demonstrated by Israel's COVID-19 vaccination strategy. Israel succeeded through a "vaccines for data" deal with Pfizer-BioNTech. In exchange for vaccine doses enabling "a vaccination rate 10 times quicker than the United States," Israel provided data from its "centralised collection of medical statistics" to help determine "whether herd immunity is achieved after reaching a certain percentage of vaccination coverage." This real-world case demonstrated how transparent data-sharing created value for both parties.

The Graduated Value Exchange Framework

This approach balances innovation and protection by creating clear value exchanges at different data sensitivity levels.

Tier 1: Basic Functionality Data

- Minimal data required for core services
- Clear necessary connection to functionality
- Strong default protections with limited options
- Retention limited to service provision

Tier 2: Enhanced Experience Data

- Additional data that improves user experience
- Clear explanation of benefits and improvements
- Meaningful control options with reasonable defaults
- Appropriate retention with periodic review

Tier 3: Innovation-Enabling Data

- Data primarily valuable for future improvements
- Transparent connection to innovation objectives
- Opt-in approaches with clear value propositions
- Appropriate protections and research safeguards

QUICK WIN: Review your data collection practices to identify opportunities to move non-essential data elements from default collection to opt-in with clear value explanation.

Real-world example: A healthcare technology company implemented a three-tier data framework for their patient management AI. Basic functionality data was collected with minimal options but strong protection guarantees. Enhanced experience data included optional elements with clear explanations of how this improved service. Innovation data was collected only with explicit opt-in and specific consent for research and development purposes. This graduated approach allowed patients to make informed choices based on their personal privacy preferences while still enabling appropriate innovation.

Managing Consent and Control Expectations

Consent models for AI systems face unique challenges due to complexity, evolving usage, and indirect impacts. Organizations need specialized approaches to address these challenges:

1. **Contextual Consent**
   - Provide information relevant to specific situations
   - Time consent requests appropriately to decisions
   - Create ongoing rather than one-time consent models
   - Design for actual attention capabilities and contexts
2. **Layered Control Systems**
   - Create graduated control options for different needs
   - Provide both simple defaults and advanced options
   - Design for different engagement levels and interests
   - Balance comprehensiveness with usability
3. **Dynamic Permission Management**
   - Build systems for evolving consent over time

Communicating AI's Societal and Ethical Values

- Create visibility into changing practices and uses
- Provide mechanisms to update preferences
- Design for relationship evolution rather than static permission

4. **Delegated and Representative Control**
   - Develop appropriate group representation mechanisms
   - Create delegation options for complex decisions
   - Build institutional safeguards where individual control is impractical
   - Design for diverse capabilities and interests

In *Making Data Work*, I present a scenario, where failures in consent and control result in barriers to future collaboration. Stakeholders describe "situations [that] are impacting our ability to build trusting relationships…we have become less and less trusting, and this makes it more difficult to get into new vendor or partner relationships."

The Consent Triangle: Bridging Three Competing Models

PATTERN RECOGNITION: Organizations that excel at consent and control tend to bridge three competing models effectively rather than choosing one approach.

AI systems encounter three different consent models with different assumptions:

Legal-Regulatory Model

- Focus on documentation and liability
- Emphasis on comprehensive disclosure
- Orientation toward compliance requirements
- Often results in unread lengthy documents

User Experience Model

- Focus on simplicity and minimal friction
- Emphasis on common use cases and patterns
- Orientation toward efficiency and satisfaction
- Often results in oversimplified binary choices

Ethical-Trust Model

- Focus on meaningful understanding and choice
- Emphasis on genuine autonomy and respect
- Orientation toward relationship-building
- Often results in complex, multi-stage approaches

Rather than choosing one model, organizations need integrative approaches that address the legitimate concerns of each perspective.

REFLECTION QUESTIONS: Which of these three consent models dominates your organization's current approach, and how could you incorporate elements from the other two to create a more balanced system?

Real-world example: A technology company reinvented their AI consent approach to bridge these competing models. The organization created a tiered system that provided simple summaries for immediate decisions, more detailed information available through clear progressive disclosure, and comprehensive documentation for regulatory compliance. Crucially, they designed the experience so each layer was genuinely useful to different stakeholder groups rather than pushing all users through the same path. This balanced approach satisfied regulatory requirements while dramatically improving both user experience and measured understanding of data practices.

Turning Privacy into a Competitive Advantage

Far from being merely a compliance requirement, privacy can become a genuine differentiator when approached strategically.

1. **Privacy-Enhancing Technologies**
   - Implement technical approaches that protect data while enabling functionality
   - Develop federated learning and edge-computing capabilities
   - Create differential privacy implementations where appropriate
   - Build privacy by design into system architecture
2. **Trust-Based Positioning**
   - Develop marketing approaches that emphasize privacy values
   - Create clear comparison points with less protective alternatives
   - Build privacy into broader brand positioning
   - Connect data practices to organizational values
3. **Transparent Leadership**
   - Take visible positions on privacy issues and standards
   - Contribute to developing improved industry approaches
   - Demonstrate commitment to best practices beyond minimum requirements
   - Build privacy into organizational identity
4. **User-Centred Innovation**
   - Develop new capabilities that enhance protection while innovating
   - Create user experiences that make privacy manageable
   - Build features that give meaningful control
   - Design for privacy and functionality together

ACTION STEP: Identify one area where investing in privacy-enhancing technologies could create a meaningful competitive advantage for your organization's AI implementation.

The Privacy Abundance Mindset

KEY INSIGHT: Organizations that treat privacy and innovation as complementary rather than competing priorities often discover unexpected opportunities for creating value in privacy-respecting ways.

Many organizations operate from a "privacy scarcity mindset" that views privacy and data utilization as fundamentally opposed, creating zero-sum trade-offs where either innovation or protection must be sacrificed.

A counterintuitive alternative is the "privacy abundance mindset" that seeks creative approaches to advance both protection and innovation simultaneously.

This perspective:

- Invests in technologies that enable both goals
- Creates nuanced approaches rather than binary choices
- Views user trust as enabling rather than constraining
- Focuses on value creation through responsible models

Real-world example: A financial services company transformed the approach to transaction data from a compliance-focused model to a competitive advantage. Rather than simply meeting regulatory requirements, "Privacy Choice" capabilities were developed that gave customers unprecedented control over how their transaction data was used, with whom it was shared, and how long it was retained. This created both simple presets and advanced controls, with clear explanations of the benefits and limitations of different choices. The approach increased voluntary data-sharing for personalization while simultaneously building trust through visible respect for customer preferences, demonstrating that privacy and data utilization can be complements rather than competing factors.

ETHICAL AI AS A STRATEGIC ADVANTAGE

Beyond compliance requirements, ethical AI practices create strategic advantages through trust, differentiation, and sustainable innovation. Organizations can realize these advantages through systematic approaches.

How Responsible AI Builds Trust and Loyalty

Trust has tangible business value, particularly in AI implementation. As Beena Ammanath, Executive Director of the Global Deloitte AI Institute, states in a recent study: "By adopting procedures designed to promote responsibility and safeguard trust, leaders can establish a culture of integrity and innovation that enables them to effectively harness the power of AI, while also advancing equity and driving impact."

Organizations can build this trust capital through:

1. **Consistent Value Alignment**
   - Demonstrate genuine connection between values and practices
   - Show how AI decisions reflect organizational principles
   - Create visible manifestations of values in user experience
   - Build values into governance and development processes
2. **Expectation Management**
   - Set realistic expectations about capabilities and limitations
   - Communicate honestly about performance and constraints
   - Address concerns directly rather than minimizing them
   - Create aligned understanding across stakeholders
3. **Visible Accountability**
   - Establish clear responsibility for AI outcomes
   - Create appropriate oversight and review mechanisms
   - Provide recourse when issues arise
   - Demonstrate commitment to addressing problems
4. **Ongoing Relationship Investment**
   - Treat trust as a dynamic relationship rather than static state
   - Create continuous dialogue with key stakeholders
   - Invest in trust before experiencing problems
   - Recognize and respond to evolving concerns

PATTERN RECOGNITION: Organizations that build the strongest trust capital treat it as an essential business asset to be systematically developed rather than a nice-to-have or PR exercise.

The Trust ROI Framework

This model helps organizations quantify the business value of trust investments.

Direct Trust Returns

- Increased adoption and usage rates
- Higher retention and relationship longevity

- Premium pricing potential
- Lower marketing and acquisition costs
- Reduced oversight and compliance friction

Indirect Trust Returns

- Reputation enhancement and protection
- Market access advantages
- Regulatory relationship improvements
- Employee engagement and retention
- Partnership and ecosystem benefits

ACTION STEP: Implement a "trust impact analysis" for your AI initiatives to systematically track how your responsible AI practices affect specific business metrics.

Real-world example: A healthcare technology company implemented a "Trust Impact Analysis" for AI initiatives, systematically tracking how responsible AI practices affected specific business metrics. They documented 18% higher clinician adoption rates, 23% faster implementation timelines, and 31% lower regulatory compliance costs compared to industry benchmarks for similar technologies. This quantification helped justify investments in trust-building practices that might otherwise have been viewed as optional costs rather than strategic investments.

Creating Customer Preference Through Ethical Practices

Beyond building trust, ethical AI practices can create genuine preference for products and services through:

1. **Values-Based Differentiation**
   - Develop clear ethical positions that distinguish your approach
   - Create visible manifestations in product and service experience
   - Connect practices to broader organizational purpose
   - Build ethics into brand identity and positioning
2. **Empowerment-Focused Design**
   - Create user experiences that enhance agency and control
   - Design for complementary rather than replacement relationships
   - Build capabilities that augment human strengths
   - Develop interfaces that encourage autonomy and understanding
3. **Transparent Value Exchange**
   - Clearly communicate the benefits of AI capabilities
   - Explain what's given and received in the relationship
   - Make trade-offs explicit rather than hidden
   - Create genuine choice about participation levels
4. **Inclusive Development**
   - Design for diverse needs and considerations
   - Test with representative user populations
   - Address potential disparate impacts proactively
   - Create products that work well across different contexts

My research additionally highlights how problems can arise when such practices are not part of the approach. For instance, a University of Pennsylvania study found that "AI-enabled recruitment systems often perpetuate racial inequities, with 30% of Black professionals receiving job alerts below their skill level, and 40% being recommended based on identities rather than abilities."

Communicating AI's Societal and Ethical Values

QUICK WIN: Identify one feature of your AI system that could be redesigned to give users more meaningful choice, control, or agency, enhancing their experience while building trust.

THE SHORT-TERM VERSUS LONG-TERM VALUE BALANCE

Organizations often face tension between short-term optimization and long-term relationship-building in AI implementation. This creates difficult decisions about practices that might increase immediate metrics while potentially undermining long-term trust.

The Sustainable Value Hierarchy

This framework helps resolve this tension by creating a value assessment hierarchy.

Level 1: Immediate Transaction Value

- Direct revenue and conversion impacts
- Short-term engagement metrics
- Immediate cost-savings
- Current period performance

Level 2: Relationship Value

- Customer lifetime value
- Brand perception and preference
- Recommendation and advocacy
- Relationship durability

Level 3: Ecosystem Value

- Regulatory relationship quality
- Partnership and collaboration potential
- Community and stakeholder relationships
- System-level reputation and influence

PATTERN RECOGNITION: Organizations that excel at sustainable AI value optimize across all three levels of the hierarchy rather than maximizing only immediate metrics.

Real-world example: An e-commerce company developed an "Ethical AI Calculator" for recommendation systems, assessing potential changes across immediate performance, relationship strength, and ecosystem impact. When evaluating a change that would increase short-term conversion but use less transparent data practices, this broader analysis revealed potential long-term reputation and regulatory costs that outweighed the immediate gains. This approach helped the organization optimize for sustainable value rather than just quarterly results.

ATTRACTING AND RETAINING TALENT THROUGH RESPONSIBLE APPROACHES

In competitive talent markets, ethical AI practices create significant advantages in recruitment and retention.

1. **Purpose Alignment**
    - Connect AI work to meaningful impact
    - Demonstrate how ethics informs technical decisions
    - Show how values manifest in actual practices
    - Create opportunities for purpose-driven contributions

2. **Professional Pride**
   - Build development approaches that create quality pride
   - Create visibility into positive impacts and outcomes
   - Establish strong professional standards and practices
   - Recognize and celebrate responsible innovation
3. **Growth and Influence**
   - Provide opportunities to shape ethical direction
   - Create career paths that include ethical leadership
   - Develop skills in responsible implementation
   - Position team members as thought leaders
4. **Cultural Environment**
   - Build ethical discussion into normal work processes
   - Create psychological safety for raising concerns
   - Recognize and reward ethical consideration
   - Demonstrate leadership commitment to values

My research emphasizes the importance of sustainable rhythms for maintaining quality of work life. I've personally seen how effective a "life integration framework with non-negotiables" – like daily energy practices, weekly recovery time, monthly perspective days, and quarterly renewal periods – can be, especially when these building blocks are important enough to be included in calendars. As more and more people prioritize ethical, sustainable work, such considerations are becoming increasingly important for organizations looking to attract and retain top talent.

REFLECTION QUESTIONS: How does your organization currently recognize and reward ethical considerations in AI development, and what additional approaches could strengthen this aspect of your culture?

The Ethical Talent Lifecycle
This framework integrates ethics throughout the talent journey.

Attraction Phase

- Clear values communication in recruitment
- Discussion of ethical approach in interviews
- Examples of how values manifest in practice
- Transparent sharing of governance approaches

Development Phase

- Ethics training integrated with technical development
- Mentorship that includes ethical dimensions
- Exposure to diverse stakeholder perspectives
- Opportunities to contribute to ethical approaches

Retention Phase

- Ongoing alignment of work with personal values
- Recognition for ethical contributions
- Influence over evolving practices
- Career advancement that includes ethical leadership

ACTION STEP: Integrate explicit questions and criteria related to ethical AI development into your hiring processes to attract talent aligned with your values.

Communicating AI's Societal and Ethical Values

Real-world example: A technology company facing intense competition for AI talent developed an "Ethics by Design" programme that became a significant recruitment and retention advantage. The successful approach included integrating ethical considerations into their development process, creating ethics advisors who worked directly with technical teams, and establishing a recognized certification programme for responsible AI implementation. Exit interviews revealed that this ethical approach was among the top three reasons technical talent chose to join and remain with the company, alongside compensation and technical challenge.

Getting Ahead of Regulatory Requirements

Proactive approaches to emerging regulatory requirements create competitive advantages through reduced compliance costs, faster time to market, and greater implementation flexibility.

1. **Anticipatory Compliance**
   - Monitor regulatory signals and direction
   - Implement likely requirements before mandated
   - Build adaptable systems that can evolve with regulation
   - Create documentation aligned with emerging standards
2. **Standards Leadership**
   - Participate in developing industry standards
   - Contribute expertise to regulatory consultations
   - Pilot approaches that could become benchmarks
   - Share implementation learnings with the field
3. **Relationship Development**
   - Build constructive relationships with regulatory bodies
   - Create dialogue before compliance issues arise
   - Demonstrate commitment to regulatory objectives
   - Establish credibility through responsible practice
4. **Values-Based Self-Regulation**
   - Develop governance best practices beyond minimum requirements
   - Create transparent accountability mechanisms
   - Implement meaningful constraints on high-risk practices
   - Demonstrate commitment to ethical principles

Our research on Executive Leadership and AI Adoption included the Holistic AI Team's 2025 article, which emphasizes that understanding elements of the EU's AI Act is essential for CEOs operating in the EU, "not only for legal compliance but also for maintaining a competitive edge in a market increasingly driven by ethical standards."

KEY INSIGHT: Organizations that develop robust ethical governance frameworks based on their values often find they've already addressed many requirements when regulations are eventually formalized.

The Regulatory Readiness System
This framework helps organizations prepare for evolving requirements.

1. **Horizon Scanning**
   - Monitor developments across relevant jurisdictions
   - Track both formal regulation and soft guidance
   - Identify patterns and direction across markets
   - Assess maturity and likelihood of different requirements

2. **Impact Assessment**
   - Evaluate potential effects on current practices
   - Identify high-priority adaptation needs
   - Assess competitive implications of changes
   - Develop response scenarios for different outcomes
3. **Proactive Implementation**
   - Prioritize changes based on likelihood and impact
   - Create staged implementation approaches
   - Develop documentation aligned with likely requirements
   - Build adaptation capability into systems and processes
4. **Strategic Engagement**
   - Determine where to lead versus follow
   - Identify opportunities to shape emerging standards
   - Develop constructive regulatory relationship strategies
   - Create appropriate transparency about approaches

QUICK WIN: Identify one emerging regulatory requirement in your jurisdiction and implement it proactively, before it becomes mandated.

Real-world example: A financial services company implemented a "Regulatory Readiness" programme for an AI credit scoring system, systematically tracking emerging requirements across jurisdictions. They identified explainability standards as a consistent pattern and implemented a comprehensive explanation system two years before it became required in their primary market. This proactive approach allowed thoughtful implementation on their timeline rather than rushed compliance, resulting in a system that both satisfied regulatory requirements and created genuine user value through its explanations. This resulted in a competitive advantage over peers who implemented minimal, compliance-focused approaches under regulatory pressure.

CONCLUSION: THE STRATEGIC IMPERATIVE OF ETHICAL COMMUNICATION

As we've explored throughout this chapter, communicating AI's societal and ethical values isn't just a compliance requirement or public relations exercise – it is a strategic imperative that directly affects an organization's freedom to operate and ability to create value.

The most sophisticated AI capabilities create zero value if they lose social license through misaligned expectations, stakeholder mistrust, or regulatory constraints. Conversely, organizations that build strong foundations of trust through transparent, responsible communication create competitive advantages in customer preference, talent attraction, regulatory relationships, and implementation speed.

Let's conclude with five key principles for strategic AI communication:

1. **Transparency Creates Resilience, Not Vulnerability:** Organizations that proactively explain their approach, acknowledge limitations, and engage with concerns build trust capital that provides protection when challenges arise.
2. **Stakeholder Engagement Improves rather than Constrains Solutions:** Meaningful dialogue with diverse perspectives leads to more robust, sustainable approaches (compared to closed development), creating both technical and relationship advantages.
3. **Value Must Be Defined Broadly and Inclusively:** AI that creates narrow organizational value while generating broader social costs will face increasing resistance. Successful implementations create value across stakeholder ecosystems.
4. **Ethics Represents a Competitive Advantage rather than a Constraint:** Organizations that view ethical considerations as opportunities for differentiation rather than compliance burdens develop more sustainable and distinctive approaches.

Communicating AI's Societal and Ethical Values 237

5. **Trust Is Earned through Practice, Not Promises:** Ultimately, AI communication is judged by its alignment with actual implementation. Organizations build credibility when their actions consistently match their words.

FINAL REFLECTION QUESTIONS: Which of these five key principles represents the greatest opportunity for your organization to strengthen its approach to AI communication, and what specific steps could you take to realize this opportunity?

As AI becomes increasingly central to organizational success, the ability to communicate effectively about its societal and ethical dimensions will separate leaders from laggards. The frameworks and approaches in this chapter provide a roadmap for turning responsible communication from a defensive necessity into a strategic advantage.

Key Tools from This Chapter

Throughout this chapter, we've introduced several practical frameworks and tools that you can apply to your AI communication efforts. Here's a quick reference guide to these resources:

The FACTS Framework: A structured approach to designing external communications covering functionality, alignment, controls, transparency, and security

The Concern Response Matrix: A graduated approach to identifying and addressing potential concerns across different levels of maturity

The Transparency Dimensions Framework: A model for developing multidimensional transparency across purpose, process, performance, people, and protection

The Community Engagement Spectrum: A guide for selecting appropriate engagement approaches from inform to collaborate based on impact level

The Regulatory Alignment Matrix: A framework for identifying core requirements across jurisdictions and developing efficient compliance approaches

The Progressive Disclosure Framework: An approach to structuring information in tiers that balance comprehensiveness with usability

The Trust Return on Investment (ROI) Framework: A model for quantifying the business value of trust investments through direct and indirect returns

The Sustainable Value Hierarchy: A framework for balancing immediate, relationship, and ecosystem value in AI implementation decisions

Each of these tools can be adapted to your specific context and integrated into your existing AI governance and communication processes.

11 Measuring What Matters

Chapter Roadmap

In this chapter, we examine why traditional metrics often fail to capture AI's true value – and how to build measurement frameworks that bridge the gap between technical performance and business impact. We look at the unique challenges that make AI measurement different from conventional IT projects, including value chain complexity, attribution difficulties, and multi-dimensional impacts that unfold over time. You will discover practical approaches for setting up measurement systems before implementation begins, industry-specific frameworks that address sector-unique challenges, and proven strategies for capturing and communicating early wins that build organizational momentum. Through real-world case studies from manufacturing, healthcare, retail, and financial services, we show how leading organizations transformed their measurement approaches to demonstrate millions in documented value while avoiding common pitfalls that undermine AI initiatives.

THE PROOF PROBLEM IN AI INVESTMENT

Have you ever attended a meeting where an AI team proudly presents technical achievements only to fall silent when asked about business impact? This is unfortunately not a rare occurrence. I've encountered such scenes repeatedly across industries – from healthcare to finance to manufacturing.

For example, a global financial institution I worked with had invested over $15 million in an advanced AI system for customer service. At a quarterly review, the technical team arrived bursting with pride about their metrics: 96% accuracy on intent recognition, 82% first-time resolution rate, and processing time reduced by 68%.

Then the CFO asked simply: "How has this affected our customer retention?"

No answer was forthcoming. Despite all the impressive technical achievements, nobody had connected these AI capabilities to this fundamental business metric. The team had measured what was easy to measure, not what actually mattered.

PATTERN RECOGNITION: Organizations typically track technical metrics like model accuracy or processing speed while neglecting business metrics like customer retention, revenue growth, or market share – creating a dangerous disconnect between AI implementation and value perception.

This measurement gap creates five critical problems:

1. AI investments become vulnerable in budget cycles
2. Technical teams optimize for metrics disconnected from business value
3. Leadership loses confidence in AI's strategic importance
4. Organizations can't distinguish truly valuable AI from impressive demos
5. Learning and improvement focus on technical metrics rather than business dimensions

WARNING SIGN: When your team can quickly provide technical performance data but struggles to articulate business impact, you've likely fallen into the AI value fog – the dangerous space where impressive capabilities exist without demonstrable business value.

Measurement challenges aren't just operational headaches – they represent existential threats to AI initiatives. Without robust value measurement frameworks, organizations operate on faith rather than evidence, hoping rather than knowing that AI investments deliver value.

In this chapter, I'll explain why traditional measurement approaches often fail for AI and provide practical frameworks to bridge the gap between technical performance and genuine business value. I'll include a comprehensive toolkit to demonstrate AI's impact in terms that resonate across the organization – from technical teams to the boardroom.

ACTION STEP: Before reading further, take two minutes to list the top metrics your organization currently uses to evaluate AI initiatives. Mark which are technical metrics and which are business metrics. This quick audit will help you identify potential measurement gaps as we proceed.

WHY TRADITIONAL APPROACHES FALL SHORT

Organizations often approach AI measurement with methods that worked well for traditional IT projects – only to discover they fall short with AI's unique characteristics. Let's examine why, and how to overcome these challenges.

Five Unique Measurement Challenges with AI

1. **Value-Chain Complexity:** AI creates value through intricate pathways that cross traditional organizational boundaries. Unlike conventional IT solutions with relatively direct impacts, AI's effects often cascade through multiple steps, creating both intended and unintended consequences along the way.
2. **Causal Ambiguity:** Multiple factors influence outcomes simultaneously, making it difficult to isolate AI's specific contribution. While traditional projects can often use clear before-and-after comparisons, AI initiatives operate in complex environments where many variables change simultaneously.
3. **Time-Lag Complexity:** Different benefits emerge over different timeframes. Some appear immediately, while others develop gradually as systems learn and processes adapt. This creates measurement challenges when trying to evaluate success using fixed timelines.
4. **Multi-Dimensional Impact:** AI affects various metrics simultaneously – often improving some while seemingly degrading others. For example, a customer service AI might reduce average call-handling time but temporarily increase the number of calls that are transferred to supervisors as AI agents learn to work with the new system and implementation stabilizes.
5. **Continuous Evolution:** Unlike static systems, AI systems improve and evolve over time. Initial performance metrics may not reflect the long-term value potential as systems learn and adapt.

KEY INSIGHT: The most common measurement mistake is treating AI like any other IT implementation. Successful organizations recognize AI's distinctive characteristics and create measurement approaches specifically designed for these complexities.

The Measurement-Value Paradox

One fundamental tension that tends to trip up most organizations relates to the spectrum of value measurement. The aspects of AI that are easiest to measure, such as technical performance and operational metrics, are often located on the spectrum furthest away from the ultimate value. Aspects closest to value, such as strategic impact and competitive advantage, are hardest to measure precisely.

This creates a natural but problematic tendency to focus on what's easy rather than on what is important – a pattern that can undermine AI's perceived value.

For a real-world example, let's look at a pharmaceutical company implementing AI for drug discovery. Initially, the team focused entirely on technical metrics like model accuracy and compound screening speed. While these showed impressive improvements, executive questions about impact on development timelines or probability of clinical success went unanswered. Only through developing a comprehensive measurement framework connecting technical metrics to R&D portfolio value did the team secure continued investment.

REFLECTION QUESTIONS: Are your AI measurement efforts concentrated on metrics that are easier to measure but further from business value, or on metrics that are harder to measure but closer to true impact?

THE QUANTITATIVE-QUALITATIVE BALANCE

Effective AI measurement requires both rigorous numbers and insightful observations. Let's examine why each approach has strengths and limitations:

Quantitative Approaches

- **Strengths:** Precision, objectivity, clear comparison, financial connection
- **Limitations:** Miss intangible impacts, reduce complexity to simplified metrics, require data history, can create false precision

Qualitative Approaches

- **Strengths:** Capture nuanced value, identify unexpected benefits, provide context, reveal stakeholder experience
- **Limitations:** Subject to bias, harder to compare systematically, difficult to connect to financial metrics, may lack credibility with some stakeholders

The most effective measurement systems integrate both approaches rather than relying exclusively on either.

QUICK WIN: Add one qualitative measurement technique (like user interviews or stakeholder experience surveys) to complement your existing quantitative metrics. This simple addition can reveal valuable insights about AI impact that numbers alone might miss.

THE PRECISION-RELEVANCE TRADE-OFF

Organizations often fall into a common trap: pursuing increasing precision in metrics that may not be relevant, while neglecting relevant dimensions because they can't be precisely measured.

Here's the counterintuitive reality: an approximate measure of something important creates more value than a precise measure of something unimportant. Organizations that accept reasonable approximation for truly relevant metrics often develop more meaningful measurement systems than those insisting on high precision regardless of relevance.

For a real-world example, let's look at a retail bank implementing AI customer service. Initially, teams focused on precise measurement of operational metrics like handle time and first-contact resolution. Despite improvements in these precise metrics, they couldn't explain stagnant customer satisfaction. When they added qualitative assessment of conversation quality and supplemented net promoter score (NPS) with short experiential interviews, they identified critical gaps the precise operational metrics had missed completely. This balanced approach led to significant improvements in both customer experience and retention.

SETTING THE FOUNDATION: PRE-IMPLEMENTATION MEASUREMENT DESIGN

Perhaps the most critical success factor in AI value measurement is establishing the right framework before implementation begins. This preparatory work enables meaningful measurement that would be difficult or impossible to create retroactively.

THE PRE-IMPLEMENTATION MEASUREMENT CHECKLIST

Before launching your AI initiative, ensure you've addressed these critical questions:

- Have we clearly articulated how this AI will create value?
- Do we have baseline measurements for our target outcomes?
- Have we identified appropriate key indicators of eventual success?
- Do we understand what other factors will influence our outcomes?
- Have we created appropriate data collection mechanisms?
- Do we have a plan for attributing changes to our intervention?
- Have we defined what success looks like across different timeframes?
- Are our measurement approaches credible to key stakeholders?

In *Making Data Work*, I've explored a line of inquiry that starts with asking, "What are we trying to achieve that is reflected in this key performance indicator (KPI)?" and then goes to the next question: "What interventions can help us improve trust and integrity as well as increase code quality by limiting errors and reducing bugs?"

THE VALUE-PATHWAY-MAPPING APPROACH

One of the most powerful pre-implementation techniques is value-pathway mapping, which serves to document the complete causal chain from AI capabilities through intermediate effects to ultimate business outcomes.

ACTION STEP: For your AI initiative, create a visual map showing each step in the value creation chain, from technical implementation through user adoption to business impact. Identify potential measurements at each stage and the assumptions connecting them.

A real-world example comes from a global manufacturing company implementing predictive maintenance AI. The team developed a comprehensive measurement framework before technical implementation began. They mapped the entire value pathway from algorithm predictions through maintenance actions to equipment performance to manufacturing output to financial results. This preparation allowed them to track value creation at each step and identify where breakdowns occurred when expected results didn't materialize. This approach demonstrated $27 million in annual value within the first year while similar initiatives elsewhere struggled to prove any value despite technical success.

QUICK WIN: Identify just one intermediate step in your AI value pathway that isn't currently being measured – and establish a baseline measurement before implementation. This bridging metric can help connect technical performance to business outcomes.

AVOIDING THE "MEASURING-THE-WRONG-THING" TRAP

A cautionary example, which I highlighted in *Making Data Work*, relates to a retail company that implemented an AI system to predict product returns but failed to establish the fundamental value metrics needed to determine success. Without analysing the historical average cost nor the overall cost of returns, it was impossible to determine if the planned 10% discount intervention would be profitable. This nearly led to an implementation that, according to stakeholders, "could actually lead to millions of dollars in negative investment returns and financial losses."

At a global insurer, developers were being discouraged from using metrics that worked against high-quality code development. The company had been tracking code bugs as a KPI but discovered this metric "acted as a disincentive for the openness it was intended to drive among engineers." Because engineers perceived reporting bugs as negative for "performance, reputation and credibility scores," they either missed or deliberately ignored such bugs.

The solution? Shifting to metrics that measured positive behaviours like testing, peer reviews, design clinic participation, design pattern reuse, code reviews and increasing automation. This created a "positive correlation with our collective purpose," according to team members.

KEY INSIGHT: Sometimes the most significant improvement doesn't come from adding new metrics but from stopping to use counterproductive ones that drive behaviours contrary to the organization's goals.

COMPREHENSIVE VALUE MEASUREMENT APPROACHES

Now that we understand the foundational challenges, let's examine specific measurement frameworks that address AI's unique characteristics.

THE TRIPLE BAT FRAMEWORK FOR LIFECYCLE MEASUREMENT

An innovative framework, which I introduced in *Value-Driven Data*, can be used to measure data and AI value across the entire implementation lifecycle. It's called the "triple BAT" framework (Baseline Alternative Theory, Baseline Alternative Testing, Baseline Alternative Tracking).

> **Step 1: Baseline Alternative Theory:** Start by establishing a clear theory of expected value change. For example, in a sales conversion optimization case, an organization might start with a baseline conversion rate of 10% and project a target of 25%. The 15% difference becomes the baseline alternative theory: a clear, testable hypothesis of value.
> **Step 2: Baseline Alternative Testing:** Validate your theory through appropriate testing methodologies. This might involve controlled experiments, parallel processing, or historical simulations depending on the context.
> **Step 3: Baseline Alternative Tracking:** Implement ongoing monitoring to ensure value persists over time, recognizing that initial gains can erode without proper attention.

The baseline alternative theory (BAT) can then be calculated as the difference between the current baseline (where we are) and the base rate (where we intend to go).

TOTAL VALUE OF AI OWNERSHIP FRAMEWORK

Traditional ROI calculations often miss substantial portions of AI's value. The total value of AI ownership (TVAO) framework provides a more comprehensive approach by measuring value across five dimensions:

1. **Direct Operational Value**
   - Immediate cost reduction
   - Process efficiency improvement
   - Quality enhancement
   - Resource optimization

2. **Revenue and Growth Value**
   - Customer acquisition improvement
   - Cross-sell and upsell enhancement
   - Retention and loyalty strengthening
   - New offering enablement
3. **Risk Mitigation Value**
   - Compliance improvement
   - Error and fraud reduction
   - Security enhancement
   - Resilience strengthening
4. **Capability-Building Value**
   - Organizational learning
   - Data asset enhancement
   - Process-intelligence development
   - Decision-quality improvement
5. **Strategic-Option Value**
   - New business model potential
   - Competitive-positioning enhancement
   - Market-entry enablement
   - Innovation-foundation creation

For a real-world example, let's consider a financial services firm. Expanding the AI investment evaluation beyond traditional ROI, the organization created a TVAO framework that explicitly accounted for benefits across the full spectrum. While traditional ROI analysis showed modest 15% returns, the comprehensive approach revealed that capability-building and strategic-option value more than doubled the effective return. This broader view helped justify continued investment despite initial ROI below usual thresholds.

QUICK WIN: Identify one "hidden value" category your organization is currently not measuring, such as risk reduction or capability-building. Create a simple assessment approach to begin capturing this dimension.

THE SECONDARY BENEFIT PARADOX: AN OVERLOOKED VALUE SOURCE

Organizations often focus measurement on primary, direct benefits while disregarding secondary benefits that may ultimately deliver significant – or even greater – value. This creates a paradoxical situation where the largest sources of value receive the least measurement attention.

The counterintuitive approach that creates better results: deliberately design measurement systems that capture secondary and tertiary effects rather than focusing exclusively on primary objectives. This broader lens often reveals value that would otherwise remain invisible.

For a real-world example, let's return to the healthcare organization implementing AI clinical decision support we previously discussed. Initially focused only on direct efficiency benefits, the team then expanded measurements to include "time reinvestment effects" (how clinicians used time saved). This led to the discovery that 78% of freed time went to additional patient interaction and consultation, creating significant quality and satisfaction improvements. These secondary benefits, measured through time allocation studies and patient experience surveys, ultimately delivered 3.7 times more value than the direct efficiency gains.

REFLECTION QUESTIONS: What secondary or tertiary effects might your AI implementation be creating that you're not currently measuring? How could these hidden effects potentially generate more value than your primary objectives?

THE MICROSOFT COPILOT SMB VALUE CASE

A Forrester Consulting research commissioned by Microsoft, shows how measuring value across multiple dimensions can justify significant AI investment. Small and medium businesses experienced 132–353% ROI over three years with Microsoft 365 Copilot, with benefits that included faster time to market, increased productivity, and improved employee satisfaction, resulting in 6% increased net revenue and 20% reduced operating costs.

This case illustrates how comprehensive measurements across both hard financial metrics and softer employee experience dimensions created a compelling value story that accelerated adoption. By quantifying both direct revenue impacts and indirect benefits like improved satisfaction, organizations could justify continued investment despite substantial implementation costs.

COMMON MEASUREMENT PITFALLS – AND HOW TO AVOID THEM

Even with the right frameworks, organizations frequently encounter predictable challenges when measuring AI value. Understanding these common pitfalls creates the opportunity to avoid them.

SEVEN DEADLY SINS OF AI MEASUREMENT

1. **The Technical Success Trap**
   - **Pitfall:** Focusing exclusively on technical performance metrics (accuracy, speed, etc.) without connecting to business outcomes
   - **Solution:** Always create explicit links between technical metrics and business value indicators, treating technical success as necessary but insufficient by itself
2. **The False Precision Problem**
   - **Pitfall:** Creating overly detailed projections and attributions despite fundamental uncertainty
   - **Solution:** Use appropriate ranges and confidence levels rather than point estimates, and match precision to actual certainty
3. **The Premature Judgement Error**
   - **Pitfall:** Evaluating long-term initiatives using short-term metrics before value has fully emerged
   - **Solution:** Create staged evaluation frameworks with appropriate metrics for different maturity phases
4. **The Isolated Measurement Mistake**
   - **Pitfall:** Measuring AI impact without accounting for other factors affecting outcomes
   - **Solution:** Implement control mechanisms and contextual analysis to isolate AI contribution
5. **The Narrow Scope Oversight**
   - **Pitfall:** Measuring only direct, intended consequences while missing broader effects
   - **Solution:** Design measurement systems that capture both intended outcomes and emergent impacts
6. **The Static Metric Failure**
   - **Pitfall:** Using fixed metrics despite evolving systems and changing value patterns
   - **Solution:** Create learning-based measurement that evolves with the system and emerging value
7. **The Stakeholder Mismatch**
   - **Pitfall:** Using metrics that don't address key stakeholder concerns and priorities
   - **Solution:** Map metrics to stakeholder value perspectives and create relevant measurements for each group

Measuring What Matters

WARNING SIGN: If you can't clearly explain how your technical metrics connect to business outcomes, you've likely fallen into the technical success trap, the most common measurement pitfall.

THE METRIC HIERARCHY FRAMEWORK: RIGHT-SIZING MEASUREMENT FOR DIFFERENT AUDIENCES

This approach creates a structured measurement system with appropriate detail at each level:

Level 1: Executive Metrics (3–5 key indicators)

- Focus on ultimate outcomes that matter to leadership
- Connect directly to strategic priorities and goals
- Integrate multiple value dimensions appropriately
- Support high-level decision-making and investment

Level 2: Management Metrics (5–10 operational indicators)

- Track key operational outcomes and performance
- Support ongoing optimization and improvement
- Provide diagnostic capability for issues
- Connect operational activity to executive metrics

Level 3: Implementation Metrics (10–20 detailed indicators)

- Monitor specific aspects of system performance
- Support detailed troubleshooting and improvement
- Track individual components of the value chain
- Provide early warning of potential issues

For a real-world example, let's look at a global bank that implemented a three-tier measurement system for an AI fraud detection initiative. For executives, they focused on three metrics: overall fraud loss reduction, customer friction change, and return on AI investment. For management, they tracked eight operational indicators, including false positive rates, investigation efficiency, and case resolution time. For the implementation team, they monitored 18 detailed metrics, covering model performance, process efficiency, and user adoption. This layered approach provided appropriate detail for each audience without overwhelming any group with unnecessary complexity.

INDUSTRY-SPECIFIC MEASUREMENT APPROACHES

Core measurement principles can be applied broadly yet different industries face unique value measurement challenges and opportunities. Let's examine industry-specific approaches that address these distinct contexts.

FINANCIAL SERVICES: BALANCING RISK AND EXPERIENCE

Financial institutions face a fundamental tension between risk management and customer experience – with AI often affecting both simultaneously:

- Stricter fraud controls reduce losses but increase customer friction
- Streamlined processes improve experience but may increase risk exposure
- Personalization enhances relevance but raises privacy considerations
- Automation increases efficiency but may reduce human judgement in complex cases

TABLE 11.1
The Financial Services Value Integration Scorecard

Value Dimension	Safety and Soundness	Operational Efficiency	Customer Value	Strategic Position
Risk Management	Loss rate reduction	Decision time improvement	Protection from harm	Risk advantage vs. competitors
Service Delivery	Policy compliance rate	Cost-to-serve reduction	Experience quality	Service differentiation
Product Offering	Product risk alignment	Delivery efficiency	Customer outcome improvement	Unique offerings
Channel Optimization	Security and compliance	Channel cost efficiency	Convenience and accessibility	Channel preference vs. alternatives

The Financial Services Value Integration Scorecard

The framework in Table 11.1 balances different value dimensions for financial AI initiatives.

Illustrating this balance is a real-world example from a retail bank that implemented an AI-based customer onboarding system, designed to simultaneously improve experience, reduce fraud, and ensure compliance. With an integrated measurement approach, the organization tracked risk metrics (fraud detection rate, false positive reduction), efficiency indicators (onboarding time, processing cost), experience measures (effort reduction, satisfaction), and strategic outcomes (competitive win rate, demographic expansion). This balanced view revealed that the implementation increased detection of genuine fraud by 41% while simultaneously reducing legitimate customer friction by 28%, creating compound value their previous siloed metrics would have missed entirely.

HEALTHCARE: INTEGRATING MEASUREMENTS

Healthcare organizations require measurement approaches that balance clinical impact, operational performance, and human experience:

1. **Clinical Outcomes**
   - Condition-specific health improvements
   - Mortality and morbidity reduction
   - Complication and readmission decrease
   - Functional status enhancement
   - Prevention effectiveness
2. **Patient Experience**
   - Satisfaction and loyalty measures
   - Perceived quality of care
   - Communication effectiveness
   - Engagement in care process
   - Trust in providers and system
3. **Clinician Experience**
   - Work satisfaction and fulfilment
   - Burnout reduction
   - Professional development
   - Team collaboration quality
   - Meaningful work proportion
4. **Cost and Efficiency**
   - Total cost of care
   - Resource utilization optimization

Measuring What Matters

- Process efficiency improvement
- Waste and duplication reduction
- Value per healthcare dollar

KEY INSIGHT: In healthcare, the most successful AI measurement approaches recognize the interconnected nature of the quadruple aim dimensions. Improvements in clinician experience often drive better patient outcomes, while enhancements in efficiency can free resources for quality improvements.

For a real-world example, let's look at an integrated health system implementing AI clinical decision support. Through creating a balanced scorecard based on the quadruple aim, they tracked clinical metrics (diagnostic timeliness, treatment appropriateness), patient experience (satisfaction, engagement), clinician experience (satisfaction, burnout), and efficiency measures (documentation time, appropriate resource use). This comprehensive approach revealed that the implementation improved diagnostic accuracy by 23% and reduced documentation time by 35%, but initially increased clinician stress due to workflow changes. Addressing these experience issues not only improved clinician satisfaction but enhanced clinical adoption, creating a virtuous cycle of increasing value.

RETAIL: THE CUSTOMER JOURNEY MEASUREMENT MODEL

This framework tracks AI impact across the complete retail value chain:

1. **Pre-Purchase Stage**
   - Targeting efficiency and effectiveness
   - Customer acquisition cost optimization
   - Browse-to-cart conversion improvement
   - Digital engagement enhancement
   - Product discovery optimization
2. **Purchase Stage**
   - Cart-to-purchase conversion
   - Transaction size and composition
   - Pricing optimization effectiveness
   - Cross-sell and upsell performance
   - Payment and checkout efficiency
3. **Fulfilment Stage**
   - Inventory position accuracy
   - Order processing efficiency
   - Delivery speed and reliability
   - Return rate and management
   - Fulfilment-cost optimization
4. **Post-Purchase Stage**
   - Customer satisfaction and advocacy
   - Repurchase rate and cadence
   - Relationship development
   - Service efficiency and effectiveness
   - Lifetime value enhancement

A real-world example comes from a specialty retailer that created an integrated measurement framework for AI initiatives that spanned marketing personalization, inventory optimization, and customer service. They mapped metrics across the full customer journey: from targeted acquisition through to post-purchase engagement, integrating operational measures like inventory efficiency with experience indicators like satisfaction. This approach revealed that the personalization AI was creating a 34% increase in first-time buyer conversion while the inventory AI reduced out-of-stocks by 28%. However, service interactions were creating friction that limited relationship development. Addressing these service gaps increased customer lifetime value by 22% beyond the initial benefits.

BREAKING DOWN SILOS WITH CROSS-FUNCTIONAL MEASUREMENT

Cross-functional measurements not only capture value more accurately – they can also help to reduce fragmentation due to organizational silos.

I've seen considerable impact when I worked with a major retail organization that attempted to measure the value of AI initiatives. The initial fundamental problem they encountered was that each department used entirely different metrics. Marketing measured engagement and conversion rates, operations focused on efficiency gains, customer service tracked resolution times, and finance monitored cost savings. These siloed measurements created an incomplete picture of AI's true impact and fostered competition rather than collaboration.

The breakthrough came when they developed a "Value Impact Matrix" – a unified measurement framework showing how each AI initiative contributed to multiple departmental goals simultaneously. This approach revealed unexpected connections: a customer service AI not only improved resolution times but also generated valuable product feedback for R&D and reduced returns costs for operations. By making these cross-functional impacts visible, the unified measurement framework transformed how departments viewed their interdependence and fostered collaborative approaches to AI development.

According to SQOR.ai research, "when organisations integrate all their KPIs into a single, unified format, they can address issues regarding silos. Unified KPIs provide real-time insights allowing leaders to assess how different metrics interact. This interconnection enhances collaboration between teams [by] revealing the hidden trends driving more informed decision-making."

THE SHARED KPI APPROACH

In the "10 ways to turn organizational silos into collaboration engines" research by Ricardo Gulko, marketing, sales, and customer experience teams demonstrated significant improvements when unified around single objectives like customer retention. Gulko's research shows how "each department bears equal responsibility for meeting the retention goals," which "lowers conflict and promotes shared ownership of achievement." This approach to shared KPIs resulted in "a more flexible, customer-focused business in addition to increased operational efficiency."

KEY INSIGHT: AI initiatives provide a unique opportunity to build cross-functional collaboration through shared KPIs, as outcomes naturally impact multiple departments simultaneously. By designing measurement systems that highlight these interconnections, organizations can use AI measurement as a tool for breaking down organizational silos.

PATTERN RECOGNITION: Organizations with siloed measurement approaches typically struggle with three key issues: conflicting success definitions across departments, attribution disputes when multiple functions contribute to outcomes, and missed opportunities to leverage synergies between departments.

BUILDING EFFECTIVE AI VALUE DASHBOARDS

Effective measurement requires not just selecting the right metrics but presenting them in ways that drive understanding and action. Well-designed AI value dashboards translate complex measurement into actionable insight.

Layered Dashboard Architecture

This approach creates connected but distinct views for different stakeholders.

1. **Value Summary Layer**
   - High-level outcomes across key dimensions
   - Overall performance against targets
   - Trend indications showing direction
   - Exception highlighting for attention areas
   - Integrated view of multiple initiatives
2. **Business Performance Layer**
   - Detailed operational metrics
   - Process-specific impact measures
   - User adoption and engagement patterns
   - Customer and stakeholder effects
   - Time-series trends and patterns
3. **Technical Performance Layer**
   - Detailed system operation metrics
   - Usage analytics and patterns
   - Data quality and coverage indicators
   - Model performance measures
   - Technical health indicators
4. **Causal Connection Layer**
   - Linkages between different metrics
   - Conversion rates between stages
   - Attribution analysis for outcomes
   - Correlation and relationship analysis
   - Value-pathway visualization

The Decision-Driven Dashboard Design Process

The ultimate purpose of measurement is to improve decisions. Effective dashboards explicitly connect metrics to specific choices and actions.

1. **Decision Inventory Development**
   - Identify key decisions influenced by the AI system
   - Document decision frequency and timing
   - Clarify decision owner roles and responsibilities
   - Understand current decision process and factors
   - Determine decision quality and confidence needs
2. **Information Requirement Analysis**
   - For each decision, identify required information
   - Determine critical metrics for each decision type
   - Clarify confidence levels needed for different choices
   - Document contextual information requirements
   - Identify comparison and benchmark needs
3. **Decision-Centric Visualization Design**
   - Organize information around specific decisions
   - Create views that directly support choice points
   - Include relevant context and supporting data
   - Design for decision timing and frequency
   - Incorporate confidence and uncertainty indicators

4. **Decision Process Integration**
   - Embed dashboard use in decision workflows
   - Establish clear action thresholds and triggers
   - Create feedback loops from decisions to outcomes
   - Develop consistent decision documentation
   - Build systemic learning from decision patterns

ACTION STEP: Identify one key decision that your AI system should inform – and design a focused dashboard view specifically for that decision. Include only the metrics that directly inform this choice, appropriate context, and clear action thresholds.

CAPTURING AND COMMUNICATING EARLY WINS

While comprehensive measurement frameworks provide long-term value, organizations also need approaches for identifying and leveraging early successes to build momentum and support.

QUICK WIN DISCOVERY PROCESS

This structured approach helps organizations identify and prioritize early value realization opportunities.

1. **Value-Pathway Mapping**
   - Document the complete chain from AI to ultimate value
   - Identify intermediate outcomes along the pathway
   - Determine which steps can deliver early partial value
   - Look for "low-hanging fruit" with minimal dependencies
2. **Stakeholder Value Analysis**
   - For key stakeholders, identify immediate pain points
   - Determine capabilities that address pressing needs
   - Find opportunities visible to influential groups
   - Identify early adopters who can showcase benefits
3. **Implementation Barrier Assessment**
   - Evaluate potential implementation challenges
   - Identify opportunities with minimal obstacles
   - Determine quick technical integration points
   - Assess organizational readiness factors
4. **Value Timing Estimation**
   - Evaluate how quickly different benefits can emerge
   - Estimate minimum viable capability for initial value
   - Identify staged value release opportunities
   - Determine appropriate measurement timing

During a retail transformation, I integrated frameworks into a continuous cycle of value creation, where the assessment process feeds directly into value delivery planning. When a critical timing risk in the organization's AI deployment strategy was discovered, this was not only documented but translated into a value opportunity. Building this element into the delivery roadmap, creating weekly measurement points – and using it to deepen executive relationships resulted in transforming a $2 million risk into a $5 million opportunity.

QUICK WIN: Identify one AI capability that could create visible value within 30 days with minimal implementation barriers. Create a focused implementation plan with clear before-and-after measurement to demonstrate early success.

THE VALUE NARRATIVE FRAMEWORK

This approach creates compelling value communication from measurement data.

1. **Impact Selection**
   - Identify the most meaningful and relatable outcomes
   - Select metrics with clear before-and-after contrast
   - Find examples with specific, concrete impact
   - Choose stories relevant to target audience
   - Balance different value dimensions
2. **Narrative Development**
   - Create clear situation-complication-resolution arc
   - Include both quantitative and qualitative elements
   - Develop specific examples and illustrative cases
   - Incorporate authentic stakeholder perspectives
   - Connect to broader purpose and mission
3. **Evidence Integration**
   - Support narrative with credible measurement
   - Balance anecdotal and systematic evidence
   - Include appropriate context and comparison
   - Address limitations and qualifications honestly
   - Provide appropriate supporting detail
4. **Audience Adaptation**
   - Tailor story to specific stakeholder concerns
   - Adjust technical detail to audience knowledge
   - Connect to relevant priorities and objectives
   - Use language and framing that resonates
   - Design appropriate visualization and format

KEY INSIGHT: Successful value communication combines both evidence and narrative. The most compelling approach provides credible measurement data within the context of a relatable story that helps stakeholders understand the human impact of the technical implementation.

THE SELF-FUNDING AI FRAMEWORK

Strategic use of early success can create self-funding dynamics that sustain AI initiatives through initial value capture.

1. **Value Capture Design**
   - Identify measurable value with financial impact
   - Create mechanisms to calculate realized value
   - Establish baseline and performance comparison
   - Develop attribution approach for shared credit
   - Create value documentation processes
2. **Reinvestment Mechanism Development**
   - Establish formulas for value-sharing
   - Create governance for reinvestment decisions
   - Develop processes for capturing available resources
   - Set appropriate reinvestment timeframes
   - Build accountability for reinvested resources
3. **Portfolio Approach Implementation**
   - Balance quick-return and longer-term initiatives

- Create explicit funding relationships between projects
- Develop stage-gate approach with value milestones
- Establish portfolio-level measurement
- Implement learning-based resource reallocation
4. **Leadership Alignment Creation**
    - Secure executive support for reinvestment approach
    - Establish appropriate governance mechanisms
    - Create transparent reporting on value and reinvestment
    - Develop explicit agreements on value capture
    - Build sustainable case for ongoing investment

For a real-world example, let's highlight a telecommunications company that created a formal "Value Reinvestment Program" for customer AI initiatives. They established clear measurement frameworks that documented cost-savings from each implementation phase, then developed agreements with finance to reinvest 60% of verified savings into subsequent phases. The remaining 40% flowed to the bottom line, creating immediate financial benefit while sustaining ongoing development. This approach created a self-funding cycle that maintained investment through a challenging budget period when other initiatives were cut, ultimately delivering significantly greater total value than would have been possible with traditional up-front funding alone.

CREATING SUSTAINABLE VALUE MEASUREMENT

Beyond early wins, organizations need approaches for sustained value measurement that evolves with maturing AI initiatives.

The Sustainable Measurement System

This framework creates measurement approaches that maintain relevance and momentum over time.

1. **Measurement Governance**
    - Establish clear ownership for ongoing measurement
    - Create regular review-and-refresh processes
    - Develop explicit stakeholder information commitments
    - Build measurement into operational rhythms
    - Create accountability for measurement quality
2. **Automation and Integration**
    - Automate data collection where possible
    - Integrate measurement into operational systems
    - Reduce manual effort for routine assessment
    - Create self-updating visualization tools
    - Build efficient exception-based reporting
3. **Evolution Mechanisms**
    - Establish processes for metric review and refinement
    - Create mechanisms to retire irrelevant measures
    - Develop approaches for adding new metrics as needed
    - Build learning loops from measurement experience
    - Create flexibility while maintaining consistency
4. **Value Communication Renewal**
    - Periodically refresh value narrative approach
    - Create new perspectives on established value
    - Develop evolving stakeholder-specific communication

Measuring What Matters

- Implement measurement insight-sharing mechanisms
- Maintain connection to strategic priorities

WARNING SIGN: When measurement becomes a burdensome "checkbox exercise" rather than a valuable business insight tool, your measurement system has likely become disconnected from actual decision-making needs.

The Portfolio Value Premium

As AI initiatives multiply within organizations, measurement must evolve from project-level to portfolio-level assessment. Organizations often measure AI initiatives as independent projects and simply sum their separate values. This approach misses a crucial reality: properly managed AI portfolios create compound value that exceeds the sum of individual initiative returns.

The counterintuitive approach that better captures value: explicitly measure interaction effects, capability leverage, and knowledge transfer across initiatives. This portfolio perspective often reveals 20–40% additional value missed by siloed measurement.

Illustrating this approach is the real-world example from a financial services organization that implemented a comprehensive portfolio measurement approach for 12 concurrent AI initiatives. Beyond initiative-specific metrics, the organization explicitly tracked cross-cutting elements, including data asset leverage, model reuse, knowledge transfer, and capability application across projects. This portfolio-level assessment revealed that approximately 35% of the total value came from these synergistic effects rather than independent project returns. This insight led them to redesign their governance to explicitly maximize these cross-initiative benefits, significantly increasing their total return on AI investment.

REFLECTION QUESTIONS: What potential synergies exist between your organization's AI initiatives that might create additional value beyond the sum of their individual impacts? How might you measure these portfolio-level benefits?

MEASUREMENT SUCCESS STORIES

Let's look at several organizations that transformed their approach to measuring AI value – and the results they achieved.

Manufacturing Company: Building a Value-Tracking System

The Situation: A global manufacturing company had invested heavily in AI maintenance and quality systems but struggled to demonstrate value. Budget reviews triggered difficult questions about ROI, and technical teams focused on algorithm performance metrics that executives found disconnected from business priorities. Implementation momentum was stalling despite promising capabilities.

The Breakthrough Approach: The company developed a comprehensive "AI Value Tracking System" with several pioneering elements:

1. **Value-Pathway Mapping:** Detailed documentation showed how AI capabilities connected to operational improvements and ultimately to financial outcomes, identifying measurable points throughout the chain.
2. **Multi-Level Metrics:** A tiered measurement framework – with executive metrics (downtime reduction, quality improvement, maintenance cost), operational metrics (failure prediction accuracy, inspection precision), and technical metrics (model performance, data quality) – was implemented.

3. **Counterfactual Analysis:** A structured approach was developed for estimating what would have happened without AI intervention, allowing credible calculation of value from avoided failures and quality issues.
4. **Staged Value Expectations:** Explicit value-projection models were created for different maturity phases, setting appropriate expectations for each stage rather than promising full value immediately.
5. **Automated Value Capture:** Systems were implemented to automatically track and document value created, reducing the burden of manual measurement and increasing credibility.

The Results: This transformed approach delivered remarkable outcomes:
- Documented $27.3 million in annual value within 18 months
- Secured additional investment expanding the program by 40%
- Accelerated adoption across previously hesitant facilities
- Created a measurement approach that became a company standard
- Transformed perception from "promising technology" to "proven value driver"

KEY INSIGHT: The manufacturing company succeeded by connecting technical metrics to operational outcomes to financial results – building a complete measurement chain that made value visible and credible to all stakeholders.

Retail Bank: Holistic Customer Onboarding Measurement

The Situation: A retail bank had implemented an AI-based customer onboarding system with competing objectives – improving customer experience while simultaneously enhancing fraud detection and regulatory compliance. Initial measurement approaches focused primarily on operational efficiency metrics, creating an incomplete picture of the system's true impact.

The Breakthrough Approach: The bank developed an integrated measurement framework that tracked four key value dimensions simultaneously:
1. **Risk Metrics:** Fraud detection rate, false positive reduction, compliance adherence
2. **Efficiency Indicators:** Onboarding time, processing cost, staff productivity
3. **Experience Measures:** Customer effort, satisfaction, completion rates
4. **Strategic Outcomes:** Competitive win rate, demographic expansion, product adoption

The Results: This holistic measurement approach revealed powerful insights that siloed metrics would have missed:
- Detection of genuine fraud increased by 41%
- Legitimate customer friction decreased by 28%
- Account opening completion rates improved by 34%
- The system created compound value across all dimensions simultaneously

What made this approach powerful was its ability to reveal interdependencies between seemingly competing goals. For example, the organization discovered that improving explainability in fraud detection actually enhanced security and customer experience simultaneously, leading to compounding benefits across dimensions.

QUICK WIN: For your AI implementation, identify one potential positive interaction between dimensions that might create compound value (e.g., how improved accuracy might enhance both operational efficiency and customer experience). Create specific metrics to track this interaction effect.

HEALTHCARE SYSTEM: THE QUADRUPLE AIM FRAMEWORK

The Situation: A healthcare system implementing AI clinical decision support struggled with fragmented, siloed measurements. The technology team was providing different metrics to different stakeholders – accuracy and safety data for clinical teams, efficiency and cost measures for administrative leaders, and high-level summaries for executives seeking overall value demonstration. But this fragmented approach created competing narratives about success and made it difficult for anyone to see the complete picture of the AI system's impact.

The Breakthrough Approach: The organization unified the measurement approach around healthcare's "Quadruple Aim" framework:

1. **Clinical Outcomes:** Diagnostic timeliness, treatment appropriateness, complication rates
2. **Patient Experience:** Satisfaction, engagement, trust measures
3. **Clinician Experience:** Satisfaction, burnout indicators, collaboration metrics
4. **Cost Efficiency:** Documentation time, resource utilization, waste reduction

The Results: This balanced approach revealed crucial insights that a single-dimension view would have missed:

- Diagnostic accuracy improved by 23%
- Documentation time reduced by 35%
- Initial implementation increased clinician stress
- Addressing clinician experience enhanced adoption and further improved outcomes

The healthcare system's quadruple aim approach provides a balanced measurement framework that evaluates how departments collectively contribute to patient outcomes rather than competing for isolated metrics.

PATTERN RECOGNITION: Organizations that create balanced, multi-dimensional measurement frameworks often discover that addressing "soft" dimensions like user experience improves "hard" metrics like efficiency and financial outcomes.

FORRESTER'S AI VALIDATION PROCESS

The challenge Michael Kasparian, Forrester CIO, faced was common: when AI systems flagged client relationships as "low value" or "at risk," business users would disagree based on their personal knowledge of those clients.

As Kasparian explains: "I think where folks struggled initially was like, 'You're saying it's low, but I think they're doing great,' so why, right? And so being able to have the proof behind it is important with any of the models."

To address this scepticism, Kasparian's team at Forrester built transparency into their AI recommendations by showing the underlying factors that drove each assessment: "Is it that they're not active with us? Is it that they're not engaged? And then ultimately: What can they do with it?"

By revealing these specific data points, stakeholders could understand why the AI reached its conclusions and either accept the assessment or provide additional context the system might have missed.

This transparent approach helped build trust in AI systems by making their reasoning processes visible and understandable to business users, demonstrating that sometimes the most important measurement isn't of outcomes but of the process that produces those outcomes.

THE VALUE-MEASUREMENT DASHBOARD TOOLKIT

To help implement the concepts we've covered, here are practical tools you can use to enhance your AI value measurement approach.

THE METRIC-TO-DECISION MAPPING TOOL

This simple but powerful tool ensures your metrics directly connect to important decisions.

1. List the key decisions influenced by your AI system
2. For each decision, identify the specific metrics that inform this choice
3. Check if you're currently measuring these decision-critical metrics
4. Identify any metrics you're tracking that don't connect to important decisions
5. Create a plan to add missing decision-critical metrics and consider eliminating unconnected ones

ACTION STEP: Complete this metric-to-decision mapping for your top three AI-influenced decisions this week. This quick exercise will immediately reveal gaps in your current measurement approach.

THE TECHNICAL-TO-BUSINESS TRANSLATION MATRIX

Use this tool to create clear connections between technical metrics and business outcomes:

1. List your key technical performance metrics vertically
2. List important business outcomes horizontally
3. In each intersection cell, document how the technical metric influences the business outcome
4. Identify gaps where connections aren't clear
5. Use this completed matrix to show stakeholders how technical performance drives business results

Example Translation

- **Technical:** "Improved model accuracy by 15%"
- **Business:** "Reduced decision risk by $3 million annually"

QUICK WIN: Create a simple one-page "technical-to-business translation guide" for your AI implementation that helps technical teams explain their metrics in business terms and enables business stakeholders to understand the relevance of technical performance.

THE STAGED VALUE EXPECTATIONS FRAMEWORK

This tool helps set appropriate expectations across different timeframes.

Immediate Horizon (0–3 months)

- Technical performance metrics achieving target levels
- Initial user adoption reaching defined thresholds
- First operational improvements becoming visible
- Early stakeholder feedback showing positive trends

Mid-Term Horizon (3–12 months)

- Process efficiency metrics showing sustained improvement
- User proficiency reaching advanced levels
- Initial financial impacts becoming measurable
- Operational integration achieving defined milestones

Long-Term Horizon (12+ months)

- Strategic positioning metrics showing meaningful change
- Financial returns meeting or exceeding projections
- Organizational capabilities demonstrating measurable enhancement
- Competitive advantage becoming evident in market metrics

DOWNLOADABLE RESOURCE: The AI Value Measurement Canvas template will be available in the online resource centre. This comprehensive worksheet helps you design a complete measurement approach for your AI initiative, from technical metrics to business outcomes.

MAKING MEASUREMENT WORK IN PRACTICE: KEY IMPLEMENTATION INSIGHTS

As we conclude, let's highlight several critical insights that can help you implement effective measurement approaches in your organization.

THE VALIDATION-DECISION BALANCE

At its foundation, the measurement challenge is about balancing validation rigor with decision-making needs. Organizations face a genuine tension between attribution rigor and actionable insights:

- Stringent attribution requirements may delay decision-making
- Oversimplified attribution can lead to incorrect conclusions
- Perfect attribution is rarely possible in complex systems
- Practical decisions require timely assessment that may be imperfect

The solution isn't choosing a single point on this spectrum, but matching validation approaches to specific decision contexts. Critical investment decisions may warrant more rigorous attribution methods, while optimization choices can often proceed with simpler validation.

WARNING SIGN: If your measurement processes are so time-consuming or complex that decisions are regularly made without waiting for results, you likely have a mismatch between your validation approach and decision-making needs.

THE CAPABILITY COMPOUNDING EFFECT

Organizations often measure capability development as a static outcome with the idea that a new ability either does or does not exist. This misses a crucial dynamic: AI-enabled capabilities compound over time, creating accelerating rather than linear returns.

The counterintuitive approach that better captures value: measure capability development as a rate of change and improvement trajectory rather than a fixed state. This dynamic measurement reveals value that static assessment tends to miss.

I've seen this in action at a manufacturing company that created a "Capability Development Index." The index tracked how AI implementation affected seven key organizational capabilities, including predictive operations, supply chain visibility, and manufacturing flexibility. Each capability was assessed quarterly on a maturity scale and tracked in both absolute levels and improvement rates. This approach revealed that the AI was creating acceleration effects: capabilities were improving at an increasing – rather than linear – rate. The compounding pattern justified continued investment despite some capabilities still being below desired absolute levels.

REFLECTION QUESTIONS: Is your organization measuring capability development as a static achievement or as a dynamic improvement trajectory? How might focusing on improvement rates rather than absolute levels change your understanding of AI's impact?

CROSS-FUNCTIONAL MEASUREMENT OPPORTUNITIES

In the "10 ways to turn organizational silos into collaboration engines" research mentioned above, we've heard from Ricardo Saltz Gulko how marketing, sales, and customer experience teams demonstrated significant improvements when unified around single objectives.

Beyond the obvious benefits, such as reducing conflict and promoting shared ownership of achievements, these insights suggest that measurement frameworks themselves can serve as collaboration tools. By creating unified, cross-functional metrics, organizations can use AI measurement as a mechanism for breaking down silos and fostering collaborative approaches to value creation.

David Haberlah explains that "shared KPIs address the issue of organizational silos by measuring how effectively teams work towards collective outcomes, driving a sense of shared responsibility. This serves to break down the tendency to focus mainly on departmental metrics and encourages adoption of a more holistic perception of organisational success."

ACTION STEP: Identify one opportunity to create a shared KPI between two departments affected by your AI implementation. Design this metric to encourage collaboration rather than competition, and measure success at the intersection of their joint work.

CONCLUSION: FROM MEASUREMENT TO ACTION

As we've explored throughout this chapter, effective value measurement for AI requires more than simply tracking metrics – it demands a comprehensive approach that connects technical performance to ultimate business outcomes, balances different value dimensions, addresses attribution challenges, and evolves as initiatives mature.

Organizations that cannot measure and demonstrate impact in terms that matter to key stakeholders will struggle to create value from even the most sophisticated AI capabilities. Conversely, even relatively straightforward implementations can deliver substantial returns when supported by measurement systems that make their value visible, attributable, and actionable.

THE FIVE PRINCIPLES OF EFFECTIVE AI VALUE MEASUREMENT

1. **Start with Value Pathways That Go beyond Metrics:** Effective measurement begins with a clear understanding of how AI creates value, mapping the complete chain from technical performance to ultimate outcomes and measuring appropriately throughout.
2. **Match Measurement to Maturity:** Different implementation stages require different measurement approaches, with metrics and expectations that evolve as initiatives progress from initial deployment to full maturity.
3. **Link Technical and Business Metrics:** The most important measurement challenge is bridging the gap between what technical teams can easily measure and what business leaders care about. Credible connections between these different worlds are required.
4. **Measure What Matters – Not Just What's Easy:** The aspects of AI value that are most easily measured are often furthest from ultimate value, while the dimensions closest to genuine impact are hardest to measure. Effective approaches prioritize relevance over convenience.
5. **Design Measurement for Decisions – Not Just Reporting:** The ultimate purpose of measurement is to improve decisions about implementation, optimization, and investment.

Effective systems explicitly connect metrics to specific choices rather than simply reporting performance.

Your Next Steps

To begin applying these principles in your organization:

1. **Map Your AI Value Pathways:** Document the complete chain from technical implementation to business impact
2. **Assess Your Measurement Maturity:** Determine if your current approach matches your implementation stage
3. **Create Technical-to-Business Translations:** Develop clear connections between technical and business metrics
4. **Audit Your Metrics for Decision Relevance:** Ensure you're measuring what matters most for key decisions
5. **Implement a Tiered Measurement Approach:** Design appropriate metrics for different stakeholder groups

By applying these principles through frameworks appropriate to your specific context, you can transform AI measurement from a technical exercise to a powerful tool for value creation, demonstrating impact in terms that build sustained support for your initiatives.

KEY INSIGHT: The ultimate goal of measurement isn't proving value – it's creating it. Effective measurement approaches don't just document AI's impact; they enhance it by improving decisions, focusing resources on high-return areas, and building organizational alignment around shared value objectives.

12 Creating Sustainable AI Value

Chapter Roadmap

In this chapter, we examine what separates sustainable AI value from temporary success and how organizations can build the foundations for long-term impact rather than fleeting wins. We look at a number of themes, including why AI value often fades after initial success as well as the crucial steps of project-to-capability transformation, integrating AI into core business processes, creating continuous improvement systems and building organizational capabilities for sustainability. Helpful insights relate to understanding and preventing value erosion, establishing funding models for long-term success and managing knowledge and IP assets.

BEYOND INITIAL SUCCESS

I was sitting in the quarterly review meeting of a Fortune 100 company's AI transformation initiative. The CIO was showcasing impressive early results: a 42% reduction in customer service handling time, 28% improvement in forecast accuracy, and several other metrics that demonstrated clear initial success.

Then the CEO asked a seemingly simple question: "What will these numbers look like a year from now?"

The room went quiet. The team had focused so intensely on launching the AI initiative and demonstrating early wins that they hadn't created a clear plan for sustaining and growing that value over time.

This scenario repeats itself across organizations. The initial excitement of AI implementation – complete with impressive demos, early adopter enthusiasm, and quick wins – often gives way to a more challenging reality: maintaining and expanding value delivery in the face of changing conditions, evolving user behaviours, and competing priorities.

PATTERN RECOGNITION: When organizations launch AI initiatives, they often celebrate immediate wins but fail to establish mechanisms to maintain value as conditions change. Watch for this missing piece in your own AI roadmap.

I've come to realize that the real challenge isn't getting the technology to work – it is keeping it delivering value over time. When a large healthcare system successfully deployed an AI diagnostic tool with impressive initial results, the team celebrated what seemed like a major victory. Six months later, usage had declined dramatically, and accuracy had degraded. What happened? The cross-functional team that had guided the implementation had disbanded, returning to their departmental roles. No persistent mechanism existed to monitor integration issues, gather user feedback, or update the system as medical practices evolved.

Research from RAND Corporation identifies this pattern clearly: 68% of AI projects that initially succeeded still failed to deliver sustained value six months after implementation. This analysis found that the primary factor in maintaining value was "institutionalized cross-functional communication channels" that persisted beyond initial project teams.

WARNING SIGN: If cross-functional collaboration teams disband after initial implementation, you're likely heading towards what RAND Corporation researchers call the "implementation cliff," where AI initiatives lose support before reaching maturity.

Creating Sustainable AI Value

It's a scenario I've seen repeated across industries and applications. Beyond the healthcare example mentioned above, there was a manufacturing company whose predictive maintenance AI initially reduced downtime dramatically but slowly lost effectiveness as equipment and processes changed. Plus, the financial institution whose fraud detection system delivered substantial early savings but gradually produced increasing false positives as criminal tactics evolved.

The common thread wasn't a technical issue but a failure to achieve sustainability – the inability to transform initial success into enduring value. The challenge wasn't creating AI capabilities but creating AI capabilities that could sustain and grow their value contribution over time.

This sustainability gap represents perhaps the most significant threat to AI's promise in most organizations. It's the difference between AI as a series of interesting but ultimately transient projects and AI as a transformative force that creates compounding value year after year.

REFLECTION QUESTIONS: Think about your organization's most successful AI implementation. What specific mechanisms exist to ensure it continues delivering value six months, one year, or three years from now?

FROM PROJECT TO PERMANENT CAPABILITY

The first critical shift in creating sustainable AI value is moving from a project mindset to a capability mindset, and this requires treating AI not as a finite initiative but as an enduring organizational capability that requires ongoing evolution.

A quote from Stela Solar, Microsoft's Global Head of Artificial Intelligence Solutions, illustrates how AI deployment requires constant attention and alignment efforts on the one hand and sustained support on the other. She says, "The data is showing signals that the organization hasn't seen before, the forecasting models have suddenly had to be rebuilt. And so it is super critical to again come to the core of it – of aligning the AI work to the core business objectives. And being really agile with it. Sponsorship from leadership allows for that agility, allows for the investment."

PROJECT-TO-CAPABILITY TRANSFORMATION

Traditional technology implementation follows a project-centric approach: define scope, build solution, deploy, declare success, and move on. Due to AI's nature and value creation patterns, this model is not suitable for maximizing benefits from AI deployment. For that, a capability model is required.

Project Model Limitations for AI

- Treats AI as a static solution rather than a learning system
- Creates artificial completion points when evolution should continue
- Allocates resources in bursts rather than sustained investment
- Measures success at deployment rather than ongoing value creation
- Enables development team dispersal just when learning should accelerate

Capability Model Advantages

- Recognizes AI as a continuously evolving system
- Creates structures for ongoing adaptation and improvement
- Establishes sustainable resource allocation models
- Measures success through continued value generation
- Maintains knowledge and expertise for compounding returns

KEY INSIGHT: UPS transformed package delivery efficiency with the Network Planning Tool. Following initial implementation, the company built "ongoing collaborative processes between operational units" that continuously refined the system as business conditions changed. The difference between UPS's success and many failed AI initiatives wasn't the initial quality but the presence of persistent mechanisms to maintain alignment as conditions evolved.

This example embodies a crucial principle: "AI doesn't fail in week one, it fades by month six" without proper sustaining mechanisms.

The Capability Transformation Framework

This structured approach helps organizations transition from project to capability mindset.

1. **Value Continuity Planning**
   - Develop explicit post-implementation evolution strategy
   - Create long-term value roadmaps beyond initial deployment
   - Establish ongoing improvement and expansion objectives
   - Build staged enhancement plans with clear milestones
   - Develop multi-year value projection models
2. **Persistent Team Structures**
   - Transition from temporary project teams to sustained capability teams
   - Create clear ongoing ownership and accountability
   - Establish appropriate skills for long-term evolution
   - Develop knowledge retention and transfer mechanisms
   - Build career paths within AI capabilities

As Jeanne Achille, Founder & CEO of Devon Group, emphasizes: "There's a talent pool out there that has AI expertise," but competition for these individuals is intense. She recommends that leaders should "run towards them and make sure you've got some of them to your team or at least have relationships with vendors that can provide you with that AI expertise" to build sustainable capabilities.

ACTION STEP: Identify which skill sets are most critical for long-term AI capability maintenance in your organization. Create a plan to either recruit these skills or develop relationships with vendors who can provide this expertise consistently.

3. **Sustainable Funding Models**
   - Shift from one-time project funding to ongoing capability investment
   - Create value-based funding mechanisms tied to outcomes
   - Develop appropriate operational versus capital allocation
   - Build financial models for continuous enhancement
   - Create reinvestment approaches from realized value
4. **Governance Evolution**
   - Transition from project governance to capability governance
   - Establish ongoing decision rights and prioritization processes
   - Create structures for balancing maintenance and enhancement
   - Develop appropriate risk management for evolving systems
   - Build stakeholder engagement models for sustained influence

Tension Point: The Innovation-Stability Balance

Organizations face a fundamental tension in the capability approach:

- Continual innovation risks disrupting established value streams

Creating Sustainable AI Value

- Excessive stability leads to stagnation and diminishing returns
- Users need predictable experience yet improving functionality
- Technical teams need both maintenance discipline and creative freedom

Our research found that one of the biggest tensions in AI sustainability occurs between the need for consistent leadership support and the reality of fluctuating executive priorities. Their study notes that "some leaders even switch up their priorities, often impacting progressing projects before they have time to demonstrate real results." This mismatch between AI development timelines and typical executive decision cycles can undermine even well-designed initiatives before they reach maturity.

Successful organizations navigate this tension by creating structured innovation rather than choosing a single point on the spectrum.

In summary, the shift from viewing AI as a project to treating it as an organizational capability requires:

- Explicit planning for ongoing evolution
- Persistent teams that maintain expertise and accountability
- Sustainable funding models that enable continuous investment
- Governance structures that balance stability with innovation

This transformation serves to counterbalance the pattern identified by RAND Corporation: 68% of initially successful AI projects fail to deliver sustained value after six months.

THE DUAL OPERATING MODEL

This framework balances stability and innovation for sustainable AI capabilities.

Stability Layer

- Focuses on reliability, performance, and consistency
- Maintains core functionality with disciplined change management
- Ensures ongoing value from established capabilities
- Provides predictable experience for users
- Operates with service-level commitments

Innovation Layer

- Focuses on enhancement, expansion, and evolution
- Develops new capabilities and approaches
- Tests improvements before main system integration
- Creates pathways for controlled experimentation
- Operates with learning and discovery metrics

By explicitly separating these functions while maintaining clear integration points, organizations can simultaneously maintain reliable operation and drive continuous improvement.

QUICK WIN: Create separate KPIs for your stability and innovation teams. Stability teams should focus on reliability metrics (uptime, accuracy, user satisfaction), while innovation teams should track learning metrics (experiments conducted, insights generated, enhancements developed).

For a real-world example, let's look at a financial services company that implemented a "dual track" operating model for their customer AI platform. A stability team was established to be responsible for reliable daily operation and a parallel innovation team focused on enhancements and new capabilities. The stability track operated with service-level agreements and operational metrics – while the innovation track used learning metrics and controlled experimentation. Clear integration processes moved proven innovations into the stable core on a regular cadence. This balanced approach maintained 99.8% reliability while still implementing significant enhancements every quarter – a combination that had eluded the organization under previous models that oscillated between excessive stability and disruptive change.

EMBEDDING AI IN CORE BUSINESS PROCESSES

Beyond the operational model, sustainable AI value requires deep integration into the organization's fundamental processes rather than existing as a peripheral addition.

Integration Depth Levels

Level 1: Supplementary Tools

- AI exists alongside existing processes
- Use is optional and additional to standard workflows
- Limited integration with core systems
- Requires explicit decision to utilize
- Value depends on active adoption

Level 2: Workflow Integration

- AI is embedded within standard processes
- Basic integration with core systems
- Use is expected but can be bypassed
- Some automation of routine decisions
- Value partially dependent on adoption

Level 3: Process Transformation

- AI fundamentally reshapes workflows
- Deep integration with enterprise systems
- Use is the default path with exceptions
- Significant automation with human oversight
- Value largely independent of individual adoption

Level 4: Business Model Integration

- AI becomes central to value proposition
- Complete integration across systems and processes
- Use is inherent in how the organization operates
- Advanced automation with strategic human roles
- Value creation fully embedded in operating model

This process of embedding AI deeply into organizational processes requires significant architectural transformation.

At a company I call the Gaming Guru in *Making Data Work*, the organization's architecture was transformed from monolithic systems to an approach inspired by microservices, where "breaking big things into small things" enabled faster adaptation to changing business needs while maintaining system integrity. The results were dramatic.

At TAG, this transformation reduced the time to implement an infrastructure change from "six months of planning, six months of trying to secure the budget, and another six months to implement"

Creating Sustainable AI Value

to "less than six hours." This remarkable acceleration didn't just improve efficiency – it fundamentally changed how the organization could respond to emerging opportunities and challenges.

PATTERN RECOGNITION: Organizations that treat AI implementations as one-time technical projects typically achieve level 1 integration at best. Those with sustainable value creation reach level 3 or 4, where AI becomes inseparable from core business operations.

THE PROCESS INTEGRATION FRAMEWORK

This approach creates methodical deepening of AI integration into core processes.

1. **Process Mapping and Opportunity Assessment**
   - Document current process in detail
   - Identify decision points and information flows
   - Assess AI applicability and value potential
   - Evaluate technical and organizational feasibility
   - Determine appropriate integration depth
2. **Integration Design**
   - Create integrated workflow incorporating AI capabilities
   - Design appropriate human-AI interaction points
   - Develop necessary system connections and data flows
   - Establish exception handling and override mechanisms
   - Build appropriate monitoring and feedback loops
3. **Organizational Alignment**
   - Adapt roles and responsibilities to new process design
   - Develop necessary skills and capabilities
   - Align incentives and performance metrics
   - Create change management and transition approach
   - Establish governance for process operation
4. **Technical Implementation**
   - Build necessary system integrations and connections
   - Implement data pipelines and flows
   - Develop API and service interfaces
   - Create monitoring and management tools
   - Establish technical support and maintenance

KEY INSIGHT: The Visibility Paradox

Organizations often design AI implementations with visibility in mind – to highlight the AI component to demonstrate innovation and change. This approach can undermine sustainable value by creating an artificial separation between AI and normal operations.

There is a counterintuitive approach that creates better results: design for invisibility, where AI becomes so integrated into normal operations that users often don't perceive it as a separate component. This seamless integration dramatically improves long-term adoption and value sustainability compared to high-visibility implementations.

A real-world example illustrates this point. A healthcare organization implementing clinical decision support initially created a highly visible "AI Assistant" interface separate from their electronic health record. Despite strong initial interest, usage declined dramatically over time as it required extra steps outside normal workflow. When they redesigned the implementation to embed the same capabilities directly into the standard electronic health record (EHR) workflow – making the AI essentially invisible – sustained adoption increased by 340% and maintained high levels for over two years. The invisible integration created far more sustainable value than the highly visible initial approach.

ACTION STEP: Audit your AI implementations for unnecessary visibility. Ask, "Does this AI component need to be identifiable as AI, or could it be seamlessly integrated into existing workflows?" Prioritize reimplementing high-visibility, low-adoption AI tools to make them more invisible and integrated.

CREATING CONTINUOUS IMPROVEMENT MECHANISMS

Sustainable AI requires systematic approaches for ongoing enhancement rather than sporadic updates. As Jeremy Foster, VP Cloud Infrastructure at Cisco, notes: "It's not going to move … as fast as you want it to. As you get better and better at it, you'll start to be able to figure out how to take these [technical, business, and operational] groups, put them together for a purpose and a reason, and actually move things along a lot faster than you could have as you mature them in position."

Improvement Domains

- **Performance Enhancement:** Increasing accuracy, speed, and reliability
- **Scope Expansion:** Extending capabilities to new areas and functions
- **Experience Refinement:** Improving user interaction and satisfaction
- **Process Optimization:** Enhancing workflow and integration
- **Value Amplification:** Increasing business impact and outcomes

THE CONTINUOUS IMPROVEMENT SYSTEM

This framework creates structured, ongoing enhancement of AI capabilities.

1. **Feedback Collection**
   - Implement systematic user feedback mechanisms
   - Create performance monitoring and alerts
   - Establish regular stakeholder input processes
   - Develop usage pattern and adoption analysis
   - Build outcome tracking and value assessment
2. **Learning Synthesis**
   - Aggregate and analyse feedback across sources
   - Identify patterns and improvement opportunities
   - Prioritize potential enhancements
   - Develop insight-driven enhancement proposals
   - Create learning documentation and knowledge base
3. **Enhancement Implementation**
   - Design improvements based on synthesized learning
   - Create appropriate testing and validation approaches
   - Implement changes through structured process
   - Monitor impact and effectiveness
   - Communicate changes to stakeholders
4. **Cycle Acceleration**
   - Reduce time between iterations
   - Implement continuous deployment capabilities
   - Create faster feedback-to-implementation loops
   - Build learning acceleration mechanisms
   - Develop compounding improvement patterns

TABLE 12.1
The Improvement Velocity Framework

Improvement Aspect	Initial State	Intermediate State	Advanced State
Feedback Cycle	Periodic formal collection	Regular structured input	Continuous real-time feedback
Analysis Approach	Manual review and synthesis	Semi-automated pattern detection	Machine learning (ML)-assisted insight generation
Implementation Process	Scheduled major releases	Regular incremental updates	Continuous deployment capability
Learning Integration	Documentation-based sharing	Knowledge management systems	Automated insight distribution
Adaptation Speed	Quarterly improvements	Monthly enhancements	Weekly or faster iteration

REFLECTION QUESTIONS: What is your organization's current improvement cycle time for AI capabilities? How might reducing this cycle from months to weeks (or weeks to days) create competitive advantage?

THE IMPROVEMENT VELOCITY FRAMEWORK

The approach shown in Table 12.1 helps organizations accelerate their improvement cycles over time.

For a real-world example, let's look at a global technology company that transformed their customer AI platform from quarterly batch updates to continuous improvement by systematically evolving each aspect of their enhancement process. They progressed from periodic satisfaction surveys to real-time feedback collection, from manual analysis to ML-assisted pattern detection, and from major releases to continuous deployment. This evolution reduced their improvement cycle time from 90+ days to under one week while significantly increasing enhancement quality through tighter feedback loops. The accelerated learning capability became a major competitive advantage as they could respond to changing conditions far faster than competitors who relied traditional update approaches.

This example demonstrates how improvement velocity itself can become a competitive advantage. By systematically accelerating learning and adaptation cycles, organizations can create sustained value advantages that competitors using traditional update approaches struggle to match.

QUICK WIN: Map your organization's current position on the Improvement velocity framework. Identify one aspect where you could advance from initial to intermediate (or intermediate to advanced) state within 90 days, and create an implementation plan.

BUILDING LASTING ORGANIZATIONAL CAPABILITIES

Truly sustainable AI value depends not just on technology but on the organization's ability to continually leverage and evolve those capabilities. As Jay Patel, CEO of Amtech, emphasizes, "Barriers are going to happen, and if barriers are going to happen, we have to actively work on breaking down those barriers so [we] can work as an organisation for one goal."

Critical Organizational Capabilities

- **AI Literacy:** Broad understanding of capabilities and applications
- **Data Fluency:** Ability to work effectively with data across functions
- **Experimental Mindset:** Comfort with hypothesis testing and learning

- **Cross-Functional Collaboration:** Effective integration across domains
- **Change Adaptability:** Capacity to evolve practices and approaches

Insights into how sustained effort is needed for truly transformational change come from Abhinav Singhal, Chief Strategy Officer of Asia Pacific's Thyssenkrupp.

He acknowledges that in his organization, "AI is still not moving the needle. It doesn't yet create the margins to really affect the P&L." However, he maintains a strategic long-term view, stating that "in five to seven years, it will be a critical game-changer in how we compete and serve our customers." Organizations must "invest in it until it reaches that pivotal point where it becomes strong enough to fuel growth on its own."

THE CAPABILITY DEVELOPMENT SYSTEM

This framework creates systematic organizational capability building.

1. **Capability Assessment**
   - Evaluate current organizational strengths and gaps
   - Identify critical capabilities for sustainable value
   - Benchmark against leading practices
   - Create capability maturity model
   - Develop prioritized capability-building plan
2. **Learning Infrastructure**
   - Implement formal training and development programs
   - Create knowledge-sharing mechanisms
   - Develop communities of practice
   - Build experiential learning opportunities
   - Establish mentorship and coaching systems
3. **Structural Enablement**
   - Align organizational structure with capability needs
   - Create appropriate roles and responsibilities
   - Develop career paths that build critical skills
   - Implement cross-functional mechanisms
   - Establish centres of excellence where appropriate
4. **Cultural Reinforcement**
   - Align recognition and rewards with capability-building efforts
   - Create leadership modelling of target capabilities
   - Implement experimentation and learning norms
   - Develop appropriate risk tolerance
   - Build continuous improvement expectations

KEY INSIGHT: The Capability Diffusion Effect

Organizations often concentrate AI capabilities in specialized teams or centres of excellence. While this creates short-term efficiency, it typically limits long-term value as this can create bottlenecks in knowledge dissemination and reduce cross-organizational learning.

The counterintuitive approach that creates better results: deliberately diffuse AI capabilities throughout the organization over time, accepting short-term inefficiency for long-term capability building. This distributed expertise dramatically improves sustainable value by embedding skills throughout the organization rather than concentrating them in isolated groups.

TAG learned this lesson – and transformed the organization's approach to resource allocation. As described in *Making Data Work*, they shifted from an 80-to-20 ratio in favour of externally-sourced

Creating Sustainable AI Value

staff to a blended-by-default strategy that reduced "third-party single points of dependency." The approach included an academy-style guild of specialist practices for capability development and an "80/20 reversal" method where new team members spent 80% of an individual's bandwidth towards skills training and development while 20% would be available for working on projects.

While this initially faced resistance due to fears about a perceived loss of bandwidth and concerns about a potential dilution of expertise, it ultimately created sustainable value by building internal capabilities rather than remaining dependent on external expertise that would "walk out the door when the contracts with the external consultants expired."

REFLECTION QUESTIONS: Where do AI capabilities currently reside in your organization? Are they concentrated in a few specialized teams, or diffused throughout business units? What would a three-year "capability-diffusion" strategy look like for your organization?

Another real-world example comes from a global consumer goods company that initially concentrated all AI capabilities in a central "Digital Center of Excellence." While this created early efficiency, it became a bottleneck as demand grew. Implementing a three-year "capability diffusion" strategy helped to systematically build AI skills across business units while maintaining the centre as a standards and governance hub. This distributed approach initially seemed less efficient but ultimately created significantly greater and more sustainable value by embedding capabilities throughout the organization, allowing far more use cases to be developed and maintained than the centralized model could support.

ACTION STEP: Create a "capability diffusion plan" that identifies:

1. Core AI skills that should remain centralized for quality and consistency
2. Business-specific AI capabilities that should be embedded in operational teams
3. A phased approach to building distributed expertise over 18–36 months
4. Knowledge-sharing mechanisms to maintain consistency across distributed teams

"AI DOESN'T FAIL IN WEEK ONE, IT FADES BY MONTH SIX"

A critical insight about AI sustainability: catastrophic failures at launch are rare, but gradual value erosion over time is common. Understanding the patterns of this decay is essential for preventing it.

THE VALUE EROSION PATTERN

Most AI initiatives follow a predictable pattern when value sustainability isn't explicitly addressed.

Phase 1: Launch Success (Weeks 1–4)

- Strong initial interest and engagement
- Quick wins and visible improvements
- Technical performance meets expectations
- Enthusiastic stakeholder support
- Positive user feedback and adoption

Phase 2: Reality Adjustment (Months 2–3)

- Initial enthusiasm stabilizes to normal levels
- Edge cases and exceptions emerge
- User behaviours settle into patterns
- Integration frictions become apparent
- Value continues but at more realistic levels

Phase 3: Subtle Drift (Months 4–6)

- Gradual performance degradation begins
- User workarounds for limitations develop
- Changes in environment affect accuracy
- New requirements emerge without being appropriately addressed
- Value begins slow but steady decline

RAND Corporation analysis, conducted by Ryseff, De Bruhl, and Newberry confirms this pattern. It shows that 68% of AI projects that initially succeeded still failed to deliver sustained value six months after implementation. The study identified that the primary factor in maintaining value was "institutionalized cross-functional communication channels" that persisted beyond initial project teams.

WARNING SIGN: If your organization celebrates AI implementation success but doesn't track key performance indicators three to six months post-launch, you're likely missing critical signals of value erosion.

Phase 4: Accelerating Decay (Months 7–12)

- Performance issues become more significant
- User adoption starts declining noticeably
- Trust erosion affects utilization
- Growing gap between needs and capabilities
- Value decline accelerates

Phase 5: Value Collapse (beyond Year 1)

- System becomes viewed as unreliable
- Widespread workarounds or abandonment
- Technical debt accumulates rapidly
- Loss of knowledge about the system
- Value drops to minimal levels or disappears

Sanghamitra Goswami, Senior Director of Data Science and Machine Learning at PagerDuty, addresses this challenge through a phased implementation approach. In an interview with LinearB, she explains: "When we have a goal and when we have a plan at each phase, then big foundational work might seem achievable. And that is very important ... I'm saying that [for] phase 1, 2, I have a goal, and I know I can go to phase 6. And although you cannot see an ROI at 1 and 2, I know I can give you that ROI back at phase 6."

TENSION POINT: THE MAINTENANCE-ENHANCEMENT BALANCE

Organizations face a fundamental tension in sustaining AI value.

- Focus too much on maintenance, and systems become stable but stagnant
- Emphasize enhancement too heavily, and stability and reliability suffer
- Technical teams prefer working on new capabilities over maintenance
- Users need both reliability and continuous improvement

This can lead to difficult decisions about resource allocation and priority setting that directly affect sustainability.

TABLE 12.2
The Value Sustainability Index

Sustainability Factor	Low Risk	Medium Risk	High Risk
Data Evolution Plan	Comprehensive data refresh strategy	Partial update approach	No systematic data renewal
Model Retraining Approach	Automated performance-triggered retraining	Scheduled periodic updates	Manual ad hoc updates
Drift Monitoring	Comprehensive performance monitoring	Basic accuracy tracking	No systematic monitoring
User Feedback Integration	Continuous feedback collection and application	Periodic feedback gathering	No structured feedback process
Environment Change Tracking	Systematic monitoring of relevant changes	Awareness of major shifts	No environmental scanning
Knowledge Management	Comprehensive documentation and knowledge transfer	Basic documentation	Undocumented tribal knowledge
Team Continuity	Stable team with shared knowledge across multiple members	Core team retention	Complete team transition post-launch

THE VALUE SUSTAINABILITY INDEX

The framework in Table 12.2 helps organizations assess and predict AI value sustainability.

According to a ESI ThoughtLab 2020 study, only 28% of AI initiatives reach widespread deployment, indicating significant challenges in scaling AI for sustainable value. The study also reveals that AI payback periods vary considerably, with 43% of projects achieving break-even in under a year, 37% taking one to two years, and 20% requiring between two and over three years.

ACTION STEP: Use the value sustainability index to assess your organization's AI initiatives. For any initiative scoring "high risk" on three or more factors, develop a focused sustainability improvement plan targeting these specific vulnerabilities.

A real-world example relates to a financial services institution that implemented a "Value Sustainability Index" to assess all their AI initiatives before and after deployment. They discovered that projects scoring "high risk" on three or more factors experienced significant value decay within nine months, requiring costly remediation. By proactively addressing these factors during design and implementation, they improved average value sustainability from 68% to 94% after 12 months across their AI portfolio, dramatically improving total return on investment.

FRAMEWORK: SIGNALS OF LONGEVITY VS. SIGNS OF SLOW FAILURE

Identifying early indicators of sustainability or decay allows proactive intervention before value erosion becomes significant.

Longevity Signals

- **Performance Stability:** Consistent accuracy and reliability over time
- **User Engagement Growth:** Increasing depth and breadth of usage
- **Feedback Integration:** Continuous improvement based on input
- **Exception Handling Refinement:** Growing capability to manage edge cases
- **Environmental Adaptation:** Successful adjustment to changing conditions

Slow Failure Signs

- **Performance Drift:** Gradual degradation in accuracy or reliability

- **Adoption Plateau or Decline:** Stalling or decreasing usage patterns
- **Feedback Disregard:** Input not influencing system evolution
- **Exception Accumulation:** Growing number of unhandled cases
- **Environmental Divergence:** Increasing mismatch with current conditions

THE EARLY WARNING SYSTEM

This framework helps organizations detect sustainability issues before significant value erosion.

1. **Performance Monitoring**
    - Track accuracy and reliability trends over time
    - Compare actual versus expected performance
    - Identify statistically significant degradation
    - Monitor precision/recall balance changes
    - Assess performance across different segments
2. **Adoption Analytics**
    - Measure usage frequency and patterns
    - Track user engagement depth and breadth
    - Monitor optional versus required usage
    - Assess cross-functional adoption differences
    - Evaluate new-versus-existing user patterns
3. **Feedback Analysis**
    - Collect and categorize user feedback
    - Identify recurring themes and issues
    - Track feedback volume and sentiment trends
    - Assess response time and resolution rate
    - Monitor feedback impact on enhancements
4. **Environmental Scanning**
    - Track relevant external and internal changes
    - Assess potential impact on system performance
    - Monitor data source modifications
    - Identify emerging requirements and needs
    - Evaluate competitive and market developments

QUICK WIN: Create a monthly "AI value dashboard" that tracks key sustainability metrics across your AI portfolio: performance stability, adoption trends, feedback response rates, and environmental alignment. Flag any area showing early warning signs for immediate intervention.

A real-world example offering valuable insights comes from a retail company implementing an "AI Sustainability Monitoring System" for a customer recommendation engine. The system identified critical early warning signs of degradation. A subtle performance drift in specific product categories was detected three months before it would have affected overall metrics. Declining usage patterns among most valuable customers was identified. And growing feedback about outdated recommendations was tracked. This early detection allowed targeted intervention that maintained performance, preventing the 30–40% value erosion the organization had experienced with previous AI initiatives that lacked such monitoring.

RITUALS THAT KEEP BELIEF ALIVE

Beyond technical sustainability, organizational belief in AI's value significantly impacts long-term outcomes. Specific practices can maintain this essential faith.

Creating Sustainable AI Value

Belief Maintenance Mechanisms

- **Value Celebration:** Showcasing positive impacts regularly
- **Progress Demonstration:** Visible improvement and evolution
- **Shared Learning:** Distribution of insights and discoveries
- **User Recognition:** Acknowledgment of adoption and contribution
- **Continuous Relevance:** Connection to evolving priorities

THE BELIEF RITUAL FRAMEWORK

This approach creates systematic practices that help maintain organizational confidence.

1. **Regular Value Reviews**
   - Conduct structured value assessment sessions
   - Share concrete examples of positive impact
   - Connect AI contribution to strategic priorities
   - Address concerns and limitations honestly
   - Create forward-looking value projections
2. **Evolution Showcases**
   - Demonstrate system improvements and enhancements
   - Highlight new capabilities and features
   - Connect changes to user feedback and needs
   - Show learning and adaptation in action
   - Create excitement about future developments
3. **Success Storytelling**
   - Identify and document compelling use cases
   - Create narrative around impact and benefits
   - Include authentic user perspectives and experiences
   - Distribute stories through multiple channels
   - Refresh and update narratives over time
4. **User Community Building**
   - Create forums for user interaction and sharing
   - Facilitate peer learning and best practices
   - Develop champions and advocates
   - Build identity around system usage
   - Create status and recognition for contributions

PATTERN RECOGNITION: Organizations that maintain strong belief in their AI initiatives use explicit rituals to make progress visible and celebrate contributions. These aren't just communication tactics – they are essential sustainability mechanisms.

KEY INSIGHT: The Honesty-Confidence Paradox

Organizations often fear that acknowledging AI limitations will undermine confidence and adoption. This leads to overpromising and hidden constraints that eventually erode trust far more severely than candour would have.

The counterintuitive approach that creates better results: be transparent about both capabilities and limitations throughout the lifecycle. This apparent risk actually builds deeper trust and more sustainable adoption than overstated claims, creating resilience when inevitable challenges arise.

Consider the lessons from the following real-world example: A healthcare organization implementing clinical decision support created a "Confidence Calibration Program" that explicitly communicated system limitations alongside capabilities. This included quarterly forums, where both successes and challenges were openly discussed – and where areas in which the system excelled and where it needed improvement were acknowledged.

The resulting transparency initially seemed risky but created significantly stronger clinician trust than a competing system at another hospital that emphasized only positive aspects. When both systems encountered similar performance issues, the transparent approach maintained 86% sustained usage while the other system experienced a drop to 37% as trust collapsed.

ACTION STEP: Implement a quarterly "AI Transparency Forum" where stakeholders review both successes and challenges of key AI initiatives. Focus on honest assessment of current capabilities and of limitations being addressed, and share a concrete plan for evolution.

Storytelling as a Sustaining Force

Beyond formal practices, the narratives that develop around AI systems significantly impact their sustained value.

Strategic Narrative Elements

- **Purpose Clarity:** Clear explanation of why the system exists
- **Impact Demonstration:** Tangible evidence of positive effects
- **Evolution Story:** Narrative of learning and improvement
- **Future Vision:** Compelling direction for continued development
- **Participant Identity:** Role of stakeholders in the ongoing story

The Narrative Sustainability Framework

This approach creates compelling and evolving stories that help maintain momentum.

1. **Core Narrative Development**
   - Create foundational story explaining purpose and vision
   - Include origin, current state, and future direction
   - Incorporate genuine organizational values and priorities
   - Develop authentic voice and perspective
   - Build flexible structure that can evolve over time
2. **Impact Story Collection**
   - Systematically gather concrete impact examples
   - Document both expected and unexpected outcomes
   - Include diverse perspectives and experiences
   - Capture challenges and how they were overcome
   - Create repository of story elements and cases
3. **Narrative Refresh Process**
   - Regularly update core narrative with new elements
   - Incorporate emerging impacts and achievements
   - Evolve future vision as capabilities mature
   - Adapt to changing organizational context
   - Maintain connection to enduring purpose
4. **Multi-Channel Distribution**
   - Create appropriate versions for different audiences
   - Utilize diverse communication channels and formats
   - Incorporate narrative elements in regular interactions

Creating Sustainable AI Value

- Enable stakeholder sharing and amplification
- Measure narrative resonance and impact

QUICK WIN: Develop a "living narrative" for your most important AI initiative. Document its origin story, current impact, and future vision in a format that can be easily updated and shared across the organization.

MEASURING TRACTION, NOT JUST PERFORMANCE

Traditional AI measurement focuses on technical performance, but sustainability requires broader assessment of organizational traction.

Key Traction Dimensions

- **Usage Depth:** Extent and sophistication of utilization
- **Capability Absorption:** Integration into skills and practices
- **Organizational Embedding:** Incorporation into standard operations
- **Decision Influence:** Impact on choices and actions
- **Evolution Momentum:** Continuing development and improvement

THE TRACTION MEASUREMENT FRAMEWORK

This approach creates comprehensive assessment of AI sustainability factors.

1. **Usage Analytics**
   - Track utilization patterns and trends over time
   - Measure breadth of adoption across functions
   - Assess depth of feature and capability usage
   - Evaluate optional versus required utilization
   - Monitor user retention and continuation
2. **Capability Assessment**
   - Measure user skill and proficiency development
   - Assess understanding of appropriate application
   - Evaluate ability to interpret and apply outputs
   - Track knowledge diffusion across the organization
   - Measure emerging use cases and applications
3. **Process Integration**
   - Assess incorporation into standard workflows
   - Measure procedural dependence on capabilities
   - Evaluate documentation and formalization
   - Track training and onboarding integration
   - Assess continuity planning and resilience
4. **Impact Evaluation**
   - Measure influence on decisions and actions
   - Assess attribution to business outcomes
   - Evaluate strategic alignment and contribution
   - Track value perception among stakeholders
   - Measure comparative advantage creation

REFLECTION QUESTIONS: Does your organization measure AI success primarily by technical performance metrics or by its traction within the organization? How would measuring traction dimensions change your understanding of which AI initiatives are truly succeeding?

TENSION POINT: THE USAGE-VALUE BALANCE

Organizations can face a fundamental tension in traction measurement.

- High usage doesn't necessarily create high value
- Valuable usage may have relatively low frequency
- Different usage patterns create different value types
- Quantitative metrics may miss qualitative impact

Effective traction measurement requires balancing these perspectives rather than optimizing for usage alone.

FROM LAUNCH TO LONGEVITY

Creating sustainable AI value requires fundamental shifts in how organizations approach implementation, operation, and evolution. Let's examine specific frameworks for making these transitions.

FUNDING AND RESOURCE MODELS FOR SUSTAINABILITY

Traditional project-based funding models fundamentally misalign with AI's evolutionary nature. Sustainable value requires new approaches to resource allocation:

Funding Model Evolution
Traditional Project Model

- One-time capital allocation
- Fixed scope and deliverables
- Success defined by on-time, on-budget delivery
- Resource disbanding after implementation
- New funding request for each enhancement

Transitional Service Model

- Operational funding for ongoing maintenance
- Basic support and minor enhancements
- Separate project funding for major changes
- Minimal evolution without new projects
- Split responsibility between operations and projects

Sustainable Capability Model

- Combined operational and development funding
- Continuous evolution and enhancement
- Success defined by ongoing value creation
- Sustained team engagement, with evolving skills
- Automatic reinvestment of portion of value

Due to the significant financial commitment required for AI deployment, attention to sustainable AI value has to be an important consideration from the start.

WARNING SIGN: If your AI initiatives require entirely new funding requests for each enhancement or update, you're likely trapped in a project-based funding model, which is fundamentally misaligned with AI's evolutionary nature.

THE SUSTAINABLE FUNDING FRAMEWORK

This approach creates aligned resource models for enduring AI value.

1. **Value-Based Funding**
   - Connect resource allocation to value creation
   - Implement value capture and reinvestment mechanisms
   - Create automatic scaling based on proven returns
   - Develop portfolio-level funding across initiatives
   - Build appropriate governance for value verification
2. **Capability Team Resourcing**
   - Establish core team with ongoing funding
   - Create appropriate skill mix for maintenance and enhancement
   - Implement flexible capacity for varying needs
   - Develop knowledge-continuity mechanisms
   - Build career paths that retain critical expertise
3. **Dual-Track Allocation**
   - Secure separate but coordinated funding for stability and innovation
   - Create appropriate balance between maintenance and enhancement
   - Implement portfolio management across tracks
   - Develop stage-gate processes for innovation incorporation
   - Build governance that maintains appropriate balance
4. **Value Capture Models**
   - Create mechanisms to identify value created
   - Implement approaches to redirect portion to continued investment
   - Develop appropriate attribution and allocation processes
   - Build reinvestment governance and decision rights
   - Establish value monitoring and verification

For a real-world example, let's look at a global financial services company that transformed AI sustainability by implementing a "Value-Based Funding Model," which fundamentally changed resource allocation. The organization established baseline operational funding for each AI capability, then created automatic reinvestment of 30% of verified value creation into continued enhancement. This self-scaling approach eliminated the need for annual budget battles while ensuring that high-value capabilities received appropriate resources for evolution. The model increased average value sustainability from 14 months to over 36 months across the AI portfolio by creating resource allocation that naturally matched value potential.

ACTION STEP: For your most valuable AI capability, design a value-based funding model that:

1. Establishes baseline operational funding
2. Creates a mechanism to quantify value created
3. Implements automatic reinvestment of a specific percentage (15–30%)
4. Includes governance for verification and allocation

MOVING BEYOND PROJECT-BASED FUNDING

Building on the funding model evolution, organizations need specific mechanisms to transition from traditional approaches.

Funding Transformation Mechanisms

- **Value Pool Creation:** Establishing dedicated resources for AI capability maintenance and enhancement
- **Reinvestment Protocols:** Developing systematic approaches for redirecting value to continued development
- **Portfolio Governance:** Creating oversight that allocates resources across multiple capabilities
- **Value-Based Scaling:** Implementing automatic resource adjustment based on demonstrated impact
- **Capability Lifecycle Management:** Developing appropriate funding across different maturity stages

I've previously explored this topic in *Value-Driven Data*, where I describe the ultimate goals for a data value profit-and-loss account as threefold: transparency, adaptability, and agency. The benefits of this approach include clear visibility into value creation, enabling flexible response to changing conditions, and empowering appropriate decision-making about resource allocation.

THE FUNDING TRANSFORMATION FRAMEWORK

This goal of this approach is to create practical transition from project to capability funding.

1. **Baseline Capability Funding**
   - Establish core operational resources for each capability
   - Create dedicated allocation for essential maintenance
   - Implement minimum enhancement funding
   - Develop appropriate service level expectations
   - Build governance for baseline resource management
2. **Value Reinvestment System**
   - Create mechanisms to capture and quantify value
   - Develop formulas for automatic reinvestment
   - Implement appropriate validation and verification
   - Build governance for reinvestment oversight
   - Establish value monitoring and reporting
3. **Portfolio Optimization**
   - Create capability portfolio view across initiatives
   - Implement comparative value assessment
   - Develop resource reallocation mechanisms
   - Build appropriate decision rights and processes
   - Establish portfolio-level metrics and objectives
4. **Funding Model Governance**
   - Create appropriate oversight and accountability
   - Develop decision criteria and processes
   - Implement performance monitoring and review
   - Build adjustment mechanisms for changing conditions
   - Establish value reporting and communication

KEY INSIGHT: The Resource-Value Disconnect

Organizations typically allocate AI resources based on projected value, creating a fundamental misalignment between resource needs and value creation potential. In this scenario, early-stage capabilities with uncertain value receive too little ongoing investment, while mature applications with diminishing returns continue receiving substantial resources.

Creating Sustainable AI Value

There is a counterintuitive approach that creates better results: allocate sustainability resources primarily based on capability maturity stage rather than projected value. This lifecycle-based funding creates appropriate support when it is most needed, dramatically improving overall portfolio returns compared to purely value-based allocation.

This insight echoes a tension highlighted by PagerDuty's Sanghamitra Goswami, who emphasizes the importance of phased implementation with clear goals. As previously mentioned, she notes that while "phase 1 and 2" might not show immediate ROI, having a roadmap to "phase 6," where ROI becomes evident, helps maintain stakeholder support during the value trough period.

A real-world example from a technology company shows that this matters. The organization transformed AI resource allocation by implementing a "Capability Maturity Funding Model," which provided different resource types and levels based on development stage rather than purely anticipated value. Early-stage capabilities received higher innovation funding despite uncertain returns, while mature applications received appropriate maintenance resources despite their established value. By providing the right resources at the right time for each capability's sustainability needs, this lifecycle-based approach increased overall AI portfolio return by 47% compared to their previous value-projection model.

CREATING SELF-SUSTAINING VALUE CYCLES

Beyond funding models, truly sustainable AI requires creating virtuous cycles where value creation enables further development.

Value Cycle Elements

- **Value Identification:** Systematic recognition of benefits created
- **Value Capture:** Mechanisms to quantify and attribute impact
- **Value Translation:** Conversion of benefits to continued resources
- **Value Amplification:** Reinvestment to increase future returns
- **Value Measurement:** Ongoing assessment of cycle effectiveness

THE VALUE CYCLE FRAMEWORK

This approach creates self-reinforcing systems for sustainable AI value:

1. **Value Stream Mapping**
   - Identify all value creation pathways
   - Document how value flows through the organization
   - Determine appropriate capture points and mechanisms
   - Assess current leakage and missed opportunities
   - Create comprehensive value flow visualization
2. **Capture Mechanism Implementation**
   - Develop appropriate value identification approaches
   - Create attribution and quantification methods
   - Implement systematic recording and tracking
   - Build verification and validation processes
   - Establish value communication and reporting
3. **Reinvestment System Development**
   - Create formulas for value capture and reinvestment
   - Implement appropriate governance and oversight
   - Develop decision processes for reinvestment allocation
   - Build measurement of reinvestment effectiveness
   - Establish feedback loops for system refinement

4. **Enhancement Prioritization**
   - Create processes for identifying highest-value improvements
   - Implement evaluation criteria for enhancement options
   - Develop appropriate decision-making mechanisms
   - Build impact assessment for implemented changes
   - Establish learning system for prioritization effectiveness

QUICK WIN: Create a visual "value stream map" for one key AI initiative, showing how benefits flow through the organization. Identify potential capture points and leakage areas, then design a simple mechanism to quantify value at these points.

For a real-world example, let's look at a manufacturing company that implemented a "value cycle system" for predictive maintenance AI and thereby created true sustainability. They developed detailed value stream maps that showed how the system prevented downtime and other losses, created automated capture mechanisms that quantified these benefits, implemented a formula redirecting 40% to continued development – and established a prioritization process for enhancements based on value potential. By creating continuous improvement funded by the value already delivered, this self-reinforcing cycle increased the system's value creation by 27% annually for three consecutive years.

PORTFOLIO MANAGEMENT FOR ONGOING INVESTMENT

As AI initiatives multiply, organizations need approaches for managing capabilities as a cohesive portfolio rather than isolated projects:

Portfolio Management Elements

- **Capability Classification:** Categorization based on type and maturity
- **Value Assessment:** Evaluation of current and potential returns
- **Risk Management:** Identification and mitigation of sustainability threats
- **Resource Allocation:** Distribution of investment across capabilities
- **Strategic Alignment:** Connection to organizational priorities

THE AI PORTFOLIO MANAGEMENT FRAMEWORK

This approach creates effective governance across multiple AI capabilities.

1. **Portfolio Categorization**
   - Classify capabilities by type, function, and domain
   - Assess maturity stage and development needs
   - Evaluate strategic importance and alignment
   - Determine risk profile and characteristics
   - Create portfolio visualization and mapping
2. **Value and Performance Assessment**
   - Implement consistent value measurement across capabilities
   - Create appropriate performance metrics by category
   - Develop comparative assessment approaches
   - Build portfolio-level value aggregation
   - Establish regular review and evaluation process
3. **Resource Allocation Optimization**
   - Develop portfolio-level investment strategy
   - Create resource distribution methodology
   - Implement appropriate balance across capability types

- Build adjustment mechanisms for changing conditions
- Establish governance for allocation decisions
4. **Risk and Sustainability Management**
   - Identify portfolio-level risks and dependencies
   - Create balanced risk profile across capabilities
   - Implement appropriate mitigation strategies
   - Develop early warning systems for sustainability issues
   - Build resilience through capability diversity

TENSION POINT: THE EXPLOITATION-EXPLORATION BALANCE

Organizations face a fundamental tension in portfolio management, as mentioned in Google Cloud's Nick Morel's insights in the "From AI to ROI" podcast. He emphasizes the need for a balance between top-down leadership and grassroots movements within organizations, with "unified C-suite support" being crucial for long-term success.

- Maximizing return from established capabilities (exploitation)
- Investing in new and unproven opportunities (exploration)
- Allocating limited resources across competing needs
- Balancing short-term results with long-term potential

Effective portfolio management requires explicit approaches to this tension rather than implicit decisions.

THE AMBIDEXTROUS PORTFOLIO MODEL

This framework (see Table 12.3) creates balanced investment across different capability types.

REFLECTION QUESTIONS: How does your current AI portfolio balance across these categories? Are you over-invested in one area at the expense of others? What would a more balanced distribution look like for your organization?

A real-world example of how an "ambidextrous portfolio framework" implementation across AI capabilities transformed sustainability comes from a financial services organization. They explicitly allocated resources across four categories: core capabilities (fraud detection, credit scoring), growth capabilities (customer personalization, retention optimization), emerging capabilities (conversational interfaces, advanced analytics), and foundation capabilities (data platforms, model management). By creating an appropriate mix of short and long-term value streams that supported sustained innovation while delivering immediate returns, this approach increased the organization's three-year AI return by 72% compared to the previous focus on individual project ROI.

TABLE 12.3
The Ambidextrous Portfolio Model

Portfolio Category	Value Certainty	Time Horizon	Resource Allocation	Governance Approach
Core Capabilities	High	Short-term	60–70%	Performance optimization
Growth Capabilities	Medium	Medium-term	20–30%	Scaling and expansion
Emerging Capabilities	Low	Long-term	10–15%	Learning and discovery
Foundation Capabilities	Indirect	Ongoing	5–10%	Platform enhancement

Crafting Value-Based Deals and Partnerships

External relationships significantly impact AI sustainability, requiring approaches that align incentives for long-term value.

Partnership Alignment Elements

- **Outcome Orientation:** Focus on value creation rather than deliverables
- **Risk-Reward Sharing:** Appropriate distribution of benefits and costs
- **Evolutionary Structure:** Support for ongoing development and change
- **Knowledge Integration:** Effective transfer and application of expertise
- **Governance Mechanisms:** Joint oversight and decision processes

The Value Partnership Framework

This approach creates sustainable external relationships for AI capabilities.

1. **Outcome-Based Structuring**
   - Create agreements focused on value outcomes
   - Develop shared success metrics and targets
   - Implement appropriate risk and reward distribution
   - Build flexibility for evolving requirements
   - Establish joint accountability for results
2. **Knowledge Transfer Systems**
   - Implement systematic capability building
   - Create appropriate documentation and training
   - Develop transition and handover processes
   - Build ongoing support and expertise access
   - Establish knowledge retention mechanisms
3. **Evolutionary Governance**
   - Create joint oversight and decision bodies
   - Implement regular review and adaptation processes
   - Develop change management approaches
   - Build appropriate decision rights and procedures
   - Establish performance monitoring and feedback
4. **Ecosystem Integration**
   - Connect partners to broader capability network
   - Create appropriate data and system integrations
   - Develop collaborative innovation mechanisms
   - Build complementary capability alignment
   - Establish appropriate intellectual property approach

KEY INSIGHT: The Vendor-Partner Inversion

Organizations typically view technology providers through a vendor lens, focusing on deliverables, fixed requirements, and cost control. This approach can lead to fundamental misalignment for AI sustainability, where ongoing evolution is essential.

The counterintuitive approach that creates better results: invert the traditional hierarchy by treating key technology providers as genuine partners with shared risk and reward – and treating internal teams as "customers" focused on value and outcomes. This inverted model creates dramatically better sustainability by aligning incentives around ongoing evolution rather than fixed deliverables.

For a real-world example of the impact of creating truly collaborative relationships focused on sustained value rather than project completion, let's look at a healthcare organization that

Creating Sustainable AI Value

transformed AI sustainability by implementing a "value alliance model" with technology providers. Rather than traditional vendor contracts focused on fixed deliverables and costs, the organization created outcome-based agreements with shared success metrics, joint governance, and aligned incentives for continuous improvement. This approach increased the average effective lifespan of their AI capabilities from 19 months to over 48 months.

ACTION STEP: For your most critical AI technology provider, create a "value alliance assessment" that evaluates:

1. How aligned are their incentives with your long-term success?
2. Do agreements focus on deliverables or outcomes?
3. Is there shared governance for ongoing evolution?
4. Are there mechanisms for knowledge transfer and capability building?
5. How might restructuring this relationship create more sustainable value?

KNOWLEDGE MANAGEMENT AND INTELLECTUAL PROPERTY

A critical but often overlooked factor in AI sustainability is how organizations manage knowledge and intellectual assets developed through their initiatives.

CAPTURING AND SHARING AI INSIGHTS AND LEARNINGS

Sustainable value requires systematic approaches for preserving and distributing knowledge.

Knowledge Management Domains

- **Technical Understanding:** System design, implementation, and operation
- **Implementation Lessons:** Successes, failures, and adaptations
- **User Insights:** Adoption patterns, usage approaches, and feedback
- **Business Impact:** Value creation mechanisms and results
- **Evolution Learning:** Enhancement approaches and outcomes

THE KNOWLEDGE MANAGEMENT SYSTEM

This framework allows for comprehensive AI knowledge capture and sharing.

1. **Knowledge Capture**
   - Implement systematic documentation processes
   - Create appropriate templates and structures
   - Develop regular capture cadence and triggers
   - Build incentives for knowledge contribution
   - Establish quality standards and review
2. **Knowledge Organization**
   - Create logical taxonomy and categorization
   - Implement appropriate metadata and tagging
   - Develop search and discovery capabilities
   - Build version control and history tracking
   - Establish context preservation mechanisms
3. **Knowledge Distribution**
   - Create appropriate access and sharing mechanisms
   - Implement push-and-pull distribution approaches
   - Develop integration with workflows and processes

- Build awareness and utilization incentives
- Establish measurement of knowledge application
4. **Knowledge Evolution**
   - Create processes for updating and refining
   - Implement regular review and verification
   - Develop obsolescence identification and handling
   - Build connection to current practices and needs
   - Establish living knowledge base approach

Tension Point: The Documentation-Utilization Balance

Organizations face a fundamental tension in knowledge management.

- Comprehensive documentation creates valuable knowledge bases
- Excessive documentation burdens knowledge creators
- Detailed information supports thorough understanding
- Simplified access encourages actual utilization

Effective knowledge management requires balancing these competing needs rather than optimizing for either extreme.

WARNING SIGN: If valuable AI knowledge primarily resides in the minds of individual team members rather than in accessible organizational systems, you're at high risk for sustainability failure due to inevitable staff transitions.

In *Making Data Work,* I highlighted the risks organizations face when knowledge isn't effectively transferred and embedded. One particular conversation comes to mind, when a manager confided, "I cannot help but feel that I am being left behind. Our legacy technologies are insidiously being replaced right under our noses, yet internally, we don't have the necessary skills or even the time to develop the new skills [needed for the new technologies]."

Building Institutional Knowledge

Beyond documented information, sustainable AI requires building enduring organizational understanding.

Institutional Knowledge Elements

- **Formal Documentation:** Captured information and knowledge
- **Shared Understanding:** Common mental models and frameworks
- **Process Integration:** Knowledge embedded in workflows and procedures
- **Cultural Elements:** Values, norms, and informal practices
- **Expertise Networks:** Interconnected individual knowledge holders

The Institutional Knowledge Framework

This approach creates enduring organizational understanding of AI capabilities.

1. **Documentation System**
   - Create appropriate artefact types and templates
   - Implement systematic capture processes
   - Develop organizational and access approaches

Creating Sustainable AI Value

- Build quality standards and governance
- Establish maintenance and evolution mechanisms

2. **Community Development**
   - Create communities of practice and interest
   - Implement knowledge-sharing forums and events
   - Develop mentorship and expertise connection
   - Build recognition for knowledge contribution
   - Establish collaborative learning approaches
3. **Process Embedding**
   - Integrate knowledge into standard workflows
   - Create procedures that incorporate best practices
   - Develop training and onboarding that transfers knowledge
   - Build decision processes that apply learnings
   - Establish governance that preserves understanding
4. **Cultural Reinforcement**
   - Create knowledge-sharing expectations and norms
   - Implement appropriate recognition and incentives
   - Develop leadership modelling of knowledge behaviours
   - Build language and frameworks for shared understanding
   - Establish learning and adaptation as cultural values

QUICK WIN: Create a monthly "AI learning forum" where teams share insights, challenges, and solutions from their AI implementations. Record these sessions and create a searchable knowledge base from the content.

Protecting and Leveraging AI Assets

Strategic management of intellectual property and proprietary assets significantly impacts sustainable value.

AI Asset Categories

- **Technical Assets:** Models, algorithms, and technical approaches
- **Data Assets:** Proprietary datasets and derived insights
- **Process Assets:** Implementation approaches and methodologies
- **Experience Assets:** Usage knowledge and best practices
- **Integration Assets:** Connection points and interoperability

The AI Asset Management Framework

This approach creates strategic protection and leverage of proprietary capabilities.

1. **Asset Identification**
   - Create comprehensive inventory of AI-related assets
   - Implement appropriate categorization and classification
   - Develop value assessment for different assets
   - Build understanding of strategic importance
   - Establish ownership and responsibility
2. **Protection Strategy**
   - Create appropriate legal protection mechanisms
   - Implement technical safeguards and controls
   - Develop access and usage governance

- Build contractual and agreement frameworks
- Establish monitoring and enforcement processes

3. **Leverage Approach**
   - Create strategic approach for asset utilization
   - Implement appropriate sharing and commercialization
   - Develop partnership and licensing models
   - Build internal use and application mechanisms
   - Establish value capture from asset leverage

4. **Evolution Management**
   - Create processes for asset enhancement and development
   - Implement portfolio approach across assets
   - Develop investment prioritization methods
   - Build connection to organizational strategy
   - Establish appropriate governance and oversight

REFLECTION QUESTIONS: What proprietary AI assets has your organization developed that could be leveraged more broadly? Are these assets being systematically identified, protected, and utilized to maximize their value?

THE LONG-TERM AI VALUE ROADMAP

Creating sustainable AI value requires explicit planning for long-term evolution rather than focusing solely on immediate implementation.

Planning for Continuous Evolution

Sustainable AI needs concrete approaches for ongoing development beyond initial deployment.

Evolution Planning Elements

- **Capability Maturation:** Development of core functionality and performance
- **Scope Expansion:** Extension to new areas and use cases
- **Integration Deepening:** Enhanced connection with systems and processes
- **Intelligence Advancement:** Increased sophistication and autonomy
- **Experience Enhancement:** Improved user interaction and satisfaction

The Evolution Roadmap Framework

This approach creates structured planning for long-term AI development.

1. **Maturity Stage Mapping**
   - Define evolutionary stages for the capability
   - Create clear characteristics for each stage
   - Develop appropriate metrics and indicators
   - Build understanding of transition requirements
   - Establish current position and progression path

2. **Multi-Horizon Planning**
   - Create short-term enhancement roadmap (0–6 months)
   - Develop medium-term evolution plan (6–18 months)
   - Establish long-term direction and vision (18+ months)
   - Build connections between different horizons
   - Create appropriate governance and review

Creating Sustainable AI Value

3. **Capability Building Alignment**
   - Identify required skills and expertise for evolution
   - Create development plans for necessary capabilities
   - Implement appropriate knowledge acquisition
   - Build organizational readiness for advancement
   - Establish learning and development approaches
4. **Value Projection and Validation**
   - Create expected value model for evolutionary stages
   - Implement appropriate measurement and verification
   - Develop feedback mechanisms and adjustment processes
   - Build connection between value and investment
   - Establish learning system for projection accuracy

Tension Point: The Stability-Evolution Balance

Organizations face a fundamental tension in evolution planning:

- Users need stability and predictability in system behaviour
- Capabilities require ongoing evolution to maintain value
- Rapid change can disrupt established usage patterns
- Slow development allows value erosion and competitive disadvantage

Effective evolution planning requires explicitly addressing this tension rather than simply accepting the trade-off.

ACTION STEP: For your key AI capabilities, create a dual-track evolution roadmap that explicitly separates stability features (requiring minimal disruption) from evolutionary features (allowing more significant change). Plan integration points for moving innovations into the stable core.

Building Adaptability into AI Systems

Beyond planning, sustainable AI requires architectural and design approaches that enable efficient evolution.

Adaptability Design Elements

- **Modular Architecture:** Independent components that can evolve separately
- **Extensible Interfaces:** Connection points that support expansion
- **Configurable Parameters:** Settings that enable adjustment without rebuilding
- **Versioning Support:** Ability to manage multiple system versions
- **Testing Infrastructure:** Capabilities for validating changes and enhancements

The Adaptable Architecture Framework

This approach creates AI systems designed for efficient evolution.

1. **Modularity Implementation**
   - Create logically separated system components
   - Implement clear interfaces and contracts
   - Develop appropriate encapsulation and isolation
   - Build independent testability and deployment
   - Establish governance for module management

2. **Extension Point Design**
   - Identify likely areas for future expansion
   - Create appropriate extension mechanisms
   - Implement plugin or add-on capabilities
   - Develop API and integration approaches
   - Build governance for extension management
3. **Configuration System Development**
   - Create comprehensive parameter management
   - Implement appropriate control and governance
   - Develop testing and validation approaches
   - Build versioning and history tracking
   - Establish impact assessment processes
4. **Change Management Infrastructure**
   - Create efficient testing and validation capabilities
   - Implement appropriate staging and deployment
   - Develop rollback and recovery approaches
   - Build monitoring and verification systems
   - Establish governance for change control

KEY INSIGHT: Strategic Technical Debt

Organizations typically view technical debt as uniformly negative, creating pressure for "perfect" implementations that paradoxically reduce adaptability through excessive complexity and coupling.

The counterintuitive approach that creates better results: strategically accept appropriate technical debt in non-critical areas to enable faster evolution, while investing heavily in adaptability for core components. This balanced approach typically delivers substantially greater sustainable value than either pursuing perfection everywhere or allowing unmanaged debt accumulation.

Managing the Lifecycle of AI Capabilities

Sustainable value requires explicit management of capabilities across their complete lifecycle.

Lifecycle Stage Characteristics

1. **Emergence Stage**
   - Initial capability development
   - Limited scope and application
   - Focused adoption with key users
   - Rapid learning and adaptation
   - Foundation establishment for expansion
2. **Growth Stage**
   - Expanding capability and scope
   - Increasing adoption and utilization
   - Performance enhancement and refinement
   - Integration with additional systems
   - Value demonstration and expansion
3. **Maturity Stage**
   - Stable core capability and performance
   - Broad adoption and integration
   - Incremental enhancement and optimization

Creating Sustainable AI Value

- Standardized processes and practices
- Balanced efficiency and effectiveness

4. **Renewal Stage**
   - Significant capability advancement
   - Architectural or approach evolution
   - New application and use expansion
   - User experience transformation
   - Value model enhancement

5. **Transition Stage**
   - Capability replacement or consolidation
   - User migration and change management
   - Knowledge transfer and preservation
   - Asset utilization and leverage
   - Strategic realignment and focus

QUICK WIN: Create a lifecycle map for your AI portfolio, categorizing each capability by its current stage (emergence, growth, maturity, renewal, or transition). Use this map to identify appropriate management strategies for each capability.

The Lifecycle Management Framework

This approach creates appropriate strategies for each development stage.

1. **Stage Assessment**
   - Create clear stage definitions and characteristics
   - Implement appropriate measurement and indicators
   - Develop regular evaluation process
   - Build understanding of transition factors
   - Establish governance for stage determination

2. **Stage-Appropriate Strategy**
   - Create differentiated approaches for each stage
   - Implement suitable investment and resourcing
   - Develop appropriate metrics and expectations
   - Build governance aligned to lifecycle position
   - Establish management practices by stage

3. **Transition Management**
   - Create early warning for stage transitions
   - Implement proactive adaptation planning
   - Develop appropriate change management
   - Build stakeholder communication approach
   - Establish governance for transition oversight

4. **Portfolio Balance**
   - Create appropriate distribution across lifecycle stages
   - Implement complementary capability development
   - Develop risk management through diversity
   - Build strategic alignment across portfolio
   - Establish governance for lifecycle balance

PATTERN RECOGNITION: Organizations often apply a single management approach to all AI capabilities, regardless of lifecycle stage. Recognizing that different stages require different strategies is a hallmark of sustainable AI management.

Creating a Culture of Ongoing Innovation

Beyond technology and processes, sustainable AI value requires cultural elements that support continuous evolution.

Innovation Culture Elements

- **Learning Orientation:** Emphasis on knowledge acquisition and application
- **Experimentation Comfort:** Acceptance of appropriate testing and discovery
- **Improvement Expectation:** Assumption of ongoing enhancement rather than stasis
- **Collaboration Emphasis:** Focus on cross-functional knowledge-sharing
- **User Centricity:** Continuous connection to user needs and experiences

The Innovation Culture Framework

This approach creates an organizational environment supporting sustainable AI evolution.

1. **Leadership Modelling**
   - Create visible executive commitment to innovation
   - Implement leadership behaviours that demonstrate values
   - Develop appropriate risk tolerance and support
   - Build recognition for innovation contribution
   - Establish consistent messaging and communication
2. **Structural Enablement**
   - Create appropriate organizational structures and roles
   - Implement resources and time allocation
   - Develop collaboration mechanisms and spaces
   - Build incentive and recognition systems
   - Establish governance supporting innovation
3. **Process Integration**
   - Create innovation elements in standard workflows
   - Implement idea generation and capture mechanisms
   - Develop appropriate evaluation and selection
   - Build efficient implementation pathways
   - Establish feedback and learning loops
4. **Skill Development**
   - Create innovation-capability-building programs
   - Implement appropriate training and development
   - Develop mentoring and knowledge-sharing
   - Build recruitment for innovation mindset
   - Establish continuous learning approaches

KEY INSIGHT: Structured Innovation

Organizations often view innovation culture as inherently unstructured, creating environments with unlimited freedom but inadequate direction and discipline to translate ideas into sustainable value.

The counterintuitive approach that creates better results: implement "structured innovation" with clear processes, governance, and expectations that channel creativity towards sustainable value rather than unbounded exploration. This balanced approach typically delivers substantially greater implemented innovation than either excessive structure or complete freedom.

ACTION STEP: Create a "structured innovation protocol" for AI enhancement that includes:

1. Clear innovation domains aligned with strategic priorities
2. Specific processes for idea development and evaluation
3. Resource allocation mechanisms for promising innovations
4. Stage-gate approach for moving from concept to implementation
5. Success metrics that balance learning with value creation

CONCLUSION: CREATING TRULY SUSTAINABLE AI VALUE

As we've explored throughout this chapter, creating sustainable AI value requires fundamental shifts in how organizations approach implementation, operation, and evolution. Even the most sophisticated AI tends to stall over the long term if it can't be maintained and enhanced over time. Conversely, even relatively straightforward implementations can deliver transformative results when they are supported by appropriate sustainability practices.

Stela Solar, Microsoft's Global Head of Artificial Intelligence Solutions, touches on another important element for sustained success: leadership. She says, "All of this really is paved by a strong leader at the front or leadership group at the front, that is spotlighting what technology can do to help an organisation be agile and thrive in the current environment."

Let's conclude with five key principles for creating truly sustainable AI value:

SHIFT FROM PROJECTS TO PERSISTENT CAPABILITIES

Sustainable value requires moving beyond the traditional project model to establish ongoing capabilities with continuous evolution, appropriate resources, and clear ownership across the complete lifecycle. As the RAND Corporation study found, 68% of AI projects that initially succeeded still failed to deliver sustained value six months after implementation, with the critical factor being "institutionalized cross-functional communication channels" that persisted beyond initial project teams.

CREATE SELF-REINFORCING VALUE CYCLES

Rather than one-time investments, establish mechanisms that capture a portion of created value and automatically redirect it to continued enhancement, creating virtuous cycles of increasing returns. The financial services organization that implemented a value-based funding model with automatic reinvestment of 30% of verified value creation saw their average AI initiative lifespan increase from 14 months to over 36 months.

BUILD ADAPTABILITY BY DESIGN

Create both technical architectures and organizational structures specifically designed for efficient evolution, balancing appropriate stability with the ability to adapt to changing conditions and requirements. The Gaming Guru achieved this by transforming monolithic systems into an approach inspired by microservices, where "breaking big things into small things" enabled faster adaptation to changing business needs while maintaining system integrity.

DEVELOP TRUE INSTITUTIONAL KNOWLEDGE

Implement comprehensive approaches for capturing, preserving, and applying knowledge across the organization, establishing expertise that transcends individual team members and survives inevitable transitions. This allows organizations to counteract the conundrum highlighted by one manager, who said, "I cannot help but feel that I am being left behind [because] we don't have the necessary skills or even the time to develop the new skills [needed for new technologies]."

CONSIDER THE COMPLETE LIFECYCLE

Recognize that different development stages require different strategies, resources, and governance, creating appropriate approaches for capabilities from emergence through maturity to renewal or transition. As Abhinav Singhal of Thyssenkrupp notes, "AI is still not moving the needle. It doesn't yet create the margins to really affect the P&L. But in five to seven years, it will be a critical game-changer in how we compete and serve our customers. You have to invest in it until it reaches that pivotal point where it becomes strong enough to fuel growth on its own."

FINAL REFLECTION QUESTIONS: Which of these five principles represents the biggest opportunity for increasing AI value sustainability in your organization? What specific actions could you take in the next 30 days to begin implementing this principle?

By applying these principles, organizations can transform AI from an endless series of promising but ultimately transient projects into enduring capabilities that create compounding value year after year – the true measure of sustainable AI success.

13 Future-Proofing AI Value

Chapter Roadmap

In this chapter, we explore how to create lasting AI value in a rapidly changing landscape. I share practical frameworks, real-world examples, and actionable strategies to help you stay ahead of the curve. Expect to find out how to build adaptable technical and organizational foundations and implement strategies for navigating evolving ethical and regulatory challenges. Also included are approaches to balancing innovation with stability, methods for preparing for emerging business and collaboration models and a practical framework for assessing your future-readiness.

EVOLUTION NEVER STOPS

The boardroom fell silent as I clicked to my final slide. I'd just walked the executive team through the company's ambitious three-year AI roadmap, a plan we'd spent months developing.

"This all looks solid," the CEO finally said. "But what happens when everything we just planned becomes outdated in six months?"

It wasn't a cynical question. It was a genuine concern from a leader who'd watched AI capabilities evolve at breakneck speed. And he was right to worry.

Consider this: the generative AI wave caught many organizations off guard despite years of warning signs. Companies that had finally mastered predictive analytics suddenly faced a completely new paradigm of AI capabilities. Those who had just become comfortable with analysing historical data suddenly needed to understand prompt engineering, large language model (LLM) fine-tuning, and completely different return on investment (ROI) models.

KEY INSIGHT: The AI implementation you meticulously plan today will likely look dramatically different by the time it's fully deployed.

For from being meant to discourage you – this means to inspire you to shift how you approach AI value from the start. Future-proofing isn't about predicting what's coming, it is about building systems flexible enough to adapt when the inevitable change arrives.

REFLECTION QUESTIONS: How would your current AI projects fare if an entirely new AI paradigm emerged tomorrow? Which elements would remain valuable, and which might become obsolete?

THE EVOLVING AI VALUE LANDSCAPE

This following observation from a CTO I worked with perfectly captures the anxiety many leaders feel about AI investments.

"Everyone thinks their AI project is cutting-edge until the next advancement makes it look like a relic from another era," he said, reflecting the reality that technology evolves so rapidly that the goalposts for "advanced" and "valuable" constantly shift.

This state of affairs makes it all the more important to look at current AI capabilities and their value boundaries, consider generative AI's impact beyond the initial hype as well as the emerging frontier of agentic AI, and pay attention to both regional and industry-specific approaches to AI development.

Current AI Capabilities and Their Value Boundaries

The AI capabilities currently deployed in most organizations fall into several broad categories:

- **Predictive Analytics and Forecasting:** Using historical data to predict future outcomes
- **Process Automation:** Streamlining workflows and reducing manual tasks
- **Natural Language Processing:** Understanding and generating human language
- **Computer Vision:** Interpreting and analysing visual information
- **Recommendation Systems:** Suggesting products, content, or actions based on patterns

Each of these capabilities creates value within certain boundaries – with the caveat that these boundaries are being expanded and sometimes completely redrawn by emerging technologies.

Generative AI: Beyond the Initial Hype

The rise of generative AI tools marked a step change in how organizations view AI's potential value.

Beyond the initial excitement, we're now seeing more nuanced applications that address specific business challenges.

- **Content Creation and Enhancement:** Not just generating text but creating contextually relevant, brand-aligned content that reduces production time by 60–80%
- **Ideation Acceleration:** Supporting innovation processes with ten times more concept variations
- **Knowledge Synthesis:** Distilling vast information stores into actionable insights, cutting research time by up to 70%
- **Process Redesign:** Reimagining workflows by utilizing AI as a thinking partner, not just a task executor

I worked with a professional services firm that initially saw generative AI as simply a way to draft emails faster. Within six months, they had completely reimagined their knowledge management strategy around these tools, turning what was initially a cost centre into a revenue-generating capability that differentiated them in the market.

WARNING SIGN: Organizations that view generative AI merely as a productivity tool are missing its transformative potential. The value isn't in the technology but in how it transforms business processes and enables new possibilities.

Agentic AI: The Next Value Frontier

If you think generative AI created disruption, prepare for the agentic AI wave. These systems don't just respond to prompts – they take initiative, make decisions, and operate autonomously towards defined goals.

Early applications include:

- Autonomous planning systems that continuously optimize resource allocation
- Self-adjusting operations management that responds to changing conditions without human intervention
- Proactive customer engagement that anticipates needs and initiates interactions
- Continuous improvement agents that identify process inefficiencies and implement solutions

Case Study: Agentic AI in Supply Chain Management

A manufacturing client recently deployed agentic AI to enhance supply chain resilience.

Rather than simply flagging potential disruptions – similarly to how previous systems functioned – this system:

1. Identifies potential supply risks
2. Evaluates alternative suppliers
3. Initiates conversations with those suppliers
4. Negotiates preliminary terms
5. Presents human decision-makers with fully formed contingency plans

The result was a 38% reduction in supply chain disruptions and a 22% decrease in emergency procurement costs.

Agentic AI represents a fundamental shift from "AI as tool" to "AI as colleague" – a shift that requires completely different approaches to value measurement, governance, and integration.

Needless to say, this change brings tension between the opportunities and risks of increasingly autonomous AI systems, one of the most significant challenges organizations will face in maximizing future AI value.

A World Economic Forum report describes the impact like this: "While agentic AI offers efficiency and innovation, it raises serious concerns regarding labour disruption, privacy, and market volatility."

PATTERN RECOGNITION: The evolution from predictive to generative to agentic AI follows a pattern of increasing autonomy and decreasing human involvement in the moment-to-moment operation of AI systems. Each step requires rethinking governance, risk, and value measurement.

REGIONAL APPROACHES TO AI DEVELOPMENT

The global AI landscape is evolving in distinctly different ways across regions, presenting both opportunities and challenges for organizations operating internationally.

United States: Private-Sector Innovation

The US approach to AI development is primarily driven by private enterprises with government support. As Henrique Schneider explains, initiatives like the $500 billion "Stargate Project" exemplify how leadership in the private sector drives national AI advancement. The US "remains at the forefront, leveraging its robust ecosystem of private-sector innovation, academic excellence, and government investment."

Europe: Regulatory-First Framework

Europe has taken a distinctive regulatory-first approach to AI development, establishing the world's first comprehensive legal framework for AI. The EU AI Act, which entered into force on August 1, 2024, represents a risk-based regulatory framework. As described in the official EU documentation, "The AI Act assigns applications of AI to three risk categories. First, applications and systems that create an unacceptable risk, such as government-run social scoring of the type used in China, are banned. Second, high-risk applications, such as a CV-scanning tool that ranks job applicants, are subject to specific legal requirements. Lastly, applications not explicitly banned or listed as high-risk are largely left unregulated."

As the EU explains, "The EU's approach to artificial intelligence centres on excellence and trust, aiming to boost research and industrial capacity while ensuring safety and fundamental rights." This regulatory framework builds upon the foundation of General Data Protection

Regulation (GDPR), with the AI Act mentioning GDPR more than 30 times throughout its recitals and articles.

The European approach emphasizes that AI systems must be "safe, transparent, traceable, non-discriminatory and environmentally friendly," while requiring that national authorities provide companies with testing environments for AI that simulate conditions close to the real world to support innovation and start-ups.

China: State-Led Strategy

In contrast, Chinese leadership takes a centralized approach with significant government investment and strategic planning. The country's supportive regulatory environment enables rapid implementation of large-scale AI projects. Recent developments like DeepSeek's R1 model, created without relying on foreign semiconductor technology, demonstrate China's commitment to AI self-sufficiency.

India: Public-Private Partnership

India employs a collaborative model between government bodies and private enterprises. The IndiaAI Mission emphasizes ethical and inclusive AI development, with S. Krishnan, secretary at the Ministry of Electronics and Information Technology, noting that "India has an opportunity to create a trillion-dollar digital economy by 2025, benefitting all sectors and people."

Middle East: Leadership-Driven Adoption

The Middle East region shows remarkably strong executive support for AI initiatives. As Harrison Lung, group chief strategy officer at e&, notes that the MENA region is "on the brink of an AI-driven future" with the potential to "cement its position as a global leader in AI innovation." This sentiment is echoed by regional business leaders who demonstrate exceptional commitment to AI adoption.

Africa: Building Foundational Capacity

In contrast to regions with established AI capabilities, Africa faces different challenges. As Irene Phoebe Kiwia argues, "Africa risks falling behind without proactive measures to develop AI competencies among its leadership." Despite AI's potential for economic and social progress, there is a critical need for "comprehensive strategies" including "education, investment in technology infrastructure, and policies fostering innovation."

KEY INSIGHT: This regional variation creates both challenges and opportunities for global organizations looking to future-proof their AI value. Those with operations across multiple regions must develop strategies that account for these different environments.

INDUSTRY-SPECIFIC VALUE EVOLUTION

The value landscape isn't changing uniformly – it is evolving in industry-specific ways. Here are some examples:

Financial Services

- From risk scoring to proactive financial wellness management
- From fraud detection to anticipatory security postures
- From basic chatbots to comprehensive financial advisors

Healthcare

- From diagnostic assistance to continuous health monitoring and intervention
- From administrative automation to full clinical workflow integration
- From treatment recommendation to personalized care planning

Manufacturing

- From predictive maintenance to self-healing systems
- From quality control to continuous process optimization
- From supply chain visibility to fully adaptive supply networks

Retail

- From product recommendations to anticipatory commerce
- From inventory optimization to demand-creating business models
- From segmented marketing to individually responsive engagement

Case Study: AI in Humanitarian Aid

The International Rescue Committee's (IRC) Signpost project shows how AI can transform humanitarian aid. By employing AI chatbots to provide multilingual information to displaced individuals, the IRC has enhanced the reach and efficiency of aid distribution across multiple countries in 11 languages.

As Jeannie Annan, IRC's chief research and innovation officer, notes: "We're trying to really be clear about where the legitimate concerns are – but lean into the optimism of the opportunities and not allow the populations we serve to be left behind in solutions that have the potential to scale in a way that human-to-human or other technology can't."

PATTERN RECOGNITION: The common thread across industries is a shift from reactive to proactive, from isolated to integrated, and from augmenting existing processes to fundamentally reimagining them.

PREPARING FOR FUTURE VALUE OPPORTUNITIES

Now that we understand how the AI landscape is evolving, let's explore how organizations can prepare to capture future value opportunities. Measures that can enhance outcomes include building adaptable technical foundations, creating organizational flexibility, developing future-focused talent strategies, and creating space for innovation. Another useful exercise relates to scenario-planning for AI evolution.

BUILDING ADAPTABLE TECHNICAL FOUNDATIONS

The Technical Architecture Trap

I've seen too many organizations invest millions in rigid technical architectures optimized for current AI capabilities only to find themselves locked into outdated approaches when the next wave arrives.

A more sustainable approach focuses on building adaptable foundations.

- Modular data architectures that can incorporate new data types and sources
- Application programming interface (API)-first development that allows components to be replaced without disrupting the entire system
- Cloud-native approaches that scale flexibly with changing demands
- Composable AI strategies that combine and recombine capabilities based on evolving needs

ACTION STEP: Assess your technical foundation's adaptability with these key questions:

1. Can your systems incorporate entirely new data types without requiring major redesign?
2. How easily can new AI models be integrated with existing workflows?

3. Can processing capacity be scaled rapidly for more computationally intensive models?
4. How portable are your AI solutions between different infrastructure environments?

Case Study: The AI Content Layer

A global retailer I advised avoided the architecture trap by designing an "AI content layer"– a flexible middleware that separated core business systems from the AI capabilities that enhanced them. When generative AI emerged, the organization could integrate these new capabilities in weeks rather than the months it took their competitors.

Innovative Infrastructure Approaches

As AI models grow larger and more resource-intensive, organizations face escalating infrastructure costs. Our research on AI scaling reveals an innovative solution to this challenge: distributed GPU networks.

According to the research, "In response to the high costs and infrastructure demands of AI model training, innovative approaches such as leveraging underutilized GPUs from gaming PCs and university labs have emerged. This strategy creates distributed GPU networks, significantly reducing costs and democratizing AI development."

While this approach comes with challenges related to speed and security, it offers a viable alternative to centralized training infrastructures, particularly for organizations with limited resources. These creative approaches to technical foundations will become increasingly important as computational requirements continue to grow.

Creating Organizational Flexibility

Technical adaptability means little without matching organizational flexibility.

The Skills Portfolio Approach

Rather than focusing narrowly on current technical skills, forward-thinking organizations develop a balanced portfolio across three categories.

1. **Foundation Skills:** Core data literacy, statistical thinking, and problem-framing will remain valuable regardless of which AI technologies dominate.
2. **Current Technology Skills:** Capabilities needed to implement and maintain today's solutions.
3. **Learning Acceleration Skills:** Meta-capabilities that help the organization rapidly absorb and apply new technologies and approaches.

This balanced approach prevents the boom-and-bust cycle of hiring for the latest hot skills only to find them quickly outdated.

QUICK WIN: Map your organization's current skills distribution across these three categories to identify imbalances. Often, organizations over-invest in current technology skills at the expense of foundational and learning acceleration capabilities.

Decision-Making Processes That Embrace Evolution

Traditional ROI models and governance frameworks often struggle with rapidly evolving technologies. More adaptive approaches include:

- Option-based investment models that value flexibility alongside immediate returns
- Graduated governance that adjusts oversight based on both risk and technological maturity
- Rolling planning cycles that regularly reassess directions rather than locking into multi-year roadmaps
- Value stage-gating that focuses early phases on learning and later phases on scaling

Case Study: AI Value Sprints

A pharmaceutical company I worked with adopted "AI value sprints" – 12-week cycles focused on specific business challenges, with explicit learning goals alongside performance targets. This approach allowed them to continuously adapt their AI strategy without the "analysis paralysis" that plagued their competitors.

DEVELOPING FUTURE-FOCUSED TALENT STRATEGIES

The most sophisticated AI architecture is only as good as the people who use it. Future-proofing requires talent strategies that anticipate tomorrow's needs.

The Human-AI Talent Spectrum

As AI capabilities evolve, the human skills that create value shift as well. Organizations need talent across a spectrum that includes:

- AI producers who create and maintain AI systems
- AI translators who connect technical capabilities to business needs
- AI collaborators who work alongside AI in augmented workflows
- AI governors who ensure AI systems align with organizational values and requirements

The relative importance of these roles shifts as AI capabilities mature. Early-stage implementations often require more producers and translators, while more advanced implementations need skilled collaborators and governors.

The Skills Refresh Cycle

AI skills have a shorter half-life than most technical capabilities. Organizations need systematic approaches to continuous upskilling.

- Regular skills forecasting based on emerging technologies
- Learning partnerships with educational institutions and vendors
- Internal knowledge-sharing structures that spread expertise
- Career paths that reward continuous learning rather than static expertise

Case Study: Skills Futures

A technology services firm I advised implemented "skills futures" – quarterly assessments of emerging capabilities paired with targeted learning opportunities. This approach reduced their external hiring costs by 35% while increasing their ability to rapidly deploy new AI technologies.

CREATING SPACE FOR INNOVATION

Innovative organizations recognize that deliberate structures and time allocations are essential for future-proofing. As I always emphasize, "Innovation doesn't happen by accident. It requires deliberate space and systems."

Here is a tried-and-true approach to innovation that includes structured time investments:

- **Weekly Innovation Blocks:** Two-hour uninterrupted time with no client work, dedicated to pure thinking space and pattern recognition
- **Monthly Deep Dives:** Half-day strategic reviews for trend analysis and opportunity mapping
- **Quarterly Renewals:** Full-day off-site sessions for strategy refresh, market analysis, and personal development

ACTION STEP: Block dedicated innovation time in your calendar now, following the above model. Protect this time fiercely and use it specifically for exploring emerging AI capabilities and their potential applications. This structured approach ensures that innovation doesn't get crowded out by day-to-day demands, a particularly important consideration given the rapid pace of AI evolution.

SCENARIO PLANNING FOR AI EVOLUTION

While we can't predict exactly how AI will evolve, we can prepare for a range of possibilities. Effective scenario planning focuses on divergent futures that stress test your approach.

Four AI Evolution Scenarios

Consider how your AI strategy would perform across these potential futures:

1. **Incremental Advancement:** AI improves steadily but without dramatic breakthroughs
2. **Capability Explosion:** Rapid advances create dramatically more powerful but still specialized AI systems
3. **Regulatory Constraint:** Growing concerns lead to significant restrictions on AI applications
4. **Democratized Innovation:** AI development tools become accessible to non-specialists, driving innovation from unexpected sources

REFLECTION QUESTIONS: For each scenario, ask:

- Which of our current initiatives would become more or less valuable?
- What new opportunities would emerge?
- What risks would we need to manage?
- Which organizational capabilities would become critical?

This exercise isn't about picking the "right" scenario – it is about building approaches resilient enough to create value across multiple possible futures.

EMERGING ETHICAL AND REGULATORY CHALLENGES

Rapid technology advancements inevitably trigger changes in the ethical and regulatory landscape. Treating such considerations as afterthoughts rather than strategic imperatives will typically impede an organization's ability to operate and build lasting trust.

Better outcomes can be achieved when organizations develop strategies for anticipating regulatory developments, frameworks for building adaptable governance, and approaches to addressing emerging ethical questions.

ANTICIPATING REGULATORY DEVELOPMENTS

Rather than reacting to regulations after they emerge, forward-thinking organizations anticipate likely directions.

The Regulatory Horizon Map

Map potential regulatory developments across key dimensions:

- **Data Protection and Privacy:** Expanding rights and restrictions around data use
- **Transparency and Explainability:** Requirements to make AI decisions interpretable
- **Testing and Certification:** Pre-deployment validation requirements

- **Ongoing Monitoring:** Post-deployment surveillance obligations
- **Liability and Redress:** Responsibility frameworks for AI-related harms

For each dimension, track early indicators from:

- Draft legislation and regulatory guidance
- Industry standard-setting initiatives
- Legal precedents and case law
- Academic and policy research
- Public opinion trends

This mapping helps prioritize which guardrails to build now, even before requirements become mandatory.

Landmark Legal Precedents

Recent legal developments provide important signals about future regulatory directions. In February 2024, the Guangzhou Internet Court issued a landmark ruling that held an AI service provider liable for copyright infringement related to AI-generated content. This marked the first time a Chinese court addressed the copyright status of AI-generated works, establishing that the creator of the AI model – rather than the AI system itself – could be held accountable for copyright violations.

Similarly, China has implemented new regulations targeting "deep synthesis technologies," including AI-generated deepfakes. These rules mandate clear labelling of AI-generated content through watermarks or digital identifiers to combat disinformation and online fraud, with authorities imposing penalties on individuals and organizations that fail to comply.

These developments signal increasing regulatory attention to AI-generated content and attribution issues, a trend that is likely to continue globally.

QUICK WIN: Create a simple tracking system for AI regulations in your key markets. Assign team members to monitor specific jurisdictions or issues and report monthly on emerging developments.

Regulatory Engagement Strategy

Beyond compliance, consider how to constructively shape evolving regulations:

- Participate in industry associations and standards bodies
- Engage directly with regulators on technical capabilities and limitations
- Share case studies and evidence of responsible practices
- Collaborate on testbeds and sandboxes for emerging approaches

Case Study: Regulatory Horizon Review

A financial services firm I worked with established a quarterly "regulatory horizon review" that brought together legal, compliance, data science, and product teams. This cross-functional approach helped them identify regulatory trends early enough to influence both their product roadmap and, in some cases, the regulations themselves.

BUILDING ADAPTABLE GOVERNANCE FRAMEWORKS

Fixed governance models break under the strain of rapidly evolving technology. More resilient approaches include:

Principles-Based Governance

Rather than rigid rules that quickly become outdated, establish enduring principles that guide decision-making across evolving capabilities, such as:

- Maintaining meaningful human oversight appropriate to risk level
- Ensuring explainability proportional to impact on stakeholders
- Preserving data quality and representativeness regardless of model type
- Testing for bias across increasingly complex model architectures

These principles create a consistent framework even as specific implementations change.

The Human Oversight Spectrum

The research identifies three decision-making models based on risk management, each with varying levels of human oversight:

- **Human-in-the-Loop:** Human retains full control while AI provides recommendations
- **Human-out-of-the-Loop:** AI executes decisions without human oversight
- **Human-over-the-Loop:** Humans can adjust parameters during algorithm execution

As Pawel Gmyrek, senior researcher at the International Labour Organization, argues, a "'human above the loop' approach" remains "essential, with AI complementing human abilities rather than replacing the judgment and accountability vital to the sector."

The appropriate level of human involvement should be calibrated to the risk level and potential impact of the decisions being made.

ACTION STEP: Audit your current and planned AI systems and classify them according to the human oversight spectrum. Ensure the level of human involvement matches the risk level of each application.

Governance Design Patterns

Develop modular governance components that can be combined and calibrated based on use case characteristics:

- **Risk Tiering:** Adjusting oversight based on potential impact
- **Testing Protocols:** Validation approaches for different model types
- **Documentation Requirements:** Records appropriate to context
- **Review Cadences:** Oversight frequency based on deployment context
- **Stakeholder Engagement:** Input mechanisms scaled to affected populations

These patterns allow governance to scale appropriately instead of becoming either too burdensome or too lax.

LEARNING FROM GOVERNANCE FAILURES

Real-world examples provide valuable lessons about governance challenges with emerging AI capabilities.

Case Study: Apple's Face ID Vulnerability

Despite using "an advanced front-facing camera and machine learning to create a three-dimensional map of your face," Apple's sophisticated Face ID technology contained unexpected security vulnerabilities. Vietnam-based security company Bkav discovered "that by adhering 2D 'eyes' to a 3D mask, they could effectively unlock an iPhone with Face ID."

This case demonstrates how even well-designed AI systems from leading technology companies can contain unforeseen vulnerabilities, emphasizing the need for rigorous security testing and human oversight of AI systems in real-world contexts.

Future-Proofing AI Value

Case Study: DoNotPay's Legal Bot Misrepresentation

DoNotPay, a self-described "robot lawyer," faced legal challenges and regulatory scrutiny when critics argued the service "oversimplifies complex legal processes and could mislead users," highlighting limitations in "the accuracy of AI-generated advice." The controversy ultimately led DoNotPay to adjust its offerings "to balance AI innovation with ethical responsibility and legal compliance."

This case illustrates the dangers of overpromising AI capabilities in high-stakes' domains like legal services and the importance of transparency about system limitations.

Case Study: IBM Watson for Oncology

The IBM Watson for Oncology case represents a $4 billion AI investment failure, highlighting the significant financial risks of inadequate governance. This example underscores how governance isn't just about compliance, it is fundamentally tied to value creation and protection.

WARNING SIGN: The governance failures with the highest costs often stem from ambitious AI projects lacking sufficient real-world testing – or promising capabilities beyond what the technology can reliably deliver.

ADDRESSING EMERGING ETHICAL QUESTIONS

As AI capabilities expand, they create novel ethical challenges that organizations must navigate.

The Ethics Evolution Framework

Map ethical considerations across three horizons:

1. **Current Capabilities:** Issues we already face with deployed systems
   - Bias and fairness in algorithmic decisions
   - Privacy implications of data use
   - Transparency of automated processes
2. **Emerging Capabilities:** Issues becoming relevant with new applications
   - Human agency alongside agentic AI
   - Attribution and ownership of AI-generated content
   - Appropriate boundaries for emotional simulation
3. **Future Capabilities:** Issues we should begin considering now
   - Human-competitive AI and labour market impacts
   - Long-term power concentration through AI infrastructure
   - Potential for autonomous systems with misaligned goals

This framework helps organizations stay ahead of ethical questions rather than scrambling to address them after deployment.

Ethical Stress Testing

Beyond compliance checkboxes, organizations need to systematically explore ethical boundaries:

- Regular red team exercises that probe for potential misuse
- Edge-case analysis that examines extreme but plausible scenarios
- Stakeholder impact assessments that consider unintended consequences
- External ethics advisory input that challenges institutional blind spots

Case Study: Ethical Pre-Mortems

A technology company I advised created "ethical pre-mortems" – sessions where cross-functional teams imagined future headlines detailing ethical failures of their AI systems, then worked backward to identify preventive measures they could take today.

REFLECTION QUESTIONS: What's the worst possible headline your organization could face related to your AI initiatives? What preventive measures could you implement to avoid it?

BALANCING INNOVATION AND STABILITY

Organizations have to find a balance between pushing boundaries and maintaining reliable operations, a challenge that is particularly acute with rapidly evolving AI capabilities.

Taking a proactive stance can include effective governance, structured approaches to experimentation, risk management for advanced AI, and frameworks for integrating new capabilities.

The Governance Spectrum Debate

A significant tension exists in AI governance approaches across industries. While some organizations, like the US Congress, implement strict bans on AI tools due to security concerns, others like Absa bank embrace AI copilots to transform business operations.

This dichotomy represents what David Treat and Michael Klein call "the governance spectrum," where organizations must determine whether their governance frameworks should enable or restrict AI adoption, a critical consideration for leaders navigating the future of AI value.

KEY INSIGHT: The most successful organizations don't position themselves at either extreme of the governance spectrum but develop contextual approaches that apply different levels of governance to different use cases based on risk and value potential.

Creating Space for Experimentation

Innovation requires room to explore but random experimentation tends to waste resources.
   Structured approaches include:

The Innovation Sandbox Model

This format enables the establishment of contained environments where emerging capabilities can be tested without disrupting core operations.

- Clearly defined boundaries for experimentation
- Dedicated resources separate from operational budgets
- Accelerated governance appropriate to controlled environments
- Explicit learning goals beyond performance metrics
- Structured paths for successful experiments to scale

Case Study: The Future Capabilities Lab

A retail organization I worked with allocated 15% of their AI budget to a "future capabilities lab" – a team with the freedom to experiment with emerging technologies against real business challenges, but in controlled environments separate from production systems.

The Opportunity Portfolio

Rather than viewing AI innovation as a single track, this approach serves to develop a balanced portfolio across opportunity horizons.

- **Horizon 1:** Improving existing processes and products (70% of resources)
- **Horizon 2:** Extending into adjacent capabilities (20% of resources)
- **Horizon 3:** Exploring transformative possibilities (10% of resources)

Future-Proofing AI Value

This approach ensures continuous innovation without sacrificing stability.

ACTION STEP: Audit your current AI investments and map them to these three horizons. Many organizations find they're overinvested in horizon 1 and underinvested in horizons 2 and 3.

Managing Risk in Advanced AI Applications

As AI capabilities become more powerful, traditional risk-management approaches may prove insufficient. More robust frameworks include:

Graduated Deployment Strategies

Rather than binary go/no-go decisions, implement progressive deployment stages:

1. **Simulation:** Testing in fully artificial environments
2. **Shadow Mode:** Running alongside but not affecting decisions
3. **Human-in-Loop:** Operating with explicit approval gates
4. **Guardian Mode:** Functioning autonomously but with oversight
5. **Autonomous Mode:** Operating independently with periodic review

Progression through these stages is determined by performance, reliability, and risk factors.

Resilience by Design

Build protective mechanisms directly into AI systems:

- Boundary enforcement that prevents operation outside tested parameters
- Graceful degradation paths that maintain basic functionality during failures
- Continuous monitoring that detects drift from expected behaviour
- Automatic safeguards that engage when anomalies are detected

Case Study: Safety Envelopes

A healthcare organization implemented "safety envelopes" around their clinical decision support AI. These operational boundaries – which were defined by medical experts and that the system could not exceed regardless of its internal logic – created multiple layers of protection.

Integrating Emerging Capabilities into Core Systems

The ultimate test of future-proofing is how effectively new capabilities can be integrated into existing operations.

The Integration Readiness Assessment

Before attempting to integrate new AI capabilities, assess readiness across five dimensions:

1. **Technical Compatibility:** Data formats, API standards, and system requirements
2. **Process Alignment:** Workflow integration points and operational impacts
3. **Skills Availability:** Team capabilities to implement and maintain
4. **Governance Readiness:** Oversight mechanisms appropriate to the technology
5. **Value Clarity:** Well-defined success metrics and expected outcomes

This structured assessment prevents integration failures that often doom otherwise promising technologies.

The Capability Onboarding Framework

Develop a systematic process for bringing new AI technologies into the organization:

- Initial capability evaluation against specific use cases
- Controlled pilot with clear learning objectives
- Success criteria for both technical and business dimensions
- Integration pathway with identified dependencies
- Scaling roadmap with resource requirements

This framework prevents both over-enthusiastic adoption of unproven technologies and resistance to valuable innovations.

THE NEXT HORIZON OF AI VALUE

Beyond the immediate future lies what promises to become a broader transformation that will reshape how organizations create and capture value.

In this fast-evolving landscape, it pays to take a proactive approach, with helpful considerations that include cross-industry value transformation, value creation through technology convergence, new business models enabled by AI, and human-AI collaboration models. Staying ahead of societal implications and opportunities is also key for success.

CROSS-INDUSTRY VALUE TRANSFORMATION

AI is beginning to blur traditional industry boundaries as core capabilities become transferable across sectors.

- Healthcare diagnostics approaches applied to manufacturing quality control
- Financial risk modelling techniques used in supply chain management
- Retail recommendation systems adapted for talent management
- Content generation capabilities deployed across all customer touchpoints

Organizations that view their AI capabilities as transferable assets rather than industry-specific solutions can find unexpected growth opportunities.

VALUE CREATION THROUGH TECHNOLOGY CONVERGENCE

The convergence of AI with other emerging technologies is creating entirely new value opportunities beyond what AI can deliver alone.

AI and Internet of Things (IoT)

Case Study: Smart Cities Integration

In research published in the journal Sensors by Alahi and colleagues, the authors showcase how AI-IoT integration transforms urban management. Their case study demonstrates how this convergence enhances urban efficiency by optimizing traffic flow through real-time monitoring and predictive analytics, reducing congestion by 30%. The fusion of AI with IoT also improved energy management by automatically monitoring consumption patterns and optimizing distribution, resulting in measurable sustainability improvements across test cities.

Case Study: PwC's Smart Cooler

PwC developed an innovative smart cooler prototype that transforms inventory management in the beverage industry. According to the research, this AI-IoT solution uses custom-built sensors and deep learning models to provide real-time stock analysis and inventory management. The system processes image data through a cloud-based API, enabling instant inventory assessment and reducing restocking times by 40%. This scalable solution demonstrates the practical business value of AI-IoT integration in retail operations.

AI and Blockchain

Case Study: Ocean Protocol's Data Privacy Solution

In 2025 research by Akhilesh Sharma on blockchain-AI integration, Ocean Protocol is highlighted as a groundbreaking example of enhancing data privacy in AI applications. The protocol enables data owners to maintain control over datasets while allowing AI developers to access critical information without compromising security. Financial institutions implementing this technology reported a 45% reduction in data breach risks while maintaining 90% of the analytical capabilities needed for AI model training.

Case Study: Microsoft Azure Blockchain Service

Akhilesh Sharma's research presents Microsoft Azure Blockchain Service as a real-world implementation enabling businesses to create ethical frameworks for AI projects. The case study details how Azure's blockchain solution provides immutable audit trails that verify dataset integrity and ethical sourcing practices throughout the AI development lifecycle. Organizations using this service have demonstrated 70% higher regulatory compliance rates and significantly improved stakeholder trust through transparent ethical practices.

AI and Spatial Computing

Case Study: Walt Disney Entertainment

Kitty Wheeler's 2024 analysis showcases how Disney leverages AI and mixed reality to gain a competitive advantage in the global entertainment industry. According to Disney's Entertainment Co-Chairman Alan Bergman, the company is implementing mixed-reality technologies to create immersive experiences that blur physical and digital worlds. This strategic pivot has resulted in new revenue streams, and engagement metrics 35% higher than traditional entertainment formats, demonstrating the commercial viability of AI-spatial computing integration.

Case Study: Mayo Clinic's Medical Training

Amelia Saeed's research emphasizes how Mayo Clinic implemented AI-powered AR headsets to enhance surgical precision and medical training. The case study details how surgeons using these systems reported 40% improved accuracy in complex procedures due to real-time anatomical visualizations. Medical students training with VR simulations achieved competency 30% faster than traditional methods, providing quantifiable ROI for healthcare institutions investing in spatial computing technologies.

Key Technology
Convergence Tensions

The Human Augmentation vs. Replacement Narrative

While 33% of companies in Kevin Namunwa's Kaspersky study expressed concerns about job displacement due to AI and emerging technologies, Scoble and Cronin in their 2024 article "Navigating the Future: The Fusion of AI, AR and VR" directly counter this fear. They argue that "AI integration with spatial computing will not replace human creativity but merely augment it," enabling creators to develop previously unattainable experiences.

PATTERN RECOGNITION: This tension between fear of replacement and potential for augmentation creates a compelling narrative about how organizations can approach AI integration to maximize human potential while mitigating concerns about displacement.

The Privacy-Utility Trade-Off

In the research on blockchain-AI integration by Nakhoon Choi and Heeyoul Kim, the researchers identify a fundamental tension between data utility and privacy protection. Their 2025 study reveals that while organizations need vast datasets to train effective AI models, blockchain's immutability creates significant privacy risks when storing sensitive information.

They propose solutions like Zero Knowledge Proof and homomorphic encryption – but acknowledge these come with computational costs. This tension illustrates the complex balancing act organizations must perform to create sustainable AI value while respecting privacy concerns.

NEW BUSINESS MODELS ENABLED BY AI

Beyond improving existing operations, AI enables fundamentally new approaches to value creation.

Outcome-Based Models

Shifting from providing products or services to guaranteeing outcomes:

- Manufacturing equipment vendors selling guaranteed uptime rather than machines
- Software companies offering business results rather than applications
- Professional services firms providing verified outcomes rather than recommendations

AI's predictive capabilities make these models viable by quantifying and managing the associated risks.

Data Network Effects

Creating value through continuously improving AI that benefits from growing usage.

- Products that become more valuable as their user base expands
- Services that improve as they process more interactions
- Platforms that create increasing returns to scale

These models often feature winner-take-most dynamics, placing a premium on early adoption.

QUICK WIN: Identify one product or service in your organization that could benefit from data network effects. Create a simple feedback loop that captures user interactions to continuously improve the offering.

Ecosystem Orchestration

Using AI to coordinate complex networks of participants:

- Matching supply and demand across fragmented marketplaces
- Optimizing resource allocation across partner networks
- Facilitating collaboration among specialized contributors

These models create value through reduction of friction and improved utilization.

Case Study: Mobility as a Service

A transportation company I advised transitioned from selling vehicle fleets to providing "mobility as a service" – a comprehensive offering that used AI to optimize vehicle allocation, routing, maintenance, and driver assignment. This shift doubled their margins while reducing their customers' transportation costs by 23%.

Human-AI Collaboration Models

The most valuable future systems will combine human and artificial intelligence in ways that amplify their respective strengths.

The Collaboration Spectrum

Different tasks require different collaboration models:

- **AI as Tool:** Systems that extend human capabilities under direct control
- **AI as Teammate:** Systems that work alongside humans with some autonomy
- **AI as Manager:** Systems that coordinate human activities
- **AI as Advisor:** Systems that provide input to human decision-makers
- **Human as Governor:** Humans who oversee autonomous AI systems

Organizations need to match collaboration models to specific contexts rather than defaulting to a single approach.

ACTION STEP: Map your current and planned AI applications to the collaboration spectrum. For each application, consider whether a different collaboration model might create more value or reduce risks.

Designing for Effective Collaboration

Creating productive human-AI partnerships requires deliberate design.

- Clear communication of AI capabilities and limitations
- Transparent sharing of confidence levels and uncertainty
- Efficient transfer of context between humans and systems
- Appropriate allocation of tasks based on relative strengths
- Learning mechanisms that improve collaboration over time

Case Study: AI Practice Rooms

A legal services firm implemented "AI practice rooms" – environments where attorneys could practice working with their new document analysis AI before using it with client materials. This approach reduced resistance and accelerated effective adoption by 68%.

Societal Implications and Opportunities

The broader impact of advanced AI will shape public perception, regulatory responses, and the overall environment in which organizations operate.

The Labour Market Transformation Debate

Research on AI's labour market impact presents a significant tension between different perspectives. McKinsey Global Institute's research indicates that "generative AI alone could automate almost 10% of tasks in the US economy," with lower-wage workers (earning less than $38,000) being "14 times more likely to lose your job or need to transition to another occupation than those with wages in the higher range, above $58,000."

In contrast, Dr. Anna Tavis of New York University argues that "jobs is the wrong unit of measure when we talk about jobs [that] are going away. I think every job will have a component that I would say will be happily automated, we all have been in jobs where we want to free ourselves up to do the real creative work."

This conceptual framing – between viewing AI as a job destroyer versus a job transformer – creates both challenges and opportunities for organizations looking to prepare their workforces for AI integration.

The Trust Imperative

As AI becomes more pervasive, trust becomes a critical success factor:

- Transparency about AI use and limitations
- Demonstrable responsibility for outcomes
- Visible commitment to ethical principles
- Meaningful engagement with stakeholders
- Proactive management of societal impacts

Organizations that build trust early will maintain greater freedom to operate as scrutiny increases.

Addressing Workforce Transformation

The impact on jobs and skills will be profound but nuanced:

- Some roles will be displaced while others are enhanced
- New job categories will emerge around AI capabilities
- Required skill sets will evolve across all functions
- Career paths will require more frequent reinvention

Organizations have an opportunity to shape this transformation through:

- Proactive reskilling and upskilling programs
- Transparent communication about changing roles
- Thoughtful redesign of affected positions
- Support for transitions between career paths

Case Study: The Next Chapter Program

A financial institution I worked with created a "next chapter program" – a comprehensive approach to helping employees whose roles were significantly changed by AI to find new positions either within the company or elsewhere, with educational support and transition assistance.

WARNING SIGN: Organizations that fail to proactively address workforce transformation risk not only losing valuable talent but also face resistance that can undermine AI adoption and value creation.

PRACTICAL FRAMEWORK: THE FUTURE-PROOF VALUE MATRIX

Now let's put all these insights into practice with a framework you can use to assess and enhance your organization's readiness for the evolving AI landscape.

How to Use the Matrix

This tool helps evaluate your AI initiatives across two critical dimensions:

1. **Value Durability:** How likely can value creation continue as AI capabilities evolve
2. **Adaptation Capacity:** How easily can the initiative incorporate new AI capabilities

Future-Proofing AI Value

Plotting your initiatives on this matrix reveals four categories:

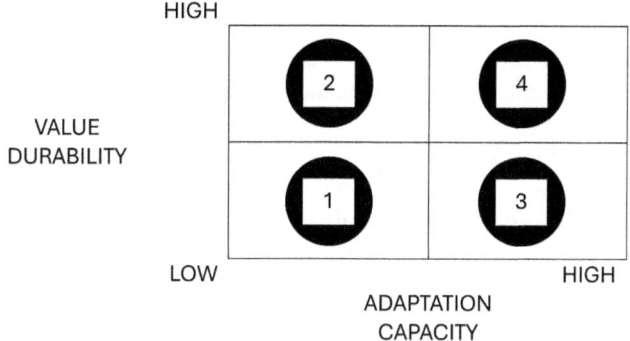

Quadrant 1: Ephemeral Opportunities (Low Durability, Low Adaptation)
- Create short-term value but will likely become obsolete
- **Example:** Implementations tied to specific model versions with rigid architectures
- **Strategy:** Extract maximum near-term value while limiting long-term investment

Quadrant 2: Rigid Foundations (High Durability, Low Adaptation)
- Create lasting value but struggle to evolve with new capabilities
- **Example:** Core process automation with inflexible workflows
- **Strategy:** Fortify value foundations while creating adaptation pathways

Quadrant 3: Flexible Experiments (Low Durability, High Adaptation)
- May not create lasting value but can easily evolve with technology
- **Example:** Innovative applications in rapidly changing domains
- **Strategy:** Use as learning vehicles and platforms for future capabilities

Quadrant 4: Future-Proof Investments (High Durability, High Adaptation)
- Create lasting value and can evolve with new capabilities
- **Example:** Modular systems addressing fundamental business needs
- **Strategy:** Prioritize for investment and scale

ACTION STEP: Applying the Future-Proof Value Matrix

1. List your current and planned AI initiatives
2. Rate each on value durability (1–10) and adaptation capacity (1–10)
3. Plot them on the matrix
4. Develop strategies for each quadrant:
   - **Quadrant 1:** Set clear time horizons and value extraction plans
   - **Quadrant 2:** Create modularization and adaptation pathways
   - **Quadrant 3:** Establish learning objectives and evolution criteria
   - **Quadrant 4:** Prioritize for investment and expansion

This matrix helps ensure your AI portfolio balances immediate value creation with long-term adaptability. You'll find a downloadable template for this assessment in our online resource centre.

PATTERN RECOGNITION: Most organizations find their AI initiatives clustered in quadrants 1 and 2, with few in the future-proof quadrant. Simply recognizing this pattern can drive more balanced portfolio decisions.

KEY TAKEAWAYS: FUTURE-PROOFING AI VALUE

Before we wrap up, let's recap the essential insights from this chapter:

1. **The Evolution Never Stops:** AI capabilities are evolving rapidly, from predictive to generative to agentic systems. Organizations must build for adaptability rather than trying to predict exact futures.
2. **Technical Foundations Matter:** Creating modular, API-first, cloud-native architectures enables rapid integration of new capabilities as they emerge.
3. **Organizational Flexibility Is Essential:** Balanced skills portfolios, adaptable decision processes, and deliberate innovation structures create the organizational capacity to absorb new technologies.
4. **Governance Must Evolve:** Principles-based governance, varying levels of human oversight, and ethical stress testing create sustainable approaches to managing risk in rapidly evolving systems.
5. **Technology Convergence Creates New Value:** The integration of AI with IoT, blockchain, and spatial computing enables entirely new business models and value creation approaches.
6. **Human-AI Collaboration Is the Future:** Finding the right collaboration model for each context maximizes value while managing risks.
7. The future-proof value matrix provides a practical way to assess and balance your AI portfolio for both immediate value and long-term adaptability.

REFLECTION QUESTIONS: Which of these seven areas represents your organization's greatest strength? Which represents your biggest gap or opportunity?

CONCLUSION: PERPETUAL READINESS

Future-proofing AI value isn't about reaching a final state of preparation. It's about creating perpetual readiness for continuous change.

The ability to thrive will not come from perfectly predicting AI's evolution – it will be the result of building technical foundations, organizational capabilities, and strategic mindsets to absorb and capitalize on whatever comes next.

As we've explored throughout this chapter, this readiness comes from:

- Understanding the evolving AI landscape
- Building adaptable technical and organizational foundations
- Anticipating ethical and regulatory developments
- Balancing innovation with stability
- Preparing for new value models and collaboration approaches

The practical frameworks provided here will help you assess your current readiness and build the capabilities needed for sustainable AI value.

NEXT STEPS:

1. Complete the future-proof value matrix assessment for your AI initiatives
2. Identify your biggest gap in adaptation capacity
3. Select one quick win from this chapter to implement in the next week
4. Share the key insights with your leadership team

14 The Data-to-Value Integration Blueprint

Chapter Roadmap

In the pages ahead, I share a five-phase integration blueprint that has helped dozens of organizations bridge the gap between AI potential and realized value. The blueprint covers: defining shared value goals, creating cross-functional alignment, establishing data foundations, capturing early wins while building for scale, and tips on how to measure, learn and evolve. This approach addresses the most common integration failures I've observed: misaligned objectives, siloed implementation, late discovery of data issues, technology-first thinking, failure to show early wins, and inability to scale beyond pilots.

THE INTEGRATION IMPERATIVE: WHERE MOST AI VALUE GETS LOST

After reviewing the organization's AI initiatives – which saw millions invested in data infrastructure, a team of talented data scientists hired, and several promising pilots launched – the healthcare CIO slumped back in his chair. He was visibly frustrated about the fact that the expected value remained maddeningly elusive.

"We have all the right pieces," he sighed, "but we're still not seeing results."

This moment captures a key challenge that derails countless AI initiatives. The missing element? Integration.

PATTERN RECOGNITION: Most organizations approach AI implementation as a collection of separate challenges – data quality, model development, technical infrastructure, process redesign, and change management – without recognizing that value emerges only when these elements work as a cohesive system.

The truth is that integration counts for more than excellence of individual elements. Even brilliantly executed components fail to deliver value when they're not woven into a cohesive whole.

This integration challenge occurs at multiple levels:

- **Technical Integration:** Connecting data, models, and business systems
- **Process Integration:** Embedding AI into operational workflows
- **Organizational Integration:** Aligning teams, incentives, and governance
- **Strategic Integration:** Connecting AI initiatives to business objectives

WARNING SIGN: Research by RAND Corporation reveals a troubling reality: while executives expect AI projects to "take weeks instead of months," technical teams struggle with "acquiring, cleaning, and exploring organization data." At the same time, leaders may "switch up their priorities often" before projects can demonstrate results.

This tension between expectation and implementation creates a value gap that can only be bridged through deliberate integration.

The Integration Success Story: Telstra's Strategic Approach

Telstra, Australia's largest telecommunications company, demonstrates the power of integration. As Kim Krogh Andersen, Telstra's group executive for product and technology, explains:

> We're looking at an agentic AI future, to completely reinvent our business and our processes, allowing our teams to work in an ecosystem of intelligent AI agents, optimizing key tasks end to end.

To realize this vision, Telstra went beyond implementing isolated AI technologies – to forming a strategic joint venture with Accenture to integrate AI across their operations. This partnership enabled them to streamline AI and data providers while accelerating their company's AI roadmap.

The lesson? Integration isn't just a technical challenge, it is a strategic imperative.

Let's dive into five key phases to discover how to avoid these pitfalls and create lasting AI value through integration.

PHASE 1: DEFINE SHARED VALUE GOALS

"We're building an AI recommendation engine." This was the answer I received when I asked a retail executive about the goals of the organization's AI project.

The response wasn't unexpected. After all, it described what they were building. However, some key elements were missing as I didn't learn why they were building it or how they would measure success.

KEY INSIGHT: Too many AI initiatives begin with a technology-focused goal rather than a value-focused objective. Without clear value targets, projects drift, resources are wasted, and results disappoint.

The AI Value Mapping Canvas in Action

The AI value mapping canvas (introduced in Chapter 4) represents your first integration tool.
Here's how to use it:

Step 1: Identify the business challenge

Begin with a clear articulation of the business problem or opportunity:

- "Customer churn in our premium segment has increased by 15% over the past year"
- "Supply chain disruptions cost us $12 million in lost sales last quarter"
- "New product development takes 18 months, putting us behind market trends"

Step 2: Define value dimensions

For each challenge, identify relevant value dimensions:

- **Financial:** Revenue growth, cost reduction, margin improvement
- **Customer:** Satisfaction, retention, lifetime value
- **Operational:** Efficiency, quality, speed, flexibility
- **Strategic:** Competitive advantage, market position, innovation
- **Employee:** Productivity, satisfaction, capability development
- **Risk:** Reduction in compliance, security, or operational risks

The Data-to-Value Integration Blueprint

ACTION STEP: For each value dimension, specify both a quantitative target and qualitative outcome you seek. This creates a multidimensional view of success that goes beyond simple financial metrics.

Step 3: Set specific, measurable targets

Transform general value dimensions into specific metrics and targets:

- "Improve customer retention"
- "Increase premium segment retention from 82% to 90% within 12 months"
- "Reduce costs"
- "Significantly decrease inventory holding costs while maintaining fulfilment rate"
- "Speed up processes"
- "Reduce new product development cycle time from 18 months to 12 months by Q4"

Step 4: Connect AI capabilities to value targets

Identify how specific AI capabilities will contribute to each value target:

- "Predictive churn models will identify at-risk customers for proactive intervention"
- "Supply chain optimization algorithms will reduce inventory while maintaining service levels"
- "Natural language processing of customer feedback will accelerate product innovation"

Step 5: Validate alignment across stakeholders

Before proceeding, ensure all key stakeholders agree on:

- The priority of each value dimension
- The specific metrics and targets
- The feasibility of the AI approach
- The timeline for value realization
- The resources required for success

REFLECTION QUESTIONS: Does your current AI initiative have clear value targets that all stakeholders understand and support? If not, which steps of the value mapping canvas would add the most clarity?

BUILDING A VALUE REALIZATION ROADMAP

With clear value goals established, create a roadmap showing how value will be realized over time:
The value staging approach:

1. **Quick Wins:** Value realized within 3–6 months
2. **Mid-Term Gains:** Value realized within 6–12 months
3. **Long-Term Transformation:** Value realized in 12+ months

For each stage, specify:

- Value metrics and targets
- Key dependencies and prerequisites
- Resource requirements

- Critical milestones
- Responsible teams

CASE STUDY: Healthcare Patient Experience Transformation

A healthcare organization developed a value roadmap for their patient experience AI initiative with three distinct horizons:

- **3-Month Horizon:** 20% reduction in appointment scheduling time
- **9-Month Horizon:** 30% improvement in preventive care compliance
- **18-Month Horizon:** 15% reduction in readmissions through better post-discharge support

This phased approach allowed them to demonstrate value quickly while working toward more substantial outcomes that required deeper integration.

QUICK WIN: Start with a 90-minute workshop using the AI value mapping canvas. Bring together business leaders, technical teams, and end users to align on value targets before diving into technical details.

Phase 1 Summary: Define Shared Value Goals

- Identify specific business challenges, not technical solutions
- Map value across multiple dimensions (financial, customer, operational, strategic)
- Set clear, measurable targets with timeframes
- Connect AI capabilities directly to value outcomes
- Create a phased roadmap for value realization
- Validate alignment across all key stakeholders

Next, we'll explore how to build the cross-functional collaboration needed to deliver on these value goals.

PHASE 2: CREATE CROSS-FUNCTIONAL ALIGNMENT

Even with clear value targets, AI initiatives often stall because responsibility is siloed within a single department, typically IT or data science.

WARNING SIGN: When you hear statements like "that's an IT project" or "the data science team is handling that," it signals dangerous fragmentation that will limit value.

THE CROSS-FUNCTIONAL IMPERATIVE: WEX's INTEGRATION SUCCESS

At financial technology company WEX, Chief Digital Officer Karen Stroup implemented a cross-functional approach to AI integration that involved close collaboration between the CTO and other executive team members.

Her perspective is revealing: "It is incredibly important for it to be a shared goal, because we win together or we lose together, and if I'm out there trying to advocate for AI but none of my peers think that it's important and they're not sold on it, I'm probably not going to be successful." Stroup emphasizes that success came from ensuring that "all [members of the executive leadership team have] shared goals around AI," creating alignment across the organization from the CEO level down through cross-functional teams.

KEY INSIGHT: Effective AI implementation requires collaboration across multiple functions: business units (domain expertise), data science (analytical capabilities), IT (infrastructure), operations (process design), HR (talent development), finance (investment prioritization), and legal/compliance (risk management).

The Data-to-Value Integration Blueprint

THE NORWEGIAN GOVERNMENT'S FIVE-STEP ALIGNMENT PROCESS

The Norwegian government's implementation of chatbots for citizen services demonstrates a methodical cross-functional approach.

A five-step process, documented by Abonamah and Abdelhamid in Discover Artificial Intelligence, enabled the transformation of their citizen services:

1. Form diverse teams with both technical and domain expertise
2. Conduct collaborative workshops to align capabilities with needs
3. Establish open communication channels for ongoing engagement
4. Train customer service representatives on AI collaboration
5. Conduct thorough post-implementation reviews to capture learnings

ACTION STEP: Assess which of these five steps is weakest in your organization's approach to AI implementation, then create a specific plan to strengthen it.

The results speak for themselves: According to the research, organizations that employed cross-functional implementation teams for AI projects were 3.2 times more likely to achieve successful deployment compared to organizations using siloed approaches.

IMPLEMENTING THE STAKEHOLDER VALUE ALIGNMENT MATRIX

The Stakeholder value alignment matrix (introduced in Chapter 5) provides a key tool for creating cross-functional alignment. Here's how to implement it:

Step 1: Identify all relevant stakeholder groups

Go beyond the obvious participants to include:

- End users who will interact with the system
- Teams whose processes will be affected
- Departments that provide necessary data
- Groups responsible for ongoing support
- Leaders who will evaluate success

Step 2: map priorities and concerns

For each stakeholder group, document:

- Primary objectives and priorities
- Specific contributions to the initiative
- Resources they control that are needed
- Potential concerns or resistance points
- How they define and measure success

Step 3: Identify points of alignment and conflict

Analyse the completed matrix to find:

- Shared priorities that can form the basis for collaboration
- Complementary objectives where stakeholders can help each other
- Potential conflicts that need to be addressed

- Resource dependencies that require coordination
- Misaligned success metrics that could create problems

Step 4: Develop specific alignment strategies

For each point of conflict or misalignment, create a specific strategy, such as:

- Redesigning incentives to reward collaboration
- Creating shared metrics across departments
- Establishing formal coordination mechanisms
- Developing compromise solutions that address multiple needs
- Escalating true conflicts for executive resolution

CASE STUDY: Financial Services Risk vs. Experience Conflict

Through this process, a financial services company discovered that their risk management team and customer experience team had fundamentally different definitions of success for an AI-powered credit approval system:

- Risk team success metrics: reduced default rates, improved fraud detection, regulatory compliance
- Customer experience team success metrics: faster approvals, higher approval rates, simplified process

By identifying this conflict early and creating shared metrics that balanced risk control and customer satisfaction, they avoided the implementation deadlock that had derailed previous initiatives.

REFLECTION QUESTIONS: Which stakeholders in your organization might have conflicting definitions of AI success? And how could you create shared metrics that all groups can agree on?

CREATING CROSS-FUNCTIONAL GOVERNANCE

Alignment must be supported by formal governance structures that enable ongoing collaboration.

Effective AI Governance Models

1. The integration team
   A dedicated cross-functional team with representatives from:
   - Business units (process owners)
   - Data science/analytics
   - IT/technical infrastructure
   - Operations
   - Compliance/risk management

   This team meets weekly to address integration issues, coordinate activities, and ensure progress toward value goals.
2. The value steering committee
   Senior leaders from key stakeholder groups who meet monthly to:
   - Review progress against value targets
   - Address strategic alignment issues
   - Resolve cross-functional conflicts
   - Allocate resources based on value potential
   - Make critical go/no-go decisions

The Data-to-Value Integration Blueprint

3. Working-level coordination
   Day-to-day coordination through:
   - Shared project management tools
   - Joint planning sessions
   - Regular cross-functional reviews
   - Create opportunities to flag urgent challenges
   - Embedded team members across functions
   - Shared physical or virtual workspaces

CASE STUDY: Manufacturing "Value Pods"

A manufacturing company created "value pods" – small cross-functional teams with representatives from operations, data science, IT, and finance who were physically co-located and collectively responsible for delivering specific value outcomes. This approach reduced handoff delays by 60% and significantly increased implementation success rates.

QUICK WIN: Create a simple RACI matrix (responsible, accountable, consulted, informed) for your AI initiative that clearly defines cross-functional roles and responsibilities.

Building Shared Accountability for Outcomes

Governance structures must be supported by shared accountability for results.
Accountability Mechanisms:

- Joint KPIs shared across departments
- Cross-functional incentives tied to overall success
- Regular value reviews with all stakeholders
- Transparent progress tracking accessible to all
- Collective problem-solving sessions when obstacles arise

CASE STUDY: Retail "Value Pairs"

A retail organization assigned "value pairs" – executives from both business and technology functions who were jointly accountable for specific value outcomes. These pairs reported together on progress, making it impossible to blame each other's departments for shortfalls.

As Sarah Fury, marketing leader at Pantheon, described in the "Speak in Flow" podcast, this type of effort can create "a common alignment on how we can manage projects better ... but also a vocabulary that we were all using. That was the same. So, when we talked about project management, we were using this language, we were all using the same language."

Phase 2 Summary: Create Cross-Functional Alignment

- Map all stakeholders involved in or affected by the AI initiative
- Identify shared priorities and potential conflicts
- Create governance structures for ongoing collaboration
- Implement shared accountability mechanisms
- Develop a common language across functions
- Build formal and informal coordination channels

With strong cross-functional alignment in place, the next critical step is ensuring you have the data foundation needed to support your AI value goals.

PHASE 3: ESTABLISH DATA FOUNDATIONS

Many organizations discover too late that they lack the data quality, accessibility, or completeness needed to achieve their AI objectives.

KEY INSIGHT: Data quality is the ceiling on AI value: no AI solution can deliver more value than its underlying data allows.

This reality is captured perfectly by Carol Reiley, an AI industry expert: "I think data labelling is undervalued. Most of the time, you give it as an interim project, but it becomes one of the most crucial processes because if your data is mislabelled or inconsistent, then no matter how awesome your model is, you're never going to get there."

THE DATA REALITY GAP: WHY ORGANIZATIONS GET THIS WRONG

There's a pervasive conundrum in AI implementation: while organizations rush to implement sophisticated AI models, they often underinvest in the foundational data work that determines success. This disconnect leads to costly rework, disappointed stakeholders, and abandoned initiatives.

According to Deloitte's 2024 report on data readiness, organizations must follow a structured five-step process to improve data readiness:

1. Analyse current results and identify data gaps
2. Determine risk tolerance for data quality issues
3. Conduct go/no-go workshops to make informed decisions
4. Develop improvement plans with clear ownership
5. Communicate progress transparently to stakeholders

ASSESSING DATA READINESS FOR AI VALUE

Before significant investment in model development, conduct a thorough data readiness assessment:

The Data Value Chain Assessment

1. **Data Existence:** Do we have the data needed to support our value goals?
2. **Data Access:** Can we retrieve and use this data when and where needed?
3. **Data Quality:** Is the data accurate, complete, and consistent?
4. **Data Integration:** Can we connect data across relevant sources?
5. **Data Governance:** Do we have appropriate controls and permissions?
6. **Data Infrastructure:** Can our systems support the required scale and performance?

For each element, rate your readiness on a scale from 1 (major gaps) to 5 (fully ready).

ACTION STEP: Use the Data value chain assessment to evaluate your current data readiness. For any element scoring below 3, develop a specific improvement plan before proceeding with model development.

CASE STUDY: Healthcare Risk Prediction Gap

A healthcare provider conducted this assessment for a patient risk prediction initiative and discovered that while they had extensive clinical data, they lacked critical "social determinants of health" data that research showed was essential for accurate predictions. This early discovery allowed them to establish data collection mechanisms before investing in model development.

The Data-to-Value Integration Blueprint

WARNING SIGN: If your team can't easily answer basic questions about data completeness, quality, and accessibility, it's a clear signal that you need to strengthen your data foundation before proceeding with sophisticated AI development.

Addressing Critical Data Gaps and Quality Issues

Based on your assessment, develop a targeted plan to address the most critical gaps:

Prioritization Framework

1. **Impact:** How essential is this data to achieving value goals?
2. **Effort:** How difficult will it be to address the gap?
3. **Timeliness:** When will this data be needed in the project timeline?

For high-impact, low-effort gaps, implement immediate solutions:

- Improving data capture processes
- Enhancing data quality validation
- Addressing system integration issues
- Implementing data governance controls
- Upgrading infrastructure for performance

For high-impact, high-effort gaps, develop parallel workstreams:

- Begin model development with available data
- Simultaneously address foundational data issues
- Create staged implementation plans based on data availability
- Consider alternative approaches that require different data

CASE STUDY: Financial Services Progressive Implementation

A financial services company discovered their customer data was fragmented across seven systems with inconsistent identifiers. Rather than delaying their personalization initiative, they created a progressive implementation plan:

- **Phase 1:** Use data from the two most complete systems (covering 60% of customers)
- **Phase 2:** Add integration with three additional systems (reaching 85% coverage)
- **Phase 3:** Complete integration with all systems (achieving 100% coverage)

This approach allowed them to demonstrate value quickly while methodically addressing the underlying data challenges.

CASE STUDY: Siemens' Predictive Maintenance Success

Siemens' approach to data foundations was equally methodical. Their predictive maintenance implementation transformed maintenance operations. It did this by deploying AI models that analyse sensor data, from machinery to predict equipment failures, before they occur.

Using multimodal frameworks processes data from various sources – including vibration, temperature, and acoustic signals – enabled a holistic view of equipment health. This comprehensive approach reduced maintenance costs by 20% and increased production uptime by 15%.

QUICK WIN: Identify one critical data quality issue affecting your AI initiative and implement a targeted improvement process within the next 30 days.

BUILDING DATA GOVERNANCE FOR AI SUCCESS

Effective AI requires appropriate data governance that balances accessibility with control:

AI-ready data governance model

- Clear data ownership with defined responsibilities
- Documented data quality standards for critical elements
- Appropriate access controls based on data sensitivity
- Transparent data lineage tracking data sources and transformations
- Consistent metadata describing data characteristics
- Automated monitoring of data quality and completeness
- Defined processes for addressing data issues

CASE STUDY: Retail "AI Data Product Teams"

A retail organization developed "AI data product teams" – groups responsible for creating and maintaining high-quality data assets specifically designed for AI use cases. These teams operated under a product management model, treating data as a product with defined users, requirements, and quality standards.

PATTERN RECOGNITION: Organizations that treat data as a product rather than a byproduct consistently achieve higher AI success rates.

CREATING SUSTAINABLE DATA CAPABILITIES

Beyond addressing immediate gaps, it's important to build lasting capabilities for ongoing data management.

Organizational Capabilities

- Data literacy programs to improve understanding across functions
- Data quality frameworks embedded in operational processes
- Data stewardship roles within business departments
- Technical debt reduction programs for legacy systems
- Data cataloguing and discovery tools for better accessibility
- Self-service capabilities for common data needs

CASE STUDY: Manufacturing "Data Excellence" Centre

A manufacturing company established a "data excellence" centre of expertise that worked across departments to enhance data capabilities. This centre didn't own data. Instead, it provided training, tools, and support to help business units improve their own data practices. This approach led to a 45% improvement in data quality within 18 months.

CASE STUDY: Gaming Company's Vision-Obstacle-Value Framework

In *Making Data Work*, I describe how a gaming company applied the "vision-obstacle-value" (or VoV) framework to align organizational efforts and overcome challenges. In this particular organizations, executives had a clear vision ("become the Google or the Facebook of the gaming industry") and value propositions ("40–50% uplift on products within a 12-month window"), but identified obstacles including "a lack of leadership, from a data point of view" and systems with "complexity and inflexibility." The resulting success story illustrates the power of a coordinated approach.

REFLECTION QUESTIONS: What obstacles stand between your organization's data vision and the value you seek to create? How could you apply the VoV framework to address them?

The Data-to-Value Integration Blueprint

Phase 3 Summary: Establish Data Foundations

- Assess data readiness across six key dimensions
- Prioritize data gaps based on impact, effort, and timeliness
- Implement progressive improvement plans
- Create appropriate governance structures
- Build sustainable data capabilities across the organization
- Apply the vision-obstacle-value framework to overcome data challenges

With solid data foundations in place, it's time to focus on execution, including balancing quick wins with building for scale.

PHASE 4: CAPTURE EARLY WINS WHILE BUILDING FOR SCALE

Many AI initiatives deliver promising pilots but fail to scale to enterprise-wide impact. This "scaling gap" is one of the most persistent challenges in AI implementation.

THE PIPELINE VS. PLATFORM TENSION

When it comes to scaling AI, organizations face a fundamental choice: Should they implement a pipeline approach: creating distinct AI tools for different functional areas? Or should they choose a platform approach: building core AI capabilities that transform the entire business?

As Philippe Rambach, Chief AI Officer at Schneider Electric, observes: "Many people actually have two processes ... one team to go from idea to proof of concept or minimum viable product, and then another team will industrialize. Guess what? Industrialization never happens."

Instead, Rambach advocates for "the team that has the mission to deliver the solution at scale, not to deliver a proof of concept or an MVP."

This represents a profound tension in organizational approaches. While Tim Fountaine of McKinsey advocates for "moving from siloed functional work to cross-functional teams where people from the business, people from analytics, IT, operations all work side by side," many companies maintain separation between initial development and scaling teams, preventing AI solutions from moving beyond the MVP stage.

KEY INSIGHT: Successful AI implementations balance immediate value creation with building the foundations for scale from day one.

IDENTIFYING AND PRIORITIZING QUICK-WIN OPPORTUNITIES

Start by identifying opportunities that can demonstrate value rapidly:

Quick-Win Criteria

- **Feasibility:** Can be implemented with available data and resources
- **Timeline:** Can deliver measurable results within 3–6 months
- **Visibility:** Impacts processes or outcomes that stakeholders care about
- **Learning Value:** Provides insights for larger-scale initiatives
- **Strategic Alignment:** Connects to broader value goals

ACTION STEP: Create a portfolio of quick-win opportunities balanced across different value dimensions: financial impact, customer experience, operational efficiency, employee productivity, and risk reduction.

CASE STUDY: Telecommunications Triple-Win Approach

A telecommunications company identified three quick-win opportunities for their customer experience AI initiative:

1. Reducing call handling time through AI-assisted agent responses
2. Improving first-call resolution through better routing
3. Increasing self-service adoption through enhanced digital assistance

Each initiative delivered measurable results within four months, creating momentum for the broader transformation.

CASE STUDY: CP AXTRA's Strategic Early Adopters

CP AXTRA in Thailand provides another excellent example of strategic quick wins. According to Microsoft's 2024 case study, CP AXTRA became Thailand's first retailer to employ Copilot with Microsoft 365 by implementing a phased adoption strategy. The organization selected 300 strategic early adopters across marketing, finance, and strategy departments. As Shaun Wong, Group Chief Transformation Officer at CP AXTRA, explained: "Our goal wasn't just to change behaviour but also to reshape mindsets. It was important to find the right change leaders who could successfully lead the charge."

QUICK WIN: Identify an AI capability that could be implemented within 90 days and would impact a highly visible, high-priority business process.

BUILDING MOMENTUM THROUGH VISIBLE SUCCESS

Quick wins must be communicated effectively to build organizational confidence.

Success Amplification Strategies

- Value spotlight sessions that showcase early results
- Executive walkthrough demonstrations of working capabilities
- User testimonials from early adopters
- Before-and-after metrics highlighting improvements
- Success story communication through internal channels

CASE STUDY: Healthcare "Impact Clinics"

A healthcare organization created "impact clinics" – interactive sessions where users could experience new AI capabilities firsthand and see the results compared to previous approaches. These sessions converted sceptics into advocates and generated demand from other departments.

CASE STUDY: Australian Government's Copilot Trial

The Australian Government's 6-month trial of Microsoft 365 Copilot provides an excellent example of building momentum through visible success. The structured approach involved distributing Copilot licenses across various agencies and establishing a systematic evaluation framework. Key findings detailed in the Australian Government's trial of Microsoft 365 Copilot Executive Summary Report revealed measurable productivity impacts, with participants reporting time savings of up to an hour per day on specific tasks. More significantly, 40% of users successfully reallocated time to higher-value strategic planning and stakeholder engagement activities, demonstrating tangible value creation through proper data-to-AI integration.

PATTERN RECOGNITION: Organizations that methodically document and communicate early wins create a virtuous cycle of support, funding, and enthusiasm for their AI initiatives.

The Data-to-Value Integration Blueprint

CREATING TECHNICAL AND ORGANIZATIONAL FOUNDATIONS FOR SCALE

Capturing quick wins can help garner appreciation and support that allows organizations to simultaneously build the foundations needed for enterprise-wide impact.

Technical Scaling Foundations

- Modular architecture that can expand to new use cases
- Standardized data pipelines for consistent processing
- Reusable component libraries to accelerate development
- Automated testing frameworks to ensure reliability
- Scalable infrastructure that grows with demand
- Monitoring and observability for operational management

Organizational Scaling Foundations

- Skills-development programs to build required capabilities
- Change-management frameworks to support adoption
- Process-redesign methodologies to fully leverage AI
- Centre-of-excellence models to share best practices
- Knowledge-management systems to capture learnings

CASE STUDY: Financial Services "AI Factory"

A financial services firm created an "AI factory" – a combination of technical platforms, organizational capabilities, and standardized methodologies that allowed them to deploy new AI use cases 4-times faster than their previous approach.

CASE STUDY: Siemens' Industrial Copilot Ecosystem

Siemens' approach to industrial copilots demonstrates an effective technical foundation for scale. As described in the "AI Spectrum" podcast hosted by Spencer Acain, Alessia Bortolotti, Customer Discovery Manager at Siemens AG, explains how Siemens developed five distinct copilots to address specific phases: design, planning, engineering, operations, and service.

The engineering copilot exemplifies data-to-value integration by enabling automation engineers to "easily and quickly generate, optimize, debug, and document PLC code," freeing expert engineers to focus on higher-value tasks.

WARNING SIGN: If your technical architecture requires significant rework to support each new use case, you're building for pilots rather than scale.

USING SUCCESS TO FUND FUTURE PHASES

Strategic use of early value creation can help secure resources for broader transformation.

The Value Reinvestment Model

1. Quantify and communicate the value of early wins
2. Dedicate a portion of realized value to fund further initiatives
3. Create a transparent process for allocating these funds
4. Prioritize investments that build scaling capabilities
5. Track and communicate compounding returns

CASE STUDY: Retail "Value Capture Fund"

A retailer implemented a "value capture fund" that allocated 30% of documented savings from AI initiatives to fund additional AI projects. This approach created a virtuous cycle of value creation that grew their AI portfolio from three initial projects to over 20 within 18 months.

CASE STUDY: Gaming Company's Proof of Value

The gaming company mentioned above provides a perfect illustration of this approach. An initial small-scale experiment yielded 13% growth in just 2 weeks, providing evidence for larger-scale transformation and building confidence in their vision of achieving 40–50% uplift on products within a 12-month window.

REFLECTION QUESTIONS: How could you design a small-scale experiment to demonstrate the value potential of your AI initiative within 4–6 weeks?

Phase 4 Summary: Capture Early Wins While Building for Scale

- Identify opportunities that can demonstrate value quickly
- Create a balanced portfolio of quick-win initiatives
- Use success amplification strategies to build momentum
- Develop technical and organizational foundations for scale
- Implement a value reinvestment model to fund future growth
- Bridge the gap between pilot and enterprise-wide implementation

With early wins creating momentum and scaling foundations in place, the final phase focuses on creating the feedback loops needed for continuous improvement.

PHASE 5: MEASURE, LEARN, AND EVOLVE

Many organizations treat AI as a one-time implementation rather than a capability that must continuously evolve. This "evolution gap" tends to limit long-term value creation.

KEY INSIGHT: Organizations that create the most value from AI are those that build systematic feedback loops to drive continuous improvement.

IMPLEMENTING THE AI VALUE DASHBOARD

Effective measurement is essential for guiding evolution. The AI value dashboard (introduced in Chapter 11) provides a comprehensive framework:

Dashboard Implementation Steps

1. Define audience-specific views for different stakeholders
2. Select relevant metrics across value dimensions
3. Establish baseline measurements before implementation
4. Implement automated data collection where possible
5. Create visualization formats that highlight key insights
6. Set regular review cadences for different audiences

Core Dashboard Components

- Value realization tracking against targets
- Leading indicators of future value
- Implementation progress against milestones
- Adoption metrics showing usage and engagement
- Technical performance indicators for AI systems
- Cost and resource utilization measures

The Data-to-Value Integration Blueprint

ACTION STEP: Design an AI value dashboard with three distinct views: an executive summary focusing on business impact, a departmental view highlighting operational metrics, and a technical view showing system performance.

CASE STUDY: Manufacturing Multilevel Dashboard

A manufacturing company developed a multilevel dashboard with daily operational metrics for front-line managers, weekly performance trends for department heads, and monthly value realization summaries for executives. This approach ensured that every level of the organization had the information needed to drive continuous improvement.

WARNING SIGN: If your organization celebrates technical deployment as "success" without measuring actual business impact, you're likely missing substantial value.

CREATING FEEDBACK LOOPS FOR CONTINUOUS IMPROVEMENT

Beyond measurement, it is crucial to establish processes for translating insights into action.

Feedback Loop Framework

1. Regular review sessions with key stakeholders
2. Structured problem-solving processes for addressing issues
3. User feedback channels for frontline insights
4. Performance improvement teams addressing specific metrics
5. Cross-functional optimization forums for system-level improvements

PATTERN RECOGNITION: Organizations that establish multiple feedback loops – technical, operational, and strategic – identify improvement opportunities more quickly and respond more effectively.

CASE STUDY: Healthcare "AI Value Rounds"

A healthcare provider implemented "AI value rounds" – weekly sessions where clinical, technical, and administrative teams reviewed performance data from the patient risk prediction system and identified improvement opportunities. These sessions followed a structured format focused on specific metrics and action planning. Similar to how medical teams conduct patient rounds to improve care, these AI value rounds became a critical ritual for continuous improvement. The cross-functional nature of these sessions ensured that technical optimizations remained connected to clinical and operational realities.

CASE STUDY: CP AXTRA's Prompt Engineering Challenge

On implementing Microsoft Copilot, CP AXTRA discovered that technical deployment was only half the battle. As Shaun Wong, Group Chief Transformation Officer at CP AXTRA, noted: "[Team members] also had to be good at creating the right prompts that would give us the answers we needed, which [isn't always easy]. It's a challenge, but we try to make it a fun challenge." By framing prompt engineering as a "fun challenge" rather than a frustration, CP AXTRA created a positive culture of experimentation and learning that accelerated adoption.

QUICK WIN: Create a simple feedback mechanism for AI users to report both successes and challenges. Review these insights weekly to identify patterns and improvement opportunities.

ADJUSTING COURSE BASED ON VALUE SIGNALS

The ability to adapt based on value signals is critical for long-term success.

Course Correction Framework

1. Early warning indicators that identify potential issues
2. Value gap analysis to quantify shortfalls against targets
3. Root cause assessment to identify underlying reasons
4. Adjustment options development with stakeholder input
5. Decision criteria for selecting appropriate responses

Common Adjustment Types

- **Scope Refinement:** Focusing on higher-value use cases
- **Process Redesign:** Improving AI integration into workflows
- **Model Enhancement:** Upgrading capabilities to address limitations
- **Data Quality Improvements:** Addressing underlying issues
- **User Experience Refinements:** Improving adoption and utilization
- **Resource Reallocation:** Shifting investment to higher-value areas

CASE STUDY: Financial Services Risk Prediction Pivot

A financial services company had initially focused their risk prediction AI on identifying high-risk customers. Value tracking assessments revealed that the greatest opportunity was in paying attention to the "medium-risk-becoming-high-risk" segment. This insight – which emerged through systematic value tracking rather than technical assessment – led to an increase in value delivery by 45% through refocused models and intervention strategies.

REFLECTION QUESTIONS: What mechanisms do you have in place to detect when your AI initiative is delivering less value than expected? How quickly can you identify root causes and adjust?

BUILDING LASTING CAPABILITIES FOR SUSTAINED SUCCESS

The ultimate goal is creating self-sustaining capabilities that continue to generate value.

Sustained Value Capabilities

- Continuous learning culture that encourages experimentation
- Knowledge management systems that preserve insights
- Skills development programs that build needed capabilities
- Innovation processes that identify new opportunities
- Governance models that evolve with changing needs
- Technology lifecycle management for ongoing relevance

CASE STUDY: Retail "AI Value Council"

A retail organization created an "AI value council" composed of business, technology, and data science leaders who met quarterly to review value realization, assess new opportunities, and allocate resources based on demonstrated impact. This institutionalized approach ensured sustained focus long after the initial excitement had faded. By embedding value review in the organization's governance structure, they avoided the common pattern of enthusiasm followed by neglect.

CASE STUDY: Makro Wholesale's Transformation Approach

As Tirayu Songvetkasem, Group Chief Digital Officer of Makro Wholesale Business, explains:

> It isn't just about using automation to minimize redundant tasks but changing the way we work and truly transforming how the business operates at its core.

The Data-to-Value Integration Blueprint

This perspective highlights the distinction between tactical implementation and transformational change – the difference between incremental improvement and sustained value creation.

KEY INSIGHT: What differentiates successful organizations is that they view AI not as a technology project but as a business transformation opportunity enabled by technology.

Phase 5 Summary: Measure, Learn, and Evolve

- Implement value measurement across multiple dimensions
- Create systematic feedback loops for continuous improvement
- Develop mechanisms to detect and address value gaps
- Build a culture of experimentation and learning
- Establish governance structures for ongoing value optimization
- Shift from implementation mindset to transformation mindset

With all five phases of the data-to-value integration blueprint now explained, let's explore how to customize this approach for your specific organizational context.

IMPLEMENTATION GUIDE AND TOOLKITS

The five-phase blueprint provides a structured approach to AI-value integration, but effective implementation requires customization for your specific organizational context.

The implementation path will vary based on your organization's context and maturity.

Maturity-Based Customization

For AI beginners:
- Focus heavily on phase 1 (define shared value goals) and phase 3 (establish data foundations)
- Select narrowly scoped quick wins with high probability of success
- Invest in basic skills development and change management
- Establish simple measurement frameworks focused on early learning

For intermediate organizations:
- Balance across all five phases with emphasis on phase 2 (cross-functional alignment)
- Develop a portfolio of initiatives with varying time horizons
- Build reusable technical components that enable faster scaling
- Implement more comprehensive value measurement approaches

For advanced organizations:
- Focus on phase 4 (building for scale) and phase 5 (measure, learn, and evolve)
- Create enterprise-wide integration approaches across multiple initiatives
- Develop sophisticated feedback loops for continuous optimization
- Build innovation processes that identify emerging value opportunities

ACTION STEP: Assess your organization's AI maturity honestly, then prioritize the phases that will add the most value in your current context.

Industry-Based Customization

Financial services:
- Emphasize governance and risk management aspects
- Incorporate regulatory compliance requirements
- Focus on both operational efficiency and customer experience
- Address data privacy and security concerns

Healthcare:
- Balance clinical outcomes with operational efficiency
- Address unique data integration challenges across care settings
- Incorporate clinical workflow considerations
- Manage change carefully in high-stakes environments

Manufacturing:
- Focus on operational metrics and efficiency gains
- Integrate with existing operational technology systems
- Address real-time data processing requirements
- Connect AI initiatives to quality and productivity metrics

Retail:
- Emphasize customer experience and personalization
- Balance centralized and store-level implementation
- Address seasonal variations in business patterns
- Connect to omnichannel customer journeys

WARNING SIGN: Beware of generic AI implementation approaches that don't address your industry's specific challenges and value opportunities.

Resources, Skills, and Capabilities Required

Successful implementation requires a combination of resources, skills, and capabilities.

Core Resource Requirements

- Executive sponsorship with sufficient authority
- Cross-functional team with dedicated time allocation
- Technical infrastructure appropriate to AI workloads
- Data access and integration capabilities
- Implementation budget for technology and services
- Change management resources for adoption support

Critical Skill Areas

- Business domain expertise in relevant processes
- Data science and analytics capabilities
- Data engineering and architecture skills
- Process design and improvement expertise
- Project and program management experience
- Change management and communication abilities

Organizational Capabilities

- Cross-functional collaboration mechanisms
- Value-measurement frameworks
- Data-governance processes
- Agile implementation methodologies
- Innovation-management approaches
- Knowledge-sharing systems

PATTERN RECOGNITION: Organizations that build balanced teams with both technical and business skills achieve higher AI value realization rates than those that over-index on technical capabilities alone.

The Data-to-Value Integration Blueprint

CASE STUDY: Pharmaceutical Capability Development Roadmap
A pharmaceutical company created a capability development roadmap that identified the specific skills, resources, and organizational capabilities needed for each phase of the AI transformation. This roadmap guided hiring, training, and organizational development efforts over a 2-year period. By mapping required capabilities to the implementation timeline, they avoided the common pitfall of building capabilities too early (wasting resources) or too late (delaying value).

COMMON CHALLENGES AND MITIGATION STRATEGIES

Based on my experience with dozens of organizations, here are the most common implementation challenges and effective mitigation strategies:

Challenge 1: Lack of Executive Alignment

- **Symptoms:** Inconsistent priorities, resource conflicts, competing initiatives
- **Mitigation:** Use the stakeholder value alignment matrix to identify and address conflicts; create joint accountability mechanisms; establish regular executive alignment sessions

Challenge 2: Data Quality and Access Issues

- **Symptoms:** Model performance problems, delayed implementation, scope reduction
- **Mitigation:** Conduct early data readiness assessments; develop parallel data-improvement workstreams; create staged implementation based on data availability

Challenge 3: Skills and Capability Gaps

- **Symptoms:** Reliance on external resources, quality issues, scaling difficulties
- **Mitigation:** Develop internal/external talent balance strategy; create knowledge-transfer mechanisms; implement training programs aligned with project needs

Challenge 4: Adoption Resistance

- **Symptoms:** Low utilization, workarounds, negative feedback
- **Mitigation:** Involve users in design process; create change champions network; develop phased rollout with feedback loops; demonstrate visible value quickly

Challenge 5: Integration Complexity

- **Symptoms:** Extended timelines, technical issues, process breakdowns
- **Mitigation:** Develop simplified initial implementations; create integration "tiger teams;" use service-oriented architectures; implement progressive integration approaches

QUICK WIN: Create a risk register for your AI initiative that identifies potential challenges, early warning indicators, and mitigation strategies before they impact implementation.

CASE STUDY: Knowledge Multiplication at Barclays Group
By focusing on knowledge multiplication rather than comprehensive documentation, Barclays Group was able to reduce the time needed for initial client engagements by 60% while realizing an increase in identified value by three times. Organizations of various sizes have now used my "knowledge multiplication framework" as a roadmap. The framework relies on documenting value creation patterns, decision frameworks, risk identification approaches, and impact measurement systems.

CASE STUDY: Retail Implementation Risk Register
A retail organization created an "implementation risk register" that identified potential challenges for each AI transformation phase – along with specific mitigation strategies. This proactive approach allowed the organization to address issues before they could derail progress.

IMPLEMENTATION TIMELINE AND MILESTONES

While specific timelines will vary based on organizational context and initiative complexity, here's a general framework for implementation planning.

Phase 1: Define Shared Value Goals

- **Timeline:** 4–6 weeks
- Key milestones:
  - Value mapping canvas completed
  - Stakeholder alignment workshop conducted
  - Value targets established and validated
  - Value-realization roadmap developed

Phase 2: Create Cross-Functional Alignment

- **Timeline:** 2–4 weeks
- Key milestones:
  - Stakeholder value alignment matrix completed
  - Governance structure established
  - Communication protocols defined
  - Shared accountability mechanisms implemented

Phase 3: Establish Data Foundations

- **Timeline:** 8–12 weeks (in parallel with other phases)
- Key milestones:
  - Data-readiness assessment completed
  - Critical gaps addressed
  - Data-governance model implemented
  - Data-quality monitoring established

Phase 4: Capture Early Wins While Building for Scale

- **Timeline:** 12–16 weeks
- Key milestones:
  - Quick-win implementations completed
  - Value results documented and communicated
  - Technical scaling foundations established
  - Organizational capabilities developed

Phase 5: Measure, Learn, and Evolve

- **Timeline:** Ongoing
- Key milestones:
  - Value dashboard implemented
  - Review cadences established
  - Feedback loops operating
  - Course corrections implemented
  - Capability development continuing

REFLECTION QUESTIONS: Which of these phases is likely to take longer in your organizational context, and how might you adjust your implementation timeline accordingly?

CASE STUDY: Healthcare Visual Roadmap

A healthcare organization created a visual roadmap showing the parallel implementation of these phases, with specific milestones, dependencies, and decision points. This roadmap served as a communication tool and progress tracker throughout their implementation.

By making the implementation plan visual and accessible, they created shared understanding across technical and clinical stakeholders with different priorities and perspectives.

CASE STUDIES: THE BLUEPRINT IN ACTION

Let's examine how organizations across industries have applied this blueprint to create substantial value through AI.

FINANCIAL SERVICES: END-TO-END TRANSFORMATION

A global bank implemented the data-to-value integration blueprint to transform their retail banking operations:

Phase 1: Define Shared Value Goals

Identification of three primary value dimensions:
- Customer experience (reduce time-to-decision by 60%)
- Risk management (improve default prediction by 25%)
- Operational efficiency (reduce processing costs by 30%)

Phase 2: Create Cross-Functional Alignment

Establishment of a transformation squad with representatives from:
- Retail banking (product owners)
- Risk management
- Operations
- Data science
- IT infrastructure
- Customer experience

Phase 3: Establish Data Foundations

Data assessment insights revealed:
- Strong internal structured data
- Limited external data integration
- Inconsistent customer identifiers across systems
- Regulatory constraints on data use

These issues were addressed through:
- Customer data platform implementation
- Regulatory-compliant external data integration
- Identity resolution framework
- Privacy-preserving data architecture

Phase 4: Capture Early Wins While Building for Scale

Quick-win portfolio included:
- Pre-approval automation for low-risk customers (80% reduction in processing time)
- Next-best-action recommendations for customer service (22% increase in product adoption)
- Anomaly detection for fraud prevention (15% reduction in false positives)

Phase 5: Measure, Learn, and Evolve

Implementation of a comprehensive value dashboard tracking:
- Customer experience metrics (time-to-decision, satisfaction scores)
- Risk performance indicators (prediction accuracy, default rates)
- Operational efficiency measures (processing costs, automation rates)
- Innovation metrics (new use cases, improvement cycle time)

Results

- 70% reduction in time-to-decision for retail loan applications
- 32% improvement in default prediction accuracy
- 45% reduction in processing costs
- $120 million annual value creation across the retail banking division

The bank has now expanded this approach to commercial banking and wealth management divisions.

RETAIL: CUSTOMER EXPERIENCE REINVENTION

A multinational retailer applied the blueprint to transform customer experience:

Phase 1: Define Shared Value Goals

Identification of primary value targets, including:
- Increase customer lifetime value by 20%
- Reduce marketing costs by 30%
- Improve inventory efficiency by 25%
- Enhance employee productivity by 15%

Phase 2: Create Cross-Functional Alignment

Creation of "customer AI pods" that combine:
- Merchandising specialists
- Marketing experts
- Store operations representatives
- Supply chain analysts
- Data scientists
- Technology specialists

Phase 3: Establish Data Foundations

Data initiatives focused on:
- Unifying online and offline customer data
- Enhancing product attribution data for better matching
- Integrating inventory and supply chain information
- Incorporating competitive and market intelligence

Phase 4: Capture Early Wins While Building for Scale

Quick wins included:
- Personalized product recommendations (18% conversion improvement)
- Dynamic pricing optimization (12% margin improvement)
- Inventory allocation optimization (15% reduction in stockouts)
- Store-specific assortment planning (8% sales increase)

The Data-to-Value Integration Blueprint

Phase 5: Measure, Learn, and Evolve

Implementation of an evolution framework that included:
- Weekly business impact reviews
- Monthly cross-functional optimization forums
- Quarterly value portfolio assessments
- Continuous customer feedback monitoring

Results

- 24% increase in customer lifetime value
- 35% reduction in marketing costs
- 28% improvement in inventory efficiency
- $215 million annual incremental value creation

The retailer is now extending these capabilities to the supplier network, creating an integrated ecosystem of AI-powered processes.

HEALTHCARE: CLINICAL AND OPERATIONAL INTEGRATION

A healthcare system implemented the blueprint to improve patient outcomes and operational efficiency:

Phase 1: Define Shared Value Goals

Definition of value targets, including:
- Reduce readmissions by 25%
- Improve clinical staff productivity by 20%
- Decrease length of stay by 15%
- Enhance patient satisfaction by 30%

Phase 2: Create Cross-Functional Alignment

Formation of a clinical AI council with:
- Physicians from key specialties
- Nursing leadership
- Quality improvement specialists
- Operations managers
- Data scientists and analysts
- IT and EHR specialists

Phase 3: Establish Data Foundations

Setting parameters for data initiatives with a focus on:
- Integrating clinical and operational data
- Incorporating social determinants of health information
- Standardizing clinical documentation
- Implementing real-time data flows from medical devices

Phase 4: Capture Early Wins While Building for Scale

Quick wins included:
- Predictive readmission risk alerts (22% reduction in high-risk readmissions)
- Clinical documentation assistance (35% reduction in documentation time)
- Patient flow optimization (18% improvement in bed utilization)
- Medication reconciliation support (42% reduction in discrepancies)

Phase 5: Measure, Learn, and Evolve

Learning system included:
- Daily operational performance reviews
- Weekly clinical outcome assessments
- Monthly cross-functional improvement forums
- Quarterly strategic value reviews

Results

- 28% reduction in readmissions for high-risk patients
- 23% improvement in clinical staff productivity
- 17% decrease in average length of stay
- $78 million annual value creation through improved outcomes and efficiency

The health system is now expanding these capabilities to ambulatory care settings and preventive health initiatives.

Manufacturing: Supply Chain Optimization

A global manufacturer implemented the blueprint to transform supply chain operations.

Phase 1: Define Shared Value Goals

Set value targets that included:
- Reduce inventory carrying costs by 25%
- Improve on-time delivery by 15%
- Decrease production downtime by 30%
- Enhance quality metrics by 20%

Phase 2: Create Cross-Functional Alignment

Formation of supply chain excellence teams combining:
- Production managers
- Supply chain planners
- Quality specialists
- Maintenance engineers
- Data scientists
- IT integration experts
- Logistics specialists
- Procurement team members

Phase 3: Establish Data Foundations

Data initiatives focused on:
- Integrating production equipment sensor data
- Connecting ERP, MES, and supplier systems
- Standardizing product and component data
- Incorporating external factors like weather and logistics data

Phase 4: Capture Early Wins While Building for Scale

Quick wins included:
- Predictive maintenance implementation (45% reduction in unplanned downtime)
- Dynamic inventory optimization (22% reduction in safety stock)

The Data-to-Value Integration Blueprint

- Quality defect prediction (35% reduction in quality escapes)
- Supplier delivery prediction (28% improvement in scheduling accuracy)

Phase 5: Measure, Learn, and Evolve

The continuous improvement system included:
- Daily performance huddles around key metrics
- Weekly cross-functional optimization sessions
- Monthly value realization reviews
- Quarterly capability development assessments

Results

- 32% reduction in inventory carrying costs
- 18% improvement in on-time delivery performance
- 41% decrease in production downtime
- $95 million annual value creation across the supply chain

This manufacturer is now extending these capabilities to product design and development processes, creating an integrated product lifecycle optimization system.

CONCLUSION: INTEGRATION AS THE VALUE MULTIPLIER

Throughout this chapter, we've explored the data-to-value integration blueprint – a comprehensive approach to connecting the many dimensions of AI implementation into a cohesive whole that delivers sustainable value.

KEY INSIGHT: Integration itself is a value multiplier. Even perfectly executed individual components – strategy, data, technology, talent, process redesign – deliver only a fraction of their potential value in isolation. When integrated effectively, they create a system where the whole is truly greater than the sum of its parts.

The five phases of the blueprint provide a structured path from concept to value. As the case studies demonstrate, this integrated approach has helped organizations across industries transform their operations and create substantial value through AI.

An additional tool that can be helpful in this context is the VoV framework as it brings in an important dimension – that of considering the obstacles that prevent the realization of visions and value.

Finally, I want to reflect back on Karen Stroup's powerful insight about shared goals:

> It is incredibly important for it to be a shared goal, because we win together or we lose together.

This perspective is at the heart of successful data-to-value integration – the recognition that AI value creation is fundamentally a collaborative endeavour that requires technical excellence, organizational alignment, and sustained focus on outcomes that matter.

ACTION STEP: Assess your organization's current AI initiatives against the five phases of the Integration Blueprint. Identify gaps in your approach and develop a plan to address them.

Key Takeaways

- Integration is the critical factor that determines whether AI projects deliver value or become expensive experiments.

- Start with clear, shared value goals that connect AI capabilities to business outcomes.
- Build cross-functional alignment through governance structures and shared accountability.
- Establish solid data foundations before investing heavily in model development.
- Balance quick wins with building the technical and organizational foundations for scale.
- Create feedback loops that enable continuous improvement and evolution.
- Customize the blueprint for your organization's industry context and AI maturity level.

15 Leading AI Value Creation

Chapter Roadmap

In this chapter, we explore the leadership capabilities that differentiate successful AI transformations from expensive experiments. We'll learn how different leadership roles contribute to AI value creation, the specific competencies needed for success, and practical models for building effective leadership at all levels of an organization.

THE LEADERSHIP BLIND SPOT

Imagine this scene: In a gleaming corporate headquarters, a data science team unveils a sophisticated AI system. The models are elegant, the code is efficient, and the demonstrations are flawless. Yet six months later, the system gathers digital dust, virtually unused by the intended stakeholders. Despite millions invested, no measurable value materializes.

A comment by the CEO of a financial services firm perfectly summed up this conundrum.

"We have the best data science team in the industry, but we're still struggling to see results," he said, capturing a painful truth I've witnessed repeatedly during my years advising organizations on AI transformation: technical excellence alone doesn't create value.

On closer investigation of this and similar scenarios, I discovered what I now call the "leadership blind spot."

In short: the primary reason many AI initiatives fail isn't technology – it is leadership.

KEY INSIGHT: In my work with hundreds of organizations, I've discovered that leadership quality – not technical sophistication – determines whether AI initiatives deliver value or become expensive experiments.

While the organizations had invested heavily in technical talent and infrastructure, they had fundamentally underinvested in the leadership capabilities needed to:

- Bridge departmental silos that prevented effective implementation
- Translate technical capabilities into language business users understood
- Navigate the human and organizational challenges of adoption
- Maintain unwavering focus on business outcomes rather than technical achievements

This blind spot isn't rare, it is endemic. And it's costing organizations billions in failed AI investments.

PATTERN RECOGNITION: Watch for these warning signs of the leadership blind spot in your organization:

- Technical excellence receiving disproportionate attention and resources
- Responsibility for AI success delegated entirely to technical teams
- Missing translation layer between technical capabilities and business outcomes
- Cross-functional collaboration barriers remaining unaddressed
- More focus on implementation activity than business value creation

THE LEADERSHIP PARADOX: WHY MOST ORGANIZATIONS GET IT WRONG

Before diving into what works, let's confront an uncomfortable truth about why organizations struggle with AI leadership: The very qualities that make leaders successful in traditional environments often work against them in AI transformation.

Consider the following three patterns I've observed across industries.

#1: THE DELEGATION TRAP

Many executives attempt to delegate AI initiatives entirely to technical teams, believing technology is a "technical problem" rather than a leadership challenge. When the European Commission embarked on its AI transformation, initial efforts stalled precisely because leadership viewed it as a purely technical project. Success only came when senior leaders recognized they needed to actively engage in shaping the strategic direction and organizational integration, not merely approve budgets.

WARNING SIGN: If your executives view AI as "an IT project" rather than a strategic business initiative requiring their direct involvement, you're likely caught in the delegation trap.

#2: THE CERTAINTY ILLUSION

Traditional leadership values certainty and clear directions. AI transformation requires comfort with ambiguity and emergent strategies. When a manufacturing executive I worked with insisted on precise ROI projections and fixed timelines before beginning a predictive maintenance initiative, months of potential value were lost while competitors moved ahead with more adaptive approaches.

REFLECTION QUESTIONS: How comfortable are your leaders with ambiguity and emergent strategies? Can they make decisions with incomplete information and adjust course as they learn?

#3: THE CONTROL CONTRADICTION

Hierarchical control structures that work in stable environments can become barriers in AI implementation. At a large global banking group, early AI initiatives struggled until leadership shifted from controlling processes to enabling cross-functional collaboration. This shift – from directing to orchestrating – proved crucial to their eventual success (see Table 15.1).

TABLE 15.1
The Directing vs. Orchestrating Distinction

Traditional Directing	AI-Era Orchestrating
Decision-making: Top-down commands with clear authority chains	**Decision-making:** Collaborative decisions with distributed expertise
Information flow: Hierarchical reporting through formal channels	**Information flow:** Open sharing across functions and levels
Problem-solving: Leaders provide solutions and direct implementation	**Problem-solving:** Leaders facilitate collaboration to find solutions
Resource allocation: Centralized control of budgets and people	**Resource allocation:** Flexible deployment based on value opportunities
Success metrics: Functional excellence and adherence to plans	**Success metrics:** Cross-functional outcomes and value creation

In practice at the bank: Instead of a CIO dictating technical specifications to business units, leadership created cross-functional teams where technical and business experts jointly designed solutions. Rather than separate IT and business budgets, they established shared incentives for new initiatives with joint accountability for outcomes. Instead of sequential handoffs between functions, they enabled continuous collaboration throughout implementation.

This orchestrating approach recognized that AI value creation requires expertise from multiple domains working together, rather than one function directing others.

> "How do we change the style we lead?" This question, by Product Leadership Coach, Tobias Freudenreich, gets to the heart of our challenge.

AI doesn't just require new technology; it demands a fundamental evolution in how we lead.

ACTION STEP: Conduct a leadership paradox assessment with your executive team. Have them rate their comfort with delegation balance, ambiguity, and collaborative leadership on a scale of 1–10.

THE NEW LEADERSHIP MODEL FOR AI VALUE

Based on my observation of hundreds of AI initiatives across industries, successful AI transformation depends on leadership capabilities across five dimensions, each contributing to successful value creation.

Let's look at how to develop these capabilities and avoid common pitfalls.

STRATEGIC VISION: SEEING BEYOND AUTOMATION

True AI value doesn't come from automating existing processes but from reimagining business models and customer relationships. The difference? Leadership vision.

A retail CEO I worked with exemplified this approach. Despite limited technical background, she transformed her organization by articulating how AI would fundamentally change their customer relationships – moving from transaction-based interactions to continuous personalized engagement throughout the customer journey. This vision guided everything from data collection to organizational design, eventually delivering a 24% increase in customer lifetime value.

Contrast this with a common mistake: Many leaders frame AI initiatives around cost reduction through automation, immediately positioning the technology as a threat to employees rather than an opportunity for enhancement. This narrow vision constrains value potential before implementation even begins.

KEY INSIGHT: Strategic vision for AI isn't about predicting the future of technology, it is about imagining new possibilities that technology can enable.

QUICK WIN: Create a "vision beyond automation" exercise for your leadership team. Ask them to describe how AI could transform customer relationships, not just reduce costs in your organization.

CROSS-FUNCTIONAL INTEGRATION: BREAKING THE SILO CURSE

Organizational boundaries pose considerable constraints for realizing AI value. Creating it requires leaders who can bridge traditional silos.

A manufacturing company I advised created "value integration teams," cross-functional leaders who were physically co-located and collectively accountable for end-to-end value delivery. This radical approach to breaking down functional barriers delivered extraordinary results: 32% reduction in inventory costs, 18% improvement in on-time delivery, and $95 million in annual value creation.

Yet as my research on cross-functional leadership reveals, effective cross-functional leadership doesn't come naturally. It requires deliberate practice and time to develop the skills needed to bring diverse groups together for a shared purpose and move initiatives forward effectively.

PATTERN RECOGNITION: The most successful AI initiatives I've observed share a common pattern: they create organizational structures that physically bring together people from different functions with shared accountability for outcomes.

The Silo-Breaking Framework

1. Identify the end-to-end value stream you're targeting
2. Create a dedicated cross-functional team with representatives from all relevant departments
3. Assign clear, collective accountability for business outcomes (not just implementation)
4. Co-locate team members physically or create strong virtual collaboration environments
5. Establish shared metrics that prevent optimization of one area at others' expense

VALUE TRANSLATION: SPEAKING MULTIPLE LANGUAGES

Effective AI leaders translate between technical and business worlds, making complex concepts understandable while ensuring business needs shape technical implementation.

Providing non-technical explanations for common AI concepts, complete with domain-specific examples, can advance cross-functional collaboration and improve implementation success rates and adoption.

This translation capability isn't merely helpful – it is essential. Tobias Freudenreich notes in his research on cross-functional leadership that without this translation: "We just burden [the] team and we leave the team alone with trying to translate in between all these different documents and stories." When translation fails, look for the common gaps shown in Table 15.2.

QUICK WIN: Create a "translation dictionary" for your organization that maps common business terms to their technical counterparts and vice versa.

CHANGE ACCELERATION: MOVING AT THE RIGHT SPEED

AI implementation requires finding the right pace: quick enough to capture value but measured enough to bring the organization along.

At Zones, a global IT solutions provider, leadership demonstrated this balance through a four-week proof-of-value assessment of Copilot for Sales. By focusing on rapid value demonstration, they achieved immediate ROI while creating the organizational momentum needed for broader adoption.

TABLE 15.2
The Value Translation Breakdown

Business Side	Technical Side
"We need better customer insights"	"What specific variables and outcomes should we predict?"
"Make it user-friendly"	"What specific workflows and tasks need support?"
"We need this ASAP"	"What are the priority features vs. nice-to-haves?"
"It needs to work perfectly"	"What accuracy level is sufficient for business decisions?"

Leading AI Value Creation

REFLECTION QUESTIONS: Is your organization's AI implementation pace driven by competitive pressure and FOMO (fear of missing out), or by a thoughtful assessment of value opportunities and organizational readiness?

The Balanced Speed Framework

- **Too Slow:** Miss market opportunities, lose competitive position, fail to attract talent
- **Too Fast:** Skip crucial foundations, create technical debt, trigger organizational resistance
- **Just Right:** Balance quick wins with sustainable architecture, match pace to change absorption capacity

CAPABILITY-BUILDING: DEVELOPING ORGANIZATIONAL MUSCLES

The difference between one-time AI projects and sustainable transformation? Organizational capability-building.

A healthcare system I advised created an "AI leadership academy" that identified high-potential leaders from both clinical and administrative functions and developed them through formal training, cross-functional assignments, and mentoring. This created a pipeline of leaders capable of driving ongoing transformation rather than relying on external expertise for each initiative.

ACTION STEP: Assess your organization's AI capability-building efforts. Are you investing primarily in technical skills, or are you balancing technical, business, and leadership capability development?

Three-Layer Capability Model

1. **Technical Layer:** Data engineering, data science, machine learning, MLOps
2. **Business Layer:** Value identification, use case development, implementation management
3. **Leadership Layer:** Strategic vision, cross-functional integration, change management

KEY LEADERSHIP ROLES IN AI TRANSFORMATION

Successful AI value creation requires multiple leadership roles working in concert. Let's examine the critical roles and how they work together:

THE EXECUTIVE SPONSOR: AUTHORITY AND AIR COVER

The executive sponsor provides the authority, resources, and organizational air cover needed for success. They must:

- Articulate how AI connects to strategic priorities
- Allocate sufficient resources for success
- Remove organizational barriers
- Hold teams accountable for value delivery
- Maintain focus through inevitable challenges

The sponsorship gap: Many organizations assign sponsorship to leaders who lack either the authority or commitment to fulfil this role effectively. This creates the appearance of executive support without the reality.

I witnessed this gap at a healthcare organization where an AI-powered clinical decision support system was technically strong but struggled with adoption. When I investigated the cause, I discovered that the nominal executive sponsor – a mid-level IT director – lacked the authority to engage clinical leadership or influence workflow changes. The initiative only gained traction when the COO stepped in as sponsor, personally convening clinical stakeholders and reinforcing the strategic importance of the project.

KEY INSIGHT: The effectiveness of an executive sponsor isn't determined by their technical knowledge, but by their organizational authority, strategic influence, and willingness to actively engage in removing barriers.

Executive sponsor effectiveness assessment: Rate your sponsor's effectiveness on these dimensions (1–5):

- Authority to allocate resources and prioritize effort
- Credibility with key stakeholders across functions
- Willingness to actively intervene when barriers arise
- Consistency in reinforcing strategic importance
- Ability to maintain focus through challenges

A score below 20 indicates a potential sponsorship gap that needs addressing.

THE VALUE ORCHESTRATOR: DAILY INTEGRATION LEADERSHIP

While the sponsor provides authority, the value orchestrator provides daily leadership to integrate technical and business elements. This role:

- Maintains focus on value creation rather than technology
- Facilitates collaboration across functions
- Translates between technical and business languages
- Identifies and addresses integration gaps
- Drives accountability for results

The orchestration challenge: Finding leaders with the right combination of business understanding, technical knowledge, and cross-functional influence is difficult but essential.

A retail organization I advised created a formal "AI value leader" role filled by a senior merchandizing executive with experience in analytics. This leader chaired weekly integration sessions with representatives from merchandizing, marketing, supply chain, store operations, and the data science team. By maintaining relentless focus on business outcomes rather than technical specifications, they delivered their personalization initiative three months ahead of schedule with 40% higher adoption than previous technology projects.

Value orchestration success patterns: The most effective value orchestrators typically have:

1. Deep understanding of the business domain
2. Sufficient technical literacy to engage with technical teams
3. Strong relationship networks across the organization
4. Track record of delivering complex initiatives
5. Ability to communicate effectively with diverse stakeholders

QUICK WIN: Identify individuals in your organization who naturally bridge technical and business worlds. These "translators" often make excellent value orchestrators with the right support and authority.

Leading AI Value Creation

THE TECHNICAL LEADER: VALUE-FOCUSED TECHNICAL DIRECTION

The technical leader guides data science, engineering, and implementation work. For value-focused AI, effective technical leaders:

- Prioritize business outcomes over technical elegance
- Translate business requirements into technical specifications
- Manage technical risks and dependencies
- Balance quick wins with sustainable architecture
- Build bridges between technical and business teams

The technical leadership trap: Many organizations select technical leaders based solely on technical expertise, without considering their ability to connect technology to business value.

WARNING SIGN: If your technical leaders talk primarily about model accuracy, architecture elegance, and technical sophistication rather than business outcomes, user adoption, and value realization, you may be caught in the technical leadership trap.

A global technology firm repeatedly struggled with AI implementation despite world-class technical talent. When we examined their leadership selection, we discovered they consistently chose technical leaders based on publication records and technical credentials rather than their ability to connect with business stakeholders or focus on practical outcomes. By shifting these selection criteria to include business acumen and communication skills, their next initiative delivered 5 times more business value in half the time.

Redefining technical leadership excellence: The most effective technical leaders have:

- Strong technical credentials that earn respect from technical teams
- Business acumen to understand value drivers
- Pragmatic problem-solving orientation
- Communication skills to explain technical concepts to non-technical audiences
- Collaborative approach to working across functions

REFLECTION QUESTIONS: Are you selecting technical leaders based primarily on technical excellence, or are you balancing technical expertise with business acumen and communication skills?

THE CHANGE LEADER: NAVIGATING THE HUMAN DIMENSION

AI initiatives often require significant changes to processes, skills, and behaviours. The change leader focuses on the human side of implementation:

- Building awareness and understanding across the organization
- Developing training and support programmes
- Addressing resistance and concerns
- Creating feedback mechanisms to capture issues
- Celebrating successes and building momentum

The change leadership gap: Many organizations underinvest in change leadership, assuming that good technology will naturally drive adoption. This assumption accounts for many AI implementation failures.

A financial services firm designated a "VP of AI transformation" who focused exclusively on change management, creating an adoption roadmap, developing role-specific training, and establishing a network of "AI champions" across business units. This dedicated change leadership approach resulted in 85% adoption within six months – nearly triple the industry average for similar initiatives.

Change Leadership in Action: The Healthcare CMO Story

A healthcare CMO demonstrated exceptional change leadership when implementing an AI-powered clinical decision support system. Rather than focusing solely on the technical implementation, he built a comprehensive change strategy that addressed physicians' concerns about autonomy, workflow impacts, and patient relationships.

His approach included:

1. Transparent communication about what the AI system could and couldn't do
2. Involving physicians in design decisions and user testing
3. Creating role-specific training that addressed both technical skills and workflow integration
4. Establishing a feedback loop for continuous improvement
5. Measuring and celebrating early adoption successes

This human-centred approach resulted in 92% adoption compared to the industry average of 35% for similar clinical decision support tools.

ACTION STEP: Assign a dedicated change leader for your AI initiatives with clear responsibility for the human aspects of implementation.

THE AI VALUE ADVISOR: EXTERNAL PERSPECTIVE AND EXPERTISE

Many successful organizations supplement internal leadership with external advisors who provide specialized expertise and independent perspective:

- Challenge internal assumptions and biases
- Share cross-industry best practices and lessons
- Provide specialized knowledge in emerging areas
- Offer objective assessments of progress and gaps
- Help navigate organizational politics from a neutral position

When IBM established its AI ethics framework, Steve Astorino, General Manager of Product Development, Data, AI & Sustainability, highlighted the importance of external perspective to ensure AI initiatives align with both organizational values and broader ethical standards. This external viewpoint helps prevent organizational blind spots from undermining value creation.

KEY INSIGHT: External advisors (see Table 15.3) can often say things internal leaders cannot, challenging the status quo without risking their organizational position. This "outsider advantage" can be crucial for addressing sensitive topics like ethical concerns, governance gaps, or organizational barriers.

A healthcare system I advised engaged an external AI value advisor to help them assess their data readiness, develop a value measurement approach, and facilitate cross-functional alignment. This outside perspective helped them avoid common pitfalls and accelerate their learning curve.

Leading AI Value Creation

TABLE 15.3
External Advisor Value Matrix

Type of Advisor	Key Value Contributions	When to Engage
Industry expert	Cross-sector patterns, competitive insights	Early strategy, market positioning
Technical specialist	Emerging capabilities, implementation approaches	Solution design, technical road-mapping
Transformation coach	Change management, leadership development	Organizational readiness, capability-building
Ethics and governance advisor	Regulatory guidance, ethical frameworks	Risk assessment, governance design

QUICK WIN: Identify one area where your AI initiative might benefit from an external perspective. Engage an advisor with relevant expertise for a short diagnostic assessment to identify blind spots and opportunities.

CRITICAL COMPETENCIES FOR EFFECTIVE VALUE LEADERSHIP

Beyond the specific roles, certain core competencies are essential for leaders driving AI value creation:

STRATEGIC THINKING WITH TECHNICAL AWARENESS

Leaders must connect AI capabilities to business strategy while understanding technical possibilities and limitations.

A manufacturing executive I worked with exemplified this balance. With limited technical background, he developed sufficient AI understanding to identify how computer vision could transform the company's quality control process. He didn't need to understand the details of convolutional neural networks, he just needed to grasp their capabilities, limitations, and value potential.

Development Approaches for this Competency

- Engage in cross-functional learning experiences
- Gain exposure to external innovations and case studies
- Invest in guided technical education focused on business applications
- Participate in strategic scenario-planning exercises that include AI possibilities

Finding the right level of technical understanding is crucial. Consider the levels shown in Table 15.4.

TABLE 15.4
The Technical Awareness Spectrum

Technical Understanding	Description	Value to Leadership
Too general	"AI is magic that solves everything"	Cannot differentiate realistic from unrealistic
Just right	Understands key capabilities, limitations, and requirements	Can guide strategic decisions with technical realism
Too specialized	Gets lost in technical details and implementation specifics	Misses strategic forest for technical trees

PATTERN RECOGNITION: The most effective non-technical leaders I've observed don't try to become technical experts. Instead, they develop targeted understanding of:

- What different AI approaches can and cannot do
- What data is required for different applications
- How long different types of AI projects typically take
- What integration challenges typically arise
- Which ethical and governance issues need consideration

VALUE TRANSLATION

Leaders must connect technical concepts to business value in both directions, expressing business problems in ways data scientists can address and translating technical concepts for business audiences.

A retail CIO I advised excelled at this competency. When presenting to the board, he translated a complex recommendation engine into "suggesting relevant products to customers based on their preferences and behaviour, improving both customer experience and sales conversion," connecting technical capabilities directly to customer experience and revenue growth.

Development Approaches for This Competency

- Practice paired problem-solving with technical and business teams
- Participate in value definition workshops that bridge perspectives
- Develop a business-friendly vocabulary for key technical concepts
- Regularly practice communicating in mixed technical/business forums

Value Translation in Action: JP Morgan's COIN Programme

Mary K. Pratt's research highlights JP Morgan's Contract Intelligence (COIN) programme as an excellent example of value translation. Rather than focusing on the technical complexity of natural language processing, JP Morgan's leadership translated the capability into concrete business terms: "evaluates commercial loan agreements in seconds instead of the 360,000 hours of lawyer work previously required annually."

This translation accomplished three crucial things:

1. Made the value immediately understandable to business stakeholders
2. Created a clear before/after comparison that quantified impact
3. Connected the technical capability to a specific business process

QUICK WIN: Practice "value translation" with your team by taking three technical features of your current AI initiative and reframing them entirely in terms of business outcomes and user benefits.

CROSS-FUNCTIONAL LEADERSHIP

AI value creation requires collaboration across traditional organizational boundaries.

The CIO of a financial services company demonstrated this competency by creating a "one-team" approach for a risk prediction initiative. Rather than separate technical and business workstreams, she established cross-functional teams with shared objectives and combined reporting lines, breaking down the traditional IT/business divide.

Development Approaches for This Competency

- Gain cross-functional project experience
- Build teams that cross organizational boundaries

- Experience multiple business functions through rotations or assignments
- Learn conflict resolution techniques for cross-functional contexts

REFLECTION QUESTIONS: How comfortable are you operating outside your functional area of expertise? Can you effectively collaborate with and influence colleagues from different disciplines?

The cross-functional paradox: Research from Tobias Freudenreich reveals a fundamental paradox in many organizations: "We still have a director product being responsible for the product managers. We have a director UX being responsible for the UX people in the team." This creates what Freudenreich calls a "leadership lag" where team structures evolve towards cross-functionality while leadership remains functionally oriented.

This paradox creates tension for both leaders and teams:

- Leaders face divided loyalties between functional reporting lines and cross-functional projects
- Teams receive mixed messages from functional and project leadership
- Resource allocation becomes contentious between functional and project priorities
- Success metrics may conflict between functional excellence and project outcomes

Breaking the cross-functional paradox: Successful organizations resolve this tension through:

1. Clearly defined decision rights between functional and project leadership
2. Unified communication from leadership to teams
3. Joint accountability for outcomes between functional and project leaders
4. Shared success metrics that balance functional and cross-functional priorities

CHANGE LEADERSHIP

Implementing AI-driven changes requires specific capabilities related to change leadership.

To successfully lead an AI transformation, leaders need to cultivate a specific set of competencies focused on change management, communication, and strategic thinking, alongside technical expertise. These include a growth mindset, empathy, strong communication skills, and the ability to manage both the technological and human aspects of change.

Development Approaches for This Competency

- Receive change leadership training
- Gain experience with previous change initiatives
- Learn from mentors who have successfully led change
- Develop stakeholder management skills
- Receive regular feedback on change leadership effectiveness

The AI adoption flywheel: Effective change leaders understand that adoption follows a predictable pattern:

1. **Awareness** → Getting on the radar
   - **Key Question:** Do stakeholders know about the AI initiative?
   - **Critical Barrier:** Information overwhelm and competing priorities
2. **Understanding** → Creating comprehension
   - **Key Question:** Do stakeholders understand what's changing and why?
   - **Critical Barrier:** Technical complexity and unclear value proposition

3. **Use** → First engagement with the system
   - **Key Question:** Can stakeholders successfully use the system?
   - **Critical Barrier:** User experience issues and workflow disruption
4. **Value** → Experiencing tangible benefits
   - **Key Question:** Do stakeholders see clear benefit from using the system?
   - **Critical Barrier:** Delayed or unclear value realization
5. **Belief** → Developing trust in the system
   - **Key Question:** Do stakeholders trust the system consistently?
   - **Critical Barrier:** Inconsistent performance or unexplained results
6. **Advocacy** → Becoming promoters within the organization
   - **Key Question:** Are stakeholders encouraging others to use the system?
   - **Critical Barrier:** Cultural resistance to change advocacy

QUICK WIN: Assess where your current AI initiative sits on the adoption flywheel. Identify the most critical barriers at your current stage and develop targeted interventions to address them.

Learning Agility

Given AI's rapid evolution, leaders must demonstrate high learning agility.

The CEO of a retail organization exhibited this competency when generative AI emerged. Rather than sticking to the original plan, she quickly assessed the new capabilities, adjusted the AI roadmap, and redirected resources to incorporate these technologies. This adaptability allowed them to gain a six-month advantage over competitors.

Development Approaches for This Competency

- Deliberately expose yourself to diverse perspectives
- Practice reflection to capture learning
- Seek feedback on adaptation to new information
- Experiment with different approaches
- Participate in regular knowledge-sharing sessions

KEY INSIGHT: In the fast-changing AI landscape, learning agility isn't just a nice-to-have quality, it is a strategic imperative. Leaders who can quickly absorb new information, identify patterns, and adapt approaches create significant competitive advantage.

The Learning Agility Challenge: Sam Altman's Talent Approach

As detailed in Beatrice Nolan's Business Insider Africa article, Sam Altman, CEO of OpenAI, demonstrates learning agility through his talent strategy. Rather than relying on conventional wisdom about experience, Altman prioritizes learning ability: "A strategy that says 'I'm only going to hire younger people,' or 'I'm only going to hire older people,' would be misguided."

Altman explains that "inexperienced does not inherently mean not valuable" and that taking chances on high-potential people with strong learning agility can deliver high returns. Simultaneously, he recognizes that experienced professionals bring essential expertise to higher-stakes tasks, noting that "what you really want is just like an extremely high talent bar of people at any age."

Three Dimensions of Learning Agility for AI Leaders

1. **Technical Learning:** Ability to grasp new AI capabilities and limitations
2. **Business Learning:** Capacity to identify new value creation opportunities
3. **Organizational Learning:** Skill at adapting implementation approaches to organizational context

Leading AI Value Creation

ACTION STEP: Create a "learning radar" process for your leadership team to regularly scan for emerging AI capabilities, assess their potential impact on your business, and adjust strategies accordingly.

CREATING THE CONDITIONS FOR AI SUCCESS

Beyond individual competencies, leaders shape the organizational context in which AI initiatives either thrive or struggle.

SETTING A COMPELLING VISION

Leaders must articulate a vision that connects AI to organizational purpose and priorities.

The CEO of a healthcare system established "AI for better care, not just efficiency" as guiding vision. This simple but powerful framing ensured that every AI initiative balanced operational benefits with clinical improvements, preventing the common trap of cost-cutting that compromises care quality.

Effective AI Visions

- Connect clearly to strategic objectives
- Specify value outcomes, not just technology goals
- Balance stakeholder interests
- Present inspirational yet credible possibilities
- Receive consistent reinforcement across communications

Vision in Action: Absa Bank's Transformation

Microsoft's case study on Absa Bank illustrates how leadership shapes AI value narratives. Naomi Parsons, Head of Digital Delivery at Absa Corporate and Investment Banking, positioned AI not as a standalone technology but as an enabler of the bank's strategic vision: "Achieving our vision of transforming human experiences and ensuring seamless delivery can be better realized through digital tools."

This strategic framing helped secure broad organizational support and positioned AI as central to the bank's competitive strategy: "The only way to remain competitive in such a dynamic banking landscape is through the right use of technology."

PATTERN RECOGNITION: The most effective AI visions share these characteristics:

- They emphasize business and customer outcomes rather than technical capabilities
- They connect to the organization's broader purpose and strategic priorities
- They balance multiple stakeholder perspectives (customers, employees, shareholders)
- They're simple enough to remember and compelling enough to inspire action
- They guide decision-making about priorities and investments

QUICK WIN: Test your current AI vision by asking 5–10 stakeholders to describe it in their own words. If they can't articulate it consistently, or if it focuses primarily on technology rather than outcomes, it's time for a vision reset.

BUILDING CROSS-FUNCTIONAL COLLABORATION

Leaders must create both the structures and incentives for collaboration.

A healthcare organization I worked with established "Care Integration Councils". These are cross-functional teams with representatives from clinical operations, IT, quality, and administration who meet weekly and share responsibility for patient outcome improvements. This collaborative structure, combined with unified success metrics, broke down the kinds of barriers we traditionally see between departments and across organizations.

Effective Collaboration Mechanisms Include

- Cross-functional teams with clear joint objectives
- Shared metrics that prevent optimization of one area at others' expense
- Joint planning and review processes
- Collaborative physical or virtual workspaces
- Effective conflict resolution processes for competing priorities

The Cultural Challenge: Navigating Diverse Decision Styles

One fascinating case, which I discussed in depth in *Making Data Work,* illustrates the cultural dimension of cross-functional collaboration. At a European multilateral organization, a project team composed of members from eight different nationalities encountered cultural clashes in decision-making styles. When a Finnish team member banged the table and demanded, "Will somebody make a decision or are we going to keep sitting here to play decision ping pong until lunchtime," it immediately revealed how "the Swede was focusing on being democratic…and was shying away from any decisions that were at risk of not finding consensus" while "the Finn was being very pragmatic … preferring to quickly make a decision, even at risk of some contention."

This cultural dimension of collaboration requires leaders to:

1. Recognize different decision-making preferences across team members
2. Create explicit decision processes that accommodate diverse styles
3. Build mutual understanding and respect for different approaches
4. Establish clear decision rights and escalation paths

REFLECTION QUESTIONS: Which collaboration model (see Table 15.5) would work best for your organization's current AI initiatives? What structural and cultural barriers might need to be addressed to implement that model effectively?

CREATING A DATA-DRIVEN CULTURE

Leaders must foster an organizational culture where decisions are based on evidence rather than intuition or hierarchy.

The CFO of a financial services firm transformed executive meetings by starting every discussion with relevant data before entertaining opinions. This simple practice shifted the culture from opinion-based debates dominated by the most senior or vocal leaders to evidence-based discussions where the quality of analysis mattered more than authority.

TABLE 15.5
The Three Collaboration Models

Model	Description	Best For
Liaison Model	Designated representatives from each function coordinate between departments	Early-stage collaboration with minimal organizational disruption
Virtual Team Model	Cross-functional team members remain in their departments but work together on AI initiatives	Balancing functional expertise with cross-functional collaboration
Dedicated Team Model	Team members are physically co-located and fully dedicated to the AI initiative	High-priority initiatives requiring deep integration and rapid progress

Leaders Build Data-driven Cultures By

- Modelling evidence-based decision-making
- Asking for data to support recommendations
- Being willing to change direction based on new information
- Distinguishing clearly between facts and interpretations
- Setting appropriate expectations about certainty and risk

From Blame to Learning: The TAG Transformation Story

I witnessed a powerful transformation example when I was working at TAG. There, concerted efforts affected a shift from a blame culture – where a feared executive maintained a "secret black book" – to an environment where "a team member who had previously agonised over a lack of empowerment" admitted to sometimes feeling "over-empowered."

This cultural transformation from blame to learning created a foundation for data-driven decision-making by:

1. Making it safe to surface and discuss problems
2. Encouraging data-driven experimentation
3. Shifting focus from "who's at fault" to "what can we learn"
4. Empowering team members to make decisions based on data

KEY INSIGHT: A data-driven culture doesn't happen by accident. It requires deliberate leadership actions that signal what behaviours are valued and rewarded in the organization.

The Data Culture Assessment: Rate your organization's data culture on these dimensions (1–5):

- **Decision-Making:** Evidence-based vs. opinion-based
- **Risk Approach:** Learning from failure vs. punishing failure
- **Information Flow:** Transparent vs. restricted
- **Authority Basis:** Analysis quality vs. organizational position
- **Time Orientation:** Fact-finding vs. quick opinions

ACTION STEP: Identify one recurring decision process in your organization that could benefit from a more data-driven approach. Redesign the process to start with data presentation before opinion discussion.

Balancing Innovation and Governance

Leaders must create an environment that encourages innovation while ensuring appropriate governance.

A healthcare CEO I worked with established "bounded innovation" – a framework that defined clear boundaries for AI experimentation based on risk level. Low-risk applications had minimal governance requirements, while high-risk applications involving clinical decisions had rigorous oversight. This balanced approach accelerated innovation in appropriate areas while protecting patient safety.

Effective Governance Balancing Includes

- Risk-based governance that scales with potential impact
- Clear ethical guidelines for AI development and use
- Streamlined approval processes for low-risk innovations
- Regular governance review and adaptation
- Transparent reporting on both benefits and risks

TABLE 15.6
Three-Tiered Governance Model for Escalating Ethical Issues

Tier	Composition	Responsibility
Tier 1	AI product managers, responsible AI champions	Routine assessments, initial screening using established tools
Tier 2	Ethics committee of diverse experts	Exploring flagged issues that require cross-functional perspective
Tier 3	Senior leadership, representatives of ethics committee	Final decisions on high-risk use cases with significant implications

Governance Model in Action: UAE's Strategic Approach

The United Arab Emirates (UAE) demonstrates forward-thinking AI governance leadership through systematic government initiatives. As detailed in research, "in 2017, the UAE appointed a Minister of State for AI, Digital Economy, and Remote Work Applications, recognizing the need for federal oversight and strategic direction."

This was followed by a comprehensive national strategy and legislative framework, including "Law 25 of 2018 [which] grants the UAE Cabinet the authority to grant interim licenses and establish a licensing regime for innovative projects involving modern technologies or AI without existing legislation."

This case illustrates how strategic government leadership can create enabling regulatory environments while maintaining governance oversight, balancing innovation and ethical considerations.

Three-Tiered Governance

Research from the GSMA report presents an effective three-tiered governance model for escalating ethical issues within organizations (see Table 15.6).

This multi-level approach balances several competing needs:

- Efficiency (routine decisions don't require senior involvement)
- Expertise (specialized knowledge applied where needed)
- Authority (senior leaders engaged for highest-risk decisions)
- Practical implementation (graduated approach based on risk level)

WARNING SIGN: If your AI governance approach applies the same level of oversight to all AI applications regardless of risk level, you're likely either stifling innovation or insufficiently protecting against high-risk applications.

THE CEO AND BOARD'S ROLE IN AI VALUE CREATION

Top leadership plays a unique and critical role in AI transformation. CEOs and boards, for example, contribute to AI value creation through strategic positioning, value accountability, organizational barrier removal, and leadership behaviour modelling.

SETTING THE STRATEGIC CONTEXT

The CEO and board establish the foundation for AI value creation through:

- **Strategic Positioning:** Defining how AI connects to organizational strategy
- **Ambition Setting:** Establishing the level of transformation sought
- **Value Prioritization:** Determining which dimensions of value take precedence
- **Resource Commitment:** Allocating appropriate funding and talent
- **Narrative Creation:** Developing the story that explains why AI matters

Leading AI Value Creation

TABLE 15.7
Strategic Positions

Strategic Positioning	Description	Example
Competitive Necessity	AI as required to maintain market position	"Our competitors are already using AI to personalize customer experiences. We must match their capabilities or risk falling behind."
Efficiency Driver	AI as engine for operational improvement	"AI will help us do more with less, reducing costs while maintaining or improving quality."
Growth Enabler	AI as creator of new value opportunities	"AI will help us identify new customer needs and develop innovative products and services to meet them."
Transformation Catalyst	AI as fundamental to business model evolution	"AI will transform how we create value for customers, moving from episodic transactions to continuous personalized relationships."

The board of a financial services company demonstrated this by making AI transformation one of their three strategic pillars, allocating 15% of capital expenditure budget to AI initiatives, and regularly dedicating board time to reviewing progress. This strategic positioning elevated AI from a technology project to a core business transformation.

The Strategic Positioning Framework

When positioned effectively by top leadership, AI becomes integrated into the organization's strategic narrative. Consider the positioning approaches in Table 15.7.

KEY INSIGHT: How the CEO and board position AI strategically sends powerful signals throughout the organization about its importance, expected impact, and relationship to core business strategy.

REFLECTION QUESTIONS: How is AI currently positioned in your organization's strategic narrative? Is it seen as a technical initiative, a competitive necessity, or a transformative opportunity?

CREATING ACCOUNTABILITY FOR VALUE

Top leaders ensure focus remains on outcomes through:

- **Value Review Cadence:** Regular assessment of impact against targets
- **Resource Reallocation:** Shifting resources based on demonstrated value
- **Consequence Management:** Addressing underperforming initiatives
- **Success Recognition:** Celebrating and rewarding value achievement
- **Learning Systems:** Creating mechanisms to capture and apply insights

A retail board established quarterly AI value reviews where initiative leaders presented actual business impact rather than implementation progress. Initiatives that couldn't demonstrate value after two quarters were restructured or discontinued, ensuring resources flowed to value-creating efforts.

The Value Accountability System

The most effective CEOs and boards implement a structured approach to ensuring AI initiatives deliver value:

1. **Clear Value Targets:** Specific, measurable outcomes expected from each initiative
2. **Regular Value Reviews:** Structured sessions focused on business impact, not technical progress

3. **Resource Flexibility:** Ability to adjust investments based on demonstrated value
4. **Consequence Framework:** Clear processes for addressing underperforming initiatives
5. **Success Amplification:** Mechanisms to scale successful approaches across the organization

Strategic Innovation at Absa Bank

Julia Carvalho, General Manager of IBM Africa Growth Markets, emphasizes this accountability role:

> It's critical that CEOs in Africa establish and implement clear and consistent standards as it concerns the utilisation of AI across all areas of strategic focus… it will determine the level of investment and, ultimately, an organization's success in a rapidly advancing digital economy.

WARNING SIGN: If your AI value reviews focus primarily on implementation milestones rather than business impact, you may be falling into the "activity trap" where progress is measured by effort rather than outcomes.

QUICK WIN: Introduce a simple "value scorecard" for your next AI initiative review, requiring presenters to lead with actual business impact metrics before discussing implementation progress.

ADDRESSING ORGANIZATIONAL BARRIERS

Top leaders have unique power to remove structural obstacles:

- **Organizational Redesign:** Creating structures that enable integration
- **Incentive Alignment:** Ensuring rewards support collaboration
- **Policy Adaptation:** Modifying policies that impede innovation
- **Cultural Evolution:** Shifting norms that block AI adoption
- **Talent Strategy:** Building organizational capabilities as required

The CEO of a manufacturing company reorganized traditional functional structures into value streams with integrated teams responsible for end-to-end processes. This structural change removed the departmental handoffs that had previously slowed AI implementation and diluted value creation.

The Barrier Removal Framework

In my experience, CEOs and boards need to systematically identify and address five types of organizational barriers to AI value (see Table 15.8).

TABLE 15.8
Barriers to AI Value

Barrier Type	Description	Removal Approach
Structural Barriers	Organizational silos, unclear decision rights, fragmented responsibilities	Organizational redesign, integration mechanisms, matrix structures
Incentive Barriers	Reward systems that discourage collaboration or innovation	Shared metrics, joint incentives, innovation recognition
Policy Barriers	Processes and rules that impede agility and experimentation	Policy exemptions, updated procedures, risk-based governance
Cultural Barriers	Norms and attitudes that block adoption and change	Leadership modelling, success stories, symbolic actions
Capability Barriers	Missing skills and expertise needed for success	Strategic hiring, training programmes, partnership arrangements

Organizational Tension in Action

At one organization I worked with, a senior executive was caught in a classic organizational tension. She was concerned about team members who "are worried that the technology is changing too quickly" and feared "that they may not be able to adapt quickly enough." While the executive wanted technological progress, she recognized the human impact, creating a leadership tension between advancement and accommodation.

CEOs and boards must directly address these tensions rather than allowing them to stall progress. Effective approaches include:

1. Acknowledging the legitimacy of concerns rather than dismissing them
2. Creating tangible support mechanisms to help people adapt
3. Balancing the pace of change with organizational absorption capacity
4. Demonstrating commitment to people alongside commitment to technology

PATTERN RECOGNITION: The most successful AI transformations I've observed share a common characteristic: Top leaders personally engage in removing the most significant organizational barriers, signalling the strategic importance of the initiative.

MODELLING LEADERSHIP BEHAVIOURS

Top leaders set the tone through their personal behaviour:

- **Data-Driven Decision Making:** Using evidence in their own decisions
- **Cross-Functional Collaboration:** Working across traditional boundaries
- **Learning Orientation:** Demonstrating curiosity and adaptability
- **Value Focus:** Consistently prioritizing outcomes over activities
- **Ethical Consideration:** Addressing the broader impacts of AI

A healthcare CEO personally participated in AI ethics discussions, demonstrating that ethical considerations weren't just compliance requirements but core elements of the approach. This visible commitment shaped how the entire organization approached AI development (see Table 15.9).

KEY INSIGHT: What leaders pay attention to and how they behave sends stronger signals than what they say. When CEOs engage personally with AI initiatives, it communicates strategic importance more powerfully than any memo or presentation.

TABLE 15.9
CEO Behaviour Impact Matrix

Leadership Behaviour	Visible Impact	Hidden Impact
Using data in decision-making	Encourages evidence-based approaches throughout the organization	Creates psychological safety for challenging opinions with facts
Engaging with technical teams	Shows importance of technical work	Builds mutual understanding between business and technical perspectives
Admitting knowledge gaps	Normalizes learning orientation	Reduces pressure to appear all-knowing
Asking value-focused questions	Maintains focus on outcomes	Discourages activity-focused status reports
Addressing ethical implications	Elevates ethical considerations	Prevents ethics from being treated as compliance checkbox

ACTION STEP: Identify one leadership behaviour that would send the strongest signal about AI's strategic importance in your organization. Create opportunities for visible demonstration of this behaviour by top leaders.

THE LEADERSHIP DEVELOPMENT IMPERATIVE

Creating sustainable AI value requires developing leadership capabilities throughout the organization. For advancing that goal, it is useful to implement specific approaches to building AI leadership literacy, foster a value-focused mindset, create leadership development pathways, and establish communities of practice.

BUILDING AI LEADERSHIP LITERACY

Leaders need sufficient AI understanding to guide value creation.

A financial services company created an "AI for leaders" programme that provided executives with the foundational knowledge needed to guide their transformation. This programme focused on business applications rather than technical details, using case studies and simulations to make concepts concrete.

Key Elements of Leadership Literacy Include

- Core AI capabilities and how they create value
- Data foundations required for effective AI
- Implementation approaches and challenges
- Value measurement and realization practices
- Ethical and governance considerations

The Leadership Literacy Challenge

According to AI and leadership expert, David De Cremer, a fundamental challenge exists because "most of today's executives were educated in a pre-AI era" and "didn't have courses on artificial intelligence or even the concept of sustainable, tech-driven business models." This creates a foundational knowledge gap that helps explain why many organizations struggle with AI implementation: their leaders lack the educational foundation necessary to effectively guide AI initiatives even when they recognize the importance.

The AI Leadership Literacy Curriculum

Based on successful programmes I've helped develop, an effective AI leadership literacy curriculum includes:

1. **AI Fundamentals for Value Creation**
   - Key AI capabilities in business-friendly language
   - Value pathways for different AI approaches
   - Data requirements and implications
   - Implementation realities and timelines
2. **Strategic Application Cases**
   - Industry-specific value examples
   - Implementation approaches and challenges
   - Success factors and failure patterns
   - Leading indicators of value and risk
3. **Leadership Implications and Responsibilities**
   - Strategic positioning considerations
   - Organizational readiness assessment

Leading AI Value Creation

- Ethical and governance approaches
- Talent and capability requirements

QUICK WIN: Create a monthly "AI leadership roundtable," where executives can discuss real-world AI applications in your industry, share insights from their own initiatives, and learn from external experts in a safe, no-jargon environment.

Developing a Value-Focused Mindset

Leaders must focus on outcomes rather than technologies.

A retail organization developed a "value-first" workshop for all leaders involved in AI initiatives. This programme taught them to begin with clear business outcomes, work backward to required capabilities, and maintain focus on value throughout implementation.

Key Mindset Elements Include

- Starting with business challenges rather than technologies
- Thinking multidimensionally about value creation
- Balancing short-term wins and long-term transformation
- Maintaining focus on end-user needs and experiences
- Considering ethical implications alongside business value

Value-First vs. Technology-First Thinking

As noted in our research, Marc Schmitt's emphasis on leaders needing a "clear strategic vision that identifies how AI can be leveraged to achieve organizational goals" perfectly captures the value-focused mindset. This approach stands in stark contrast to technology-first thinking that begins with AI capabilities and then searches for applications.

The contrast between these approaches is striking as illustrated in Table 15.10.

The Value-First Framework

1. **Start with Business Objectives**
   - What specific business outcomes are we trying to achieve?
   - How would we measure success in business terms?
   - Who are the key stakeholders and what do they value?
2. **Identify Value Gaps**
   - Where are current approaches falling short?
   - What capabilities would create breakthrough value?
   - What constraints must be addressed?
3. **Work Backward to Solutions**
   - What capabilities could address these gaps?
   - How might AI approaches enable these capabilities?
   - What other elements (process, people, technology) need to change?

TABLE 15.10
Value-First vs. Technology-First Thinking

Technology-First Thinking	Value-First Thinking
"We should implement generative AI."	"We need to improve our customer response time."
"Let's build a recommendation engine."	"We need to increase cross-sell effectiveness."
"We should use computer vision."	"We need to reduce quality defects."
"We should explore large language models."	"We need to make knowledge more accessible."

4. **Evaluate Value-to-Effort Ratio**
   - What value can we realistically expect?
   - What investment will be required?
   - What risks need to be managed?

REFLECTION QUESTIONS: When you hear about a new AI capability, is your first thought "how could we use this?" Or "what business problems could this solve?" Your instinctive response reveals your current mindset orientation.

Creating Leadership Development Pathways

Organizations need systematic approaches to building AI leadership capabilities.

One approach that I have seen deliver impact relates to the previously mentioned "leadership academy." The key is to create dedicated leadership development pathways, where high-potential team members, ideally from across functions, have access to opportunities such as formal training, cross-functional assignments, and mentoring from senior leaders. This can create a pipeline of leaders capable of driving ongoing transformation.

Effective Development Approaches Include

- Rotational assignments across functions
- Paired leadership models (technical and business)
- Mentoring relationships between experienced and emerging leaders
- Real-world learning through project responsibilities
- Formal development in critical competencies

The WellSpan Health Leadership Development Model

In Kitty Wheeler's 2025 analysis of Microsoft's Dragon Copilot in healthcare, Dr. R. Hal Baker, SVP and Chief Digital Officer and CIO at WellSpan Health, demonstrated how leadership development approaches shape AI adoption success.

Rather than focusing on isolated technical training, Dr. Baker pursued an ecosystem approach to leadership development: "a Microsoft-powered ecosystem where AI assistance extends across our organization, delivering a consistent and intelligent experience everywhere we work."

This ecosystem approach to leadership development ensures that leaders at all levels understand how their specific domain connects to the broader organizational strategy and technical infrastructure (see Table 15.11).

TABLE 15.11
The Leadership Development Continuum

Development Stage	Focus Areas	Development Methods
Awareness	Understanding possibilities, overcoming misconceptions	Executive briefings, industry cases, technology demonstrations
Application	Connecting AI to specific business areas, identifying opportunities	Function-specific workshops, value-mapping exercises, use case development
Activation	Building implementation capabilities, removing barriers	Cross-functional projects, mentoring relationships, real-world application
Advancement	Scaling impact, developing others, evolving approaches	Leadership communities, knowledge-sharing, innovation forums

ACTION STEP: Assess your organization's current AI leadership development approach. Does it address all stages of the leadership development continuum? Are there gaps that might be limiting your value creation potential?

BUILDING A LEADERSHIP COMMUNITY

Sustainable AI leadership requires creating a community of practice.

A manufacturing company established an "AI leadership network" that connected leaders across all functions and levels who were involved in AI initiatives. This community met monthly to share experiences, solve common problems, and develop collective capabilities.

Effective Community Approaches Include

- Regular knowledge-sharing forums
- Cross-functional problem-solving sessions
- Shared learning resources and tools
- Peer coaching and feedback mechanisms
- Celebration of leadership successes

Community in Action: The KPMG Model

Martin Sokalski, Principal in Advisory and Emerging Technology Risk Services at KPMG, emphasizes how community approaches strengthen leadership capabilities: "The true art of the possible for artificial intelligence will become unlocked as soon as there is more trust and transparency. This can be achieved by incorporating foundational AI program imperatives like integrity, explainability, fairness and resilience."

The research highlights how organizations can employ "RAI champions" who "foster communities of experts, connecting different geographies and business units for active learning and sharing, addressing ethical doubts within product teams and organisations."

The Leadership Community Blueprint

1. **Structure**
   - Cross-functional representation (technical, business, support functions)
   - Multi-level participation (executives to emerging leaders)
   - Clear purpose and charter
   - Regular cadence and consistent format
2. **Content Focus**
   - Value creation approaches and results
   - Implementation challenges and solutions
   - Emerging capabilities and applications
   - Ethical considerations and governance approaches
   - Skill and capability development
3. **Engagement Methods**
   - Success story sharing
   - Problem-solving clinics
   - Expert presentations
   - Hands-on workshops
   - Peer-coaching sessions

KEY INSIGHT: Leadership communities create a multiplier effect by spreading learning horizontally across the organization rather than relying solely on top-down or formal development approaches.

QUICK WIN: Start a monthly "AI leadership lunch & learn" where leaders from different functions share their experiences with AI implementation – both successes and challenges – in an informal, collaborative setting.

THE LEADERSHIP RHYTHM: BUILDING SUSTAINABLE ENERGY FOR AI VALUE

One often-overlooked aspect of effective AI leadership is establishing the right cadence for different leadership activities. In my research, I've observed that successful AI leaders consciously manage their energy and attention through specific rhythms.

THE DAILY SUCCESS PATTERN

Effective AI leaders establish daily routines that maximize their strategic contribution

Morning Power Hours (First 3 Hours)

- Strategic work only
- No reactive tasks
- Full energy focus
- Value creation priority

Mid-Day Implementation (Next 4–5 Hours)

- Team engagement
- Client delivery
- Operational excellence
- Quality assurance

Evening Review (30 Minutes)

- Value captured
- Insights gained
- Next day preparation
- Energy management

KEY INSIGHT: I've found that the more strategic our work becomes, the more important our daily rhythms are. This is particularly relevant for AI leadership, where balancing strategic vision with implementation details is essential.

The Energy-Value Connection

When leaders are faced with AI scaling challenges, many try to address the situation by working more hours. After helping dozens work through such transitions, I've found what I consider a better approach: managing energy in addition to time.

According to my research, leaders need to recognize different "energy zones" for different types of leadership work:

1. **High-Value Creation Zone:** Hours when you're at your strategic best (typically the first three hours of the day)
2. **Implementation Zone:** When you execute plans and manage operations
3. **Recovery Zone:** Time for strategic renewal (not just downtime)
4. **Growth Zone:** Dedicated time for building future capacity

WARNING SIGN: If you find yourself consistently handling strategic AI decisions during low-energy periods or reacting to implementation issues during your high-energy times, you may be unconsciously undermining your leadership effectiveness.

REFLECTION QUESTIONS: When do you currently schedule your most strategic AI leadership work? Does this align with your highest-energy periods of the day?

The Weekly Success Structure

Different days serve different leadership purposes:

Monday: Strategy and Planning

- Week planning
- Team alignment
- Strategic reviews
- Innovation time

Tuesday to Thursday: Value Delivery

- Client engagement
- Team development
- Quality assurance
- Value capture

Friday: Growth and Learning

- Week review
- Knowledge capture
- System improvement
- Capability-building

This structured rhythm ensures that leaders maintain both the strategic perspective and operational engagement needed for AI value creation.

The Innovation Engine

Another critical aspect of weekly rhythm relates to allowing deliberate space for innovation. Successful AI leaders create specific weekly patterns to ensure innovation doesn't get crowded out by implementation pressures.

Weekly Innovation Block

- 2-hour uninterrupted time
- No client work
- Pure thinking space
- Pattern recognition

Innovation in Action: Zones' Four-Week Proof-of-Value Approach

Research citing Cordova of Cloud Wars presents how Zones, a global IT solutions provider, demonstrates effective AI leadership rhythm through a four-week proof-of-value assessment of Copilot for Sales. By focusing on rapid value demonstration across a defined time period, this leadership approach yielded immediate ROI while enhancing sales processes and data insights.

This time-boxed approach creates natural rhythms for:

- Initial assessment and planning
- Implementation and testing
- Value measurement
- Learning and adaptation

ACTION STEP: Identify one element of the weekly success structure that's currently missing or inconsistent in your leadership rhythm. Block time on your calendar for the next month to establish this practice as a habit.

COMMUNICATION: THE CRITICAL LEADERSHIP SKILL

Effective communication emerges as perhaps the most critical leadership capability for AI value creation. Key competencies include being able to translate between technical and business worlds, create compelling value narratives, and manage expectations throughout the AI journey.

TRANSLATING BETWEEN TECHNICAL AND BUSINESS WORLDS

Leaders must bridge technical and business languages.

The CIO of a healthcare organization developed a "translation guide" that provided non-technical explanations for common AI concepts, complete with healthcare-specific examples. This resource helped technical teams communicate more effectively with clinical and administrative leaders, significantly improving cross-functional collaboration.

Key Translation Capabilities Include

- Explaining technical concepts without jargon
- Using appropriate analogies and examples
- Adapting detail level to audience needs
- Connecting technical elements to business outcomes
- Distinguishing essential complexities from unnecessary ones

The Translation Challenge: Bridging Two Worlds

A lack of effective translation can represent a significant barrier for advancing AI initiatives since potential miscommunication and misalignment hinder cross-functional collaboration and impede efforts to work towards a common goal.

The Technical-to-Business Translation Framework

Successful AI leaders master the translation patterns shown in Table 15.12.

KEY INSIGHT: The most effective AI translators regularly practice "translation in both directions," expressing business needs in technical terms and technical concepts in business language.

TABLE 15.12
Technical-to-Business Translation Patterns

Technical Concept	Business Translation	Example
Model accuracy	Decision confidence	"This system can give you 95% confidence in customer churn predictions, allowing you to focus retention efforts on the right customers."
Data requirements	Business information needs	"To achieve this level of prediction, we need purchase history, support interactions, and website behaviour data. Do we currently collect this?"
Technical constraints	Business limitations	"With current data quality, we can predict next purchase within a two-week window, not the daily precision you wanted."
Implementation complexity	Resource and timeline requirements	"This is a multi-phase process requiring about six months and involvement from marketing, IT, and customer service."

Translation in Action: The Retail CIO Example

A retail CIO I worked with came up with the perfect translation for his audience. When presenting to the board, he translated a complex recommendation engine into "helping customers find products they'll love before they even know they want them," connecting technical capabilities directly to customer experience and revenue growth.

This translation accomplished three things simultaneously:

1. Made the technical concept immediately understandable
2. Connected it to customer experience (a board priority)
3. Linked it to financial outcomes (revenue growth)

QUICK WIN: Create a "jargon jar" in your next cross-functional meeting. Anyone using unexplained technical terms or acronyms contributes $1, with proceeds going to a team celebration.

CREATING COMPELLING VALUE NARRATIVES

Leaders must craft stories that connect AI to meaningful outcomes.

A manufacturing executive worked with a communication coach to develop the narrative for a predictive maintenance initiative. Rather than focusing on technical capabilities, the narrative centred on eliminating unexpected downtime that frustrated operators and disappointed customers. This human-centred story created much stronger engagement than technical descriptions.

Effective Value Narratives

- Connect AI capabilities to organizational purpose
- Illustrate value through concrete examples and stories
- Find ways to create emotional engagement with abstract concepts
- Balance inspiration with realism
- Adapt to different stakeholder perspectives

The AI Narrative Framework

Based on successful AI communication approaches, I've developed this framework for creating compelling AI value narratives:

1. **The Tension:** Frame the compelling problem
   - Establish the current situation and what's at stake
   - Make it relevant to the specific audience
   - Create emotional connection through human impact
   - Highlight the consequence of inaction
2. **The Shift:** How AI changes the equation
   - Introduce the AI capability in non-technical terms
   - Explain how it creates new possibilities
   - Differentiate from previous approaches
   - Acknowledge legitimate concerns
3. **The Impact:** Tangible outcomes and metrics
   - Describe specific benefits in audience-relevant terms
   - Provide concrete examples and measures
   - Connect to strategic priorities
   - Include both immediate wins and long-term value
4. **The Ask:** Clear next steps for decision-makers
   - Specify exactly what's needed to move forward
   - Make the request actionable and concrete

- Link the ask directly to the desired outcomes
- Create urgency without artificial pressure

Narrative in Action: Microsoft's Healthcare Approach

In Kitty Wheeler's 2025 analysis, we see how Dr. R. Hal Baker from WellSpan Health frames AI value not through technical descriptions but through an ecosystem narrative: "a Microsoft-powered ecosystem where AI assistance extends across our organization, delivering a consistent and intelligent experience everywhere we work."

This narrative connects AI to workplace experience rather than technology implementation, making it immediately relevant to healthcare professionals concerned with clinical workflows and patient care.

The Narrative Contrast: Context Matters in AI Communication

Research reveals how organizational context fundamentally shapes AI value communication approaches. While Naomi Parsons of Absa Bank successfully positioned AI as a strategic enabler ("The only way to remain competitive in such a dynamic banking landscape is through the right use of technology"), government organizations face fundamentally different considerations.

As documented by Solender and Fried, the White House's Chief Administrative Officer Catherine Szpindor declared Microsoft Copilot "unauthorized for House use" due to security concerns. This difference reflects the distinct leadership challenges each context presents.

Private sector leaders like Parsons can emphasize competitive advantage and market positioning, focusing on opportunities that drive business value and shareholder returns.

Government leaders must balance innovation with public accountability, data sovereignty, security protocols, and citizen privacy – considerations that don't exist in the same way for private organizations. The House's cautious approach reflects legitimate concerns about data handling, transparency requirements, and the higher scrutiny that comes with managing public resources and citizen information.

Rather than representing a simple contrast between opportunity-focused and risk-focused thinking, these examples demonstrate how effective AI leadership requires adapting communication strategies to organizational context, stakeholder expectations, and regulatory environments. What constitutes responsible AI leadership varies significantly between contexts where competitive advantage drives decisions vs. those where public trust and accountability are paramount.

ACTION STEP: Before drafting your AI value narrative, first assess your organizational context: Are you in a competitive private sector environment where opportunity-focused messaging resonates? A regulated industry where risk mitigation is paramount? A government setting where public accountability shapes communication? Then adapt the four-part framework (tension, shift, impact, ask) to emphasize the aspects most relevant to your context – competitive advantage for private sector, citizen benefit, and responsible stewardship for government, or regulatory compliance and risk management for highly regulated industries. Test your context-appropriate narrative with stakeholders who understand your organizational constraints and expectations.

Managing Expectations throughout the Journey

Leaders must balance enthusiasm with realism.

A retail CIO built credibility by being transparent about both the successes and challenges of a customer analytics platform. When issues arose, he communicated them proactively with both explanations and mitigation plans. This transparency increased confidence in the initiative rather than undermining it.

Effective Expectation Management Includes
- Setting appropriate timelines for value realization
- Being transparent about limitations and risks
- Distinguishing between vision and current reality
- Showing progress while acknowledging challenges
- Maintaining consistency across communications

The Truth Paradox: Why Transparency Builds Trust

One counter-intuitive insight from my work with hundreds of organizations: leaders who are transparent about AI limitations and challenges typically build more trust than those who present an overly optimistic picture.

By ensuring transparency – both about possibilities and limitations – organizations can build trust and mitigate risk, ensuring better outcomes.

The Expectation Management Framework

Successful AI leaders use the following practices to manage expectations effectively:

1. **Value-Range Setting**
   - Communicate potential value as a range, not a point estimate
   - Include both optimistic and conservative scenarios
   - Clarify assumptions behind different projections
   - Update ranges as more information becomes available
2. **Timeline Realism**
   - Distinguish between technical implementation and value realization
   - Account for organizational absorption and adaptation time
   - Build in buffer for unexpected challenges
   - Create milestone-based communication plan
3. **Capability Clarity**
   - Be explicit about what the AI can and cannot do
   - Distinguish current capabilities from future possibilities
   - Provide concrete examples of both strengths and limitations
   - Address common misconceptions proactively
4. **Continuous Calibration**
   - Regularly update stakeholders on progress and challenges
   - Acknowledge when expectations need adjustment
   - Celebrate real wins while maintaining perspective
   - Create feedback mechanisms for stakeholder concerns

WARNING SIGN: If your AI communication contains only positive messages without acknowledging limitations or challenges, you may be unintentionally setting unrealistic expectations that will undermine trust when reality hits.

QUICK WIN: Create a simple "expectation calibration" section in your next AI status update that explicitly addresses: (1) What's going better than expected, (2) What's going as expected, and (3) What's proving more challenging than expected.

CASE STUDIES IN AI LEADERSHIP EXCELLENCE

Let's examine how different organizations have built effective leadership models by looking at four case studies, where measures have led to exceptional AI value creation.

Consumer Products: The Dual-Track Leadership Model

A global consumer products company achieved remarkable results through a distinctive leadership approach:

Leadership innovation: The CEO established a dual-track structure with a Value Transformation Office focused on business outcomes and an AI Centre of Excellence focused on technical capabilities. These groups were connected through a monthly AI Leadership Council that he personally chaired.

Key Leadership Actions

- Linked AI initiatives to three priority value areas aligned with strategy
- Created a dedicated transformation fund outside normal budgeting
- Established cross-functional AI teams with clear value accountability
- Implemented an AI leadership academy to develop internal capabilities
- Created a multi-level value dashboard for consistent measurement

Results

- $350 million in documented annual value creation
- 40% reduction in product development cycle time
- 28% improvement in marketing effectiveness
- Transformation of competitive position in key markets

Leadership insight: The consistent focus on business outcomes rather than technology prevented the initiative from becoming a technical showcase without business impact.

The Leadership Balance: Technical Excellence with Business Focus

What made this dual-track model particularly effective was the careful balance of technical excellence and business focus. While the AI Centre of Excellence pursued technical innovation and capability-building, the Value Transformation Office maintained relentless focus on business outcomes.

The monthly AI Leadership Council, chaired by the CEO, provided the integration mechanism that prevented these tracks from diverging. This council:

1. Reviewed progress against value targets
2. Resolved conflicts between technical and business priorities
3. Adjusted resource allocation based on demonstrated value
4. Identified new opportunities for AI application
5. Shared learning across initiatives

PATTERN RECOGNITION: The dual-track model works particularly well in organizations with strong functional orientation and significant technical complexity. By creating dedicated structures for both technical excellence and business value, this prevents the common problem of technical priorities overshadowing business outcomes.

ACTION STEP: Assess whether your organization's current structure effectively balances technical excellence with business value focus. If these responsibilities are currently merged without clear accountability for business outcomes, consider creating distinct but tightly integrated roles – such as dedicated value orchestrators who work closely with technical teams, or regular integration forums where technical and business leaders jointly review progress against value targets.

HEALTHCARE: THE CLINICAL LEADERSHIP MODEL

A healthcare system created a distinctive clinical leadership approach to AI transformation:

Leadership innovation: The organization established a three-tier leadership model with a clinical AI council setting direction, care delivery innovation teams implementing solutions, and a patient insight panel providing feedback and ideas.

Key Leadership Actions

- Identified physician champions for each initiative with protected time
- Created care-focused success metrics that aligned with clinical values
- Established rapid feedback and improvement cycles for continuous learning
- Developed mechanisms to distribute successful innovations across the system
- Invested in clinical workflow integration expertise

Results

- 28% reduction in avoidable readmissions
- 35% improvement in preventive care compliance
- 42% decrease in administrative burden on clinicians
- $85 million annual value creation across the system

Leadership insight: The investment in physician time through protected hours created the clinical engagement that many healthcare AI initiatives lack.

The Protected Time Breakthrough

The most striking innovation in this leadership model was the commitment to protected time for physician champions. While many healthcare organizations expect clinical leaders to participate in technology initiatives on top of their regular clinical responsibilities, this organization took a different approach:

1. **Formal Role Definition:** Physician champions had clearly defined responsibilities and expectations.
2. **Protected Time:** 20% of their schedule was dedicated to AI implementation.
3. **Recognition:** Leadership contribution was factored into performance evaluations.
4. **Development Support:** Champions received leadership training and coaching.
5. **Continuity Commitment:** Champions remained involved from design through implementation.

This approach recognized that effective clinical leadership for AI requires more than occasional input, it requires sustained engagement throughout the development and implementation process.

KEY INSIGHT: The most successful healthcare AI implementations I've observed share this common characteristic: they provide protected time for clinical leadership rather than treating it as an "add-on" responsibility.

Case Connection: WellSpan Health's Approach

This case resonates with Dr. R. Hal Baker's approach at WellSpan Health. Instead of viewing AI as a technology to be implemented, Dr. Baker created a comprehensive ecosystem approach that centred on clinician experience and patient care. This leadership model, which puts clinical outcomes – rather than technology implementation – at the centre, creates stronger engagement and more sustainable value.

FINANCIAL SERVICES: THE GOVERNANCE LEADERSHIP MODEL

A global financial institution created a novel governance approach led from the board level.

Leadership innovation: The board established a multi-level governance structure with a digital transformation committee providing oversight, an AI value council driving implementation, an ethics and risk panel guiding responsible use, and value realization teams delivering specific use cases.

Key Leadership Actions

- Created a dedicated $200 million transformation fund with multi-year commitment
- Established an AI leadership recruiting strategy focusing on hybrid skills
- Developed a balanced approach to innovation and control
- Built a comprehensive value dashboard connecting activities to outcomes
- Created an AI capability academy for ongoing leadership development

Results

- $420 million annual value creation through efficiency and growth
- 32% reduction in regulatory compliance costs
- 45% improvement in risk prediction accuracy
- Transformation of competitive position in digital banking

Leadership insight: The board's creation of a dedicated governance committee with appropriate expertise sent a powerful signal about the strategic importance of AI and the need to balance innovation with responsible use.

The Multi-Level Governance Innovation

What made this financial institution's approach particularly effective was its multi-level governance model (see Table 15.13) that scaled oversight based on risk and strategic importance.

This tiered approach accomplished several critical things:

1. Sent a clear signal about strategic importance through board-level involvement
2. Created appropriate oversight for different types of decisions
3. Balanced innovation needs with risk management
4. Connected governance to value realization
5. Built governance expertise throughout the organization

TABLE 15.13
Multi-Level Governance Model

Governance Level	Composition	Responsibility	Meeting Cadence
Digital Transformation Committee	Board members with technology and transformation expertise	Strategic direction, investment approval, ethical oversight	Quarterly
AI Value Council	C-suite executives plus AI leadership	Implementation oversight, value tracking, cross-functional coordination	Monthly
Ethics and Risk Panel	Cross-functional experts in ethics, risk, compliance, technology	Ethical framework development, risk assessment, policy recommendations	Bi-weekly
Value Realization Teams	Cross-functional teams focused on specific use cases	Implementation, value delivery, stakeholder engagement	Weekly

Manufacturing: The Integration Leadership Model

A global manufacturer developed a distinctive integration-focused leadership approach:

Leadership innovation: The company created "value integration teams" composed of cross-functional leaders who were physically co-located and collectively accountable for end-to-end value delivery from specific AI initiatives.

Key Leadership Actions

- Reorganized traditional functional structures into value streams
- Implemented shared performance metrics across functions
- Created an "AI rotation program" for leadership development
- Established daily integration huddles focused on removing barriers
- Developed a learning system to capture and spread insights

Results

- 32% reduction in inventory carrying costs
- 18% improvement in on-time delivery performance
- 41% decrease in production downtime
- $95 million annual value creation across the supply chain

Leadership insight: This radical approach to physical co-location – moving leaders from different functions to the same workspace – created unprecedented collaboration that overcame decades of functional silos.

The Co-Location Revolution

The most dramatic element of this manufacturer's approach was the commitment to physical co-location. While many organizations create virtual cross-functional teams that meet periodically, this company took a more radical approach:

1. **Physical Space Redesign:** Created dedicated spaces for value integration teams
2. **Full-Time Dedication:** Team members were 100% dedicated to their value stream
3. **Leadership Proximity:** Leaders from different functions worked side-by-side daily
4. **Visual Management:** Used physical information radiators to track value and barriers
5. **Daily Integration Rituals:** Structured routines to ensure continuous alignment

By creating a physical environment that forced continuous integration, the manufacturing firm dramatically accelerated the maturation of cross-functional collaboration.

PATTERN RECOGNITION: While full co-location isn't practical for all organizations, the principle of increasing interaction frequency is universally applicable. The more frequently cross-functional team members interact, the faster they develop shared understanding and collaborative capability.

QUICK WIN: Even if full co-location isn't possible, try creating a "collaboration day" where cross-functional team members work in the same physical or virtual space one day per week, focusing exclusively on integration challenges and opportunities.

CULTURAL LEADERSHIP: THE HIDDEN SUCCESS FACTOR

Beyond structures and processes, cultural leadership emerged as a critical success factor in my research.

While most organizations focus on creating the right organizational charts and governance processes, they often overlook how cultural norms and values shape AI implementation success.

At one organization I advised, the AI initiatives stalled not because of technical challenges, but because of unspoken cultural barriers. The engineering team valued precision and thorough analysis, often spending weeks perfecting models before sharing results.

Meanwhile, the business team operated in a fast-paced environment where quick "good enough" decisions often trumped perfect solutions delivered late.

These conflicting cultural approaches to decision-making and risk created an invisible barrier that no amount of structural reorganization could overcome.

The cultural dimension of AI leadership requires conscious attention to:

- **Decision-making styles** and how they vary across teams
- **Psychological safety** that enables honest communication
- **Blame culture vs. learning culture** and its impact on innovation
- **Empowerment balance** that provides appropriate autonomy

THE CULTURAL LEADERSHIP FRAMEWORK

Based on my observations across organizations, there are four cultural dimensions that leaders must address for AI success (see Table 15.14).

TAG's Cultural Transformation

Leadership at TAG was able to transform organizational culture by addressing these dimensions directly through implementing the following deliberate leadership actions:

1. **Decision Rights Clarification:** Making explicit who had authority for different types of decisions
2. **Failure Response Shift:** Moving from blame to learning when problems occurred
3. **Information Democratization:** Making more data and context available to team members
4. **Experimental Encouragement:** Creating space for appropriate risk-taking

This cultural transformation created the foundation for data-driven decision-making by ensuring team members felt safe to surface problems, experiment with solutions, and learn from results.

KEY INSIGHT: Cultural transformation doesn't happen by accident – it requires consistent leadership behaviours that demonstrate new values in action. What leaders pay attention to, measure, reward, and model has far more impact than what they say.

TABLE 15.14
Cultural Leadership Framework

Cultural Dimension	Inhibiting Culture	Enabling Culture	Leadership Actions
Decision Approach	Consensus paralysis or top-down dictates	Balanced decision rights with appropriate involvement	Define clear decision processes with explicit roles and rights
Learning Orientation	Blame-focused with punishment for failure	Learning-focused with emphasis on growth	Model openness and learning from mistakes
Information Flow	Restricted information shared on "need to know" basis	Transparent sharing with appropriate context	Make information widely available with context for interpretation
Innovation Approach	Risk-averse with high barriers to new ideas	Balanced approach supporting appropriate experimentation	Create "safe spaces" for innovation with clear boundaries

Leading AI Value Creation

REFLECTION QUESTIONS: Which cultural dimension in your organization might be creating the biggest barrier to AI value? What specific leadership behaviours might help shift that dimension?

THE FUTURE OF AI LEADERSHIP: EMERGING CHALLENGES

As AI capabilities evolve, leadership approaches must adapt as well. Three emerging challenges deserve particular attention.

Navigating Leadership in the Generative AI Era

Generative AI's rapid emergence creates new leadership challenges that traditional AI implementation approaches cannot address. Unlike previous AI technologies that typically served specific, narrow functions, generative AI has the potential to transform virtually every business process simultaneously. This creates unprecedented coordination challenges for leaders.

A Fortune 500 retail company recently faced this challenge when different departments began experimenting with generative AI independently. Marketing launched a content generation pilot, customer service tested an AI chatbot, and operations explored automated report writing.

Within months, they had created a fragmented landscape of incompatible tools, duplicate investments, and inconsistent user experiences. The leadership team realized they needed a fundamentally different approach – one that balanced innovation speed with enterprise coherence.

This ecosystem approach requires leaders to think beyond individual use cases to comprehensive transformation.

The Generative AI Leadership Challenge

The rise of generative AI creates several distinct leadership challenges:

1. **Expectation Management:** Generative AI's impressive demos often create unrealistic expectations about capabilities and implementation timelines. Leaders must manage the gap between demos and production reality.
2. **Integration Complexity:** Generative AI touches virtually every aspect of organizational operations. Leaders must coordinate across functions to create coherent implementation approaches.
3. **Ethical Considerations:** The potential for misinformation, bias, and misuse requires leaders to implement appropriate governance without stifling innovation.
4. **Workforce Transformation:** Generative AI will transform virtually every job. Leaders must guide this transformation while maintaining workforce engagement and addressing legitimate concerns.

Two Leadership Approaches: Comprehensive vs. Incremental

Research into generative AI adoption reveals a tension between comprehensive and incremental approaches, which Table 15.15 illustrates.

Most organizations will need to balance these approaches – and create an enterprise framework while enabling agile experimentation within it.

WARNING SIGN: If your organization is pursuing generative AI without a cross-functional coordination mechanism, you risk creating fragmented approaches, duplicate investments, and inconsistent user experiences.

Leading with Ethical Responsibility

As AI systems become more powerful, leadership responsibility for ethical oversight increases. The UAE demonstrated this understanding by appointing a Minister of State for AI, Digital Economy, and

TABLE 15.15
Comprehensive Approach vs. Incremental Approach

Comprehensive Approach	Incremental Approach
Strengths:	**Strengths:**
• Creates coherent ecosystem	• Faster time to initial value
• Prevents fragmentation	• Lower initial investment
• Enables synergies across uses	• More manageable scope
• Builds enterprise-wide capabilities	• Easier to secure initial buy-in
Challenges:	**Challenges:**
• Higher initial investment	• Potential for fragmentation
• More complex coordination	• Duplicate investments
• Longer time to first value	• Missed synergies
• Risk of analysis paralysis	• Inconsistent governance

Remote Work Applications in 2017 and creating comprehensive governance frameworks. This proactive approach to ethical leadership will become increasingly important as AI capabilities expand.

The Ethics Leadership Imperative

The increasing power of AI systems creates new ethical leadership responsibilities that go beyond traditional corporate governance. As AI capabilities expand from narrow applications to systems that can generate content, make autonomous decisions, and influence human behaviour at scale, leaders must grapple with questions that didn't exist even five years ago.

Unlike previous technology implementations, AI ethics isn't just about compliance – it is about maintaining organizational legitimacy and social license to operate. The tension we see between different organizational approaches – from comprehensive ethics councils to distributed champion networks – reflects the fact that there's no one-size-fits-all solution to ethical AI leadership.

A shift from reactive to proactive ethical leadership is essential and reflects a broader recognition that AI systems fundamentally can create unintended consequences that are difficult to reverse once deployed at scale. Leaders who wait for ethical issues to emerge often find themselves in damage-control mode, whereas those who build ethical considerations into their leadership approach from the beginning create sustainable competitive advantage through stakeholder trust.

Three Models for Ethical AI Leadership

Research reveals three distinct approaches to ethical AI leadership:

1. **Centralized Ethics Council Model**
    - **Approach:** Formal ethics council with dedicated members
    - **Strength:** Consistent standards and oversight
    - **Challenge:** Potential disconnect from business realities
    - **Example:** Orange's ethics council
2. **Ethics Champions Network Model**
    - **Approach:** "RAI champions" embedded across business units
    - **Strength:** Domain-specific expertise and implementation knowledge
    - **Challenge:** Potential inconsistency across units
    - **Example:** Tech companies using "responsible AI champions"
3. **Multi-Tiered Escalation Model**
    - **Approach:** Three-level system from routine assessment to executive decisions
    - **Strength:** Balances efficiency with appropriate oversight
    - **Challenge:** Potential bottlenecks at higher levels
    - **Example:** GSMA's three-tiered governance model

The UAE's Legislative Approach

The UAE demonstrates how government leadership can create enabling regulatory environments while maintaining governance oversight. As detailed in research, "in 2017, the UAE appointed a Minister of State for AI, Digital Economy, and Remote Work Applications, recognizing the need for federal oversight and strategic direction."

This was followed by a comprehensive national strategy and legislative framework, including "Law 25 of 2018 [which] grants the UAE Cabinet the authority to grant interim licenses and establish a licensing regime for innovative projects involving modern technologies or AI without existing legislation."

This case illustrates how strategic government leadership can balance innovation and ethical considerations through:

1. Clear leadership accountability at the highest level
2. Comprehensive strategy development
3. Flexible regulatory frameworks that enable innovation
4. Appropriate oversight mechanisms

KEY INSIGHT: As AI systems become more capable and autonomous, ethical leadership must evolve from reactive to proactive, addressing potential issues before they arise rather than responding after problems occur.

ACTION STEP: Review your organization's current approach to AI ethics and governance. Does it reflect the increasing power and autonomy of AI systems? Are you proactively addressing ethical considerations or primarily reacting to issues as they arise?

BALANCING SPEED AND RESPONSIBLE IMPLEMENTATION

The competing priorities of rapid implementation and appropriate safeguards represent a core leadership challenge. Future AI leaders will need to navigate this tension more effectively, balancing speed with responsibility.

This tension appears in multiple contexts throughout the research:

1. **Executive Pressure vs. Foundational Needs**
   AI implementation is widely regarded as a strategic necessity and competitive advantage, causing urgency in driving deployment at the expense of meeting foundational needs.
2. **Technical Integration vs. People Readiness**
   The accelerating pace of technological progress can mean that human adaptation lags behind, creating a critical disconnect.
3. **Innovation Freedom vs. Governance Control**
   The research on AI governance highlights the tension between centralized and decentralized governance approaches. The Infocomm Media Development Authority (IMDA) and Personal Data Protection Commission (PDPC) report acknowledges this tension, noting that "organisations should have robust internal governance structures and measures to oversee their use of AI. These structures can be adapted or new if necessary… De-centralised governance mechanisms can be considered when centralised governance is not optimal."

The Balanced Implementation Framework

Based on my observations of successful AI implementations, leaders can resolve these tensions through a balanced implementation approach (see Table 15.16).

TABLE 15.16
Balanced Implementation Framework

Dimension	Too Slow	Too Fast	Balanced Approach
Strategic	Missing market opportunities	Insufficient strategic alignment	Phased implementation aligned with strategy
Technical	Over-engineering before value	Technical debt from shortcuts	Value-first architecture that enables scaling
People	Excessive change preparation	Change shock from rapid implementation	Change absorption matched to organizational capacity
Governance	Innovation paralysis from excessive control	Unmanaged risks from insufficient oversight	Risk-based governance scaled to potential impact

Leadership Questions for Balanced Implementation

To find the right balance between speed and responsibility, leaders should ask:

1. What is the minimum viable foundation needed before we start?
2. How can we phase implementation to deliver early value while building for the future?
3. What governance mechanisms provide appropriate oversight without creating unnecessary barriers?
4. How can we match our implementation pace to our organization's ability to absorb change?

PATTERN RECOGNITION: The most successful AI implementations typically start small but think big, balancing a focus on quick wins that create momentum with building foundations for long-term transformation.

WARNING SIGN: If your AI implementation approach lacks either speed metrics (value delivery timeline) or responsibility metrics (risk management, quality assurance, change absorption), you may be overemphasizing one dimension at the expense of the other.

THE LEADERSHIP IMPERATIVE: A CALL TO ACTION

As we conclude this exploration of AI leadership, I offer three critical insights from my work with hundreds of organizations on their AI journeys:

1. **Leadership as a Key Differentiator:** The organizations that achieve transformative value from AI aren't necessarily those with the most advanced technology or the largest data science teams. They're the ones with leaders who can connect AI to strategy, build cross-functional collaboration, translate between technical and business worlds, and maintain relentless focus on value realization.
2. **Leadership Must Evolve:** Traditional leadership approaches are insufficient for AI transformation. The rapid pace of technological change, the cross-functional nature of implementation, and the profound organizational implications require new leadership capabilities and models.
3. **Leadership Development Is Essential:** Building the leadership capabilities needed for AI value creation requires deliberate investment. Organizations must develop both individual competencies and collective leadership systems that enable sustainable transformation.

THE AI LEADERSHIP DEVELOPMENT GAP

There is an educational gap affecting AI leadership, due, in part, to a lack of "courses on artificial intelligence or even the concept of sustainable, tech-driven business models," as defined by David

Leading AI Value Creation

De Cremer in his research (mentioned above). Organizations have to address the resulting leadership readiness challenge directly.

The gap exists at multiple levels:

- **C-Suite:** Limited understanding of AI capabilities and strategic implications
- **Middle Management:** Insufficient skills to lead cross-functional implementation
- **Team Leaders:** Inadequate preparation for human-AI collaboration

Key Takeaways on AI Leadership

Throughout this chapter, we've explored how effective leadership drives AI value creation:

1. **Leadership Roles Must Be Clearly Defined and Filled Effectively:**
   - **Executive Sponsor:** Authority and air cover
   - **Value Orchestrator:** Daily integration leadership
   - **Technical Leader:** Value-focused technical direction
   - **Change Leader:** Navigating the human dimension
   - **AI Value Advisor:** External perspective
2. **Critical Leadership Competencies Must Be Developed Intentionally:**
   - Strategic thinking with technical awareness
   - Value translation between technical and business worlds
   - Cross-functional leadership across organizational boundaries
   - Change leadership through the transformation journey
   - Learning agility in a rapidly evolving landscape
3. **Leaders Must Create Organizational Conditions for AI Success:**
   - Setting a compelling vision that connects to strategy
   - Building cross-functional collaboration structures
   - Creating a data-driven culture where evidence trumps opinion
   - Balancing innovation and governance appropriately
4. **Top Leadership Plays a Unique and Critical Role:**
   - Setting the strategic context for AI initiatives
   - Creating accountability for value realization
   - Addressing organizational barriers to success
   - Modelling leadership behaviours that drive success
5. **Leadership Development Requires Systematic Investment:**
   - Building AI leadership literacy throughout the organization
   - Developing a value-focused mindset that starts with outcomes
   - Creating leadership development pathways for sustained capability
   - Building a leadership community for collective learning

YOUR LEADERSHIP ACTION PLAN

To translate these insights into action, consider these five steps:

1. **Assess your current AI leadership capabilities** across the five dimensions we've explored
2. **Identify your most critical leadership gaps** that could undermine value creation
3. **Develop a leadership capability-building plan** focused on high-priority areas
4. **Create the organizational conditions** that enable AI value through structures, processes, and culture
5. **Establish your AI leadership rhythm** that balances strategic perspective with operational engagement

TABLE 15.17
AI Leadership Assessment Framework

Leadership Dimension	Developing (1–3)	Competent (4–7)	Advanced (8–10)	Your Score
Strategic Vision	AI viewed primarily as technology initiative	AI connected to specific business outcomes	AI integrated into strategic transformation	
Cross-functional Integration	Siloed implementation with limited coordination	Formal coordination mechanisms between functions	Integrated teams with shared accountability	
Value Translation	Limited connection between technical and business concepts	Basic translation capabilities in key roles	Fluent translation at all levels	
Change Acceleration	Ad hoc change management without cohesive approach	Structured change approach for individual initiatives	Comprehensive change capability for transformation	
Capability-Building	Limited focus on building sustainable capabilities	Targeted capability development for key roles	Systematic capability-building throughout the organization	

THE AI LEADERSHIP ASSESSMENT FRAMEWORK:

Use the framework in Table 15.17 to evaluate your organization's current AI leadership capabilities across five critical dimensions.

ACTION STEP: Complete this assessment with your leadership team to identify your organization's strengths and gaps. Focus your development efforts on the areas with the largest gaps and greatest strategic importance.

LEADERSHIP DEVELOPMENT PRIORITIES

Based on your assessment, consider which leadership capabilities should be your highest development priorities:

1. **For Low Strategic Vision Score:**
   - Engage executives in AI education focused on strategic possibilities
   - Create strategic scenario-planning sessions incorporating AI capabilities
   - Develop a clear AI value narrative connected to organizational strategy
2. **For Low Cross-Functional Integration Score:**
   - Create formal integration mechanisms across functions
   - Implement shared metrics that drive collaborative behaviour
   - Consider structural changes to reduce organizational barriers
3. **For Low Value Translation Score:**
   - Develop translation tools for common AI concepts in your business
   - Create paired leadership models connecting technical and business perspectives
   - Implement communication practices that bridge technical and business languages
4. **For Low Change Acceleration Score:**
   - Invest in formal change-leadership capabilities
   - Develop a structured change approach for AI initiatives
   - Create feedback mechanisms to continuously improve change effectiveness

5. **For Low Capability-Building Score:**
   - Create formal leadership-development programmes for AI
   - Implement experiential-learning approaches through real projects
   - Establish communities of practice to share learning across the organization

The tools, templates, and frameworks included in this book and on the website (www.valuesofai.com) provide practical resources for implementation. But remember that it is the quality of your leadership – rather than the sophistication of your tools – that will ultimately determine your success in creating and capturing the multidimensional values of artificial intelligence.

CONCLUSION: THE LEADERSHIP FACTOR

As we close this exploration of AI leadership, it's worth reflecting on the fundamental insight that has emerged throughout this chapter: leadership is the decisive factor in AI value creation.

The most sophisticated AI technologies, the largest data science teams, and the most generous budgets cannot compensate for inadequate leadership. Conversely, even organizations with modest technical resources can achieve remarkable results when their leaders excel at connecting AI to strategy, orchestrating cross-functional collaboration, translating between technical and business worlds, and maintaining unwavering focus on outcomes.

Key Takeaways

1. **Leadership Transformation Precedes Technological Transformation:** Organizations that invest in developing AI leadership capabilities before or alongside technical capabilities consistently achieve better results than those focused primarily on technical excellence.
2. **New Leadership Approaches Are Required:** Traditional leadership models that worked in stable, hierarchical environments often fail in the cross-functional, rapidly evolving context of AI implementation. Leaders must develop new capabilities and adopt new approaches specifically suited to AI's unique challenges.
3. **Balance Is Essential:** The most effective AI leaders strike a delicate balance between seemingly opposing forces: technical expertise and business acumen, strategic vision and operational execution, innovation freedom and appropriate governance, speed and responsibility.
4. **Multiple Leadership Roles Working in Concert Drive Success:** No single leader can provide all the capabilities needed for AI value creation. Successful initiatives depend on multiple complementary roles – from executive sponsors to value orchestrators to technical leaders to change leaders – working together effectively.
5. **Leadership Development Is a Strategic Investment:** Building the leadership capabilities needed for AI value creation requires deliberate, systematic investment. This investment yields returns not just in current AI initiatives but in creating sustainable capabilities for ongoing transformation.

As you move forward on your AI value journey, I encourage you to make leadership development a central focus of your strategy. While AI technologies will continue to evolve at a rapid pace, the fundamental leadership principles we've explored in this chapter will remain relevant regardless of specific technical approaches.

By investing in leadership capabilities, creating the organizational conditions for success, and establishing the right rhythms and processes, you can position yourself and your organization to capture the full multidimensional value of artificial intelligence, not just today but for years to come.

16 From Chaos to Clarity
Your AI Value Creation Toolkit

Chapter Roadmap

The following pages are dedicated to resources that are designed to unlock AI value during different project stages and for various kinds of organizations. The goal for sharing the tools, templates and framework that I have found useful in my work is to remove barriers to getting started – and encourage meaningful action.

THE MONDAY MORNING PROBLEM

Asking the right questions at the right time is key. As one RAND study found, "leaders often instruct the data science team to solve the wrong problem with AI," resulting in teams "working hard for months to deliver a trained AI model that will have minimal impact on the business."

Here's the staggering scale of this misalignment: while over 65% of organizations are already regularly using generative AI – nearly double from just ten months prior – and over 45% of new code is being generated by AI, most leaders can't quantify the value they're creating. We're witnessing the fastest technology adoption in business history, yet it's being built on a foundation of fundamental confusion about what problems AI should actually solve.

The contradiction gets deeper. Research shows developers complete tasks 55% faster using GitHub Copilot and report 75% greater job satisfaction. Yet when Uplevel studied 800 developers using the same tool, they "found no significant improvements in pull request cycle time or PR throughput." Two studies, same technology, completely opposite conclusions about value creation.

This isn't just a measurement problem – it is a judgement problem. As AI industry expert Carol Reiley warns: "I think if a human can't tell, then it's going to be hard for AI to tell." When we're asking artificial intelligence to solve business challenges we haven't clearly defined, we end up wondering why the results feel disconnected from our strategic priorities.

Consider the extremes: JP Morgan's COIN programme revolutionized legal document processing, handling in seconds what previously required 360,000 hours of lawyer work annually. Meanwhile, the United States Postal Service struggles with about 950 applications that "aren't well documented" because they "lost the first developer on the project." Same technology category, radically different value outcomes.

The difference isn't due to technical sophistication, it comes down to whether organizations start with clear value objectives or get seduced by AI's technical possibilities. Again and again, I've seen leaders approach AI like they're shopping for a solution to a problem they haven't properly diagnosed. The result? Brilliant technical execution that misses the strategic mark entirely.

This gap between AI's potential and practical value creation isn't a technical problem waiting for better algorithms. It's a leadership challenge that requires better tools for connecting artificial intelligence to authentic organizational value. The Monday morning question isn't "What *can* AI do?" It is "What problems *should* AI be solving?"

This creates a fundamental tension for leaders building AI value toolkits: sophisticated frameworks may look impressive in presentations, but they rarely get used consistently in the messy reality of organizational life. The truth is that simple tools that people actually use consistently deliver more value than perfect tools that sit unused.

Yet individual frameworks, no matter how good, are often not sufficient for creating sustained success. The real power comes from how they work together.

Consider the experience of a global retailer, where teams initially tried using a comprehensive project management framework with 47 different checkpoints and dependencies. The complexity paralyzed implementation. When they switched to a simplified four-phase approach with clear decision points, progress accelerated dramatically.

THE GAME THEORY SOLUTION: INCENTIVIZING COLLABORATION THROUGH DESIGN

Here is where the research reveals that something more is needed than just "keeping it simple."

As previously mentioned, Dr. Thuc Vu, entrepreneur and co-founder of OhmniLabs and Kambria, proposes a different lens. "If you think of collaboration in business as a game, there [are] some very interesting game theoretical approaches, [where] we can not only incentivize but [actively] force collaboration between the players in the game."

That's an insight that resonates since I have found a strong correlation between the success of AI implementation and the collaborative capabilities of an organization. This means effective AI value tools go beyond providing methods – to creating incentive structures that naturally promote the collaboration AI initiatives require. Research on gamification shows that techniques incorporating "rewards, storytelling, and healthy competition" provide structured frameworks for punching through the wall of departmental silos, leading to "enhanced productivity, increased trust, and an overall supportive culture."

The Norwegian government proved this works in practice. A five-step cross-functional framework was structured to enhance success by being simple and designed to incentivize collaboration: (1) forming diverse teams including AI/ML experts, software developers, and customer support representatives; (2) conducting collaborative workshops to align technical capabilities with practical needs; (3) establishing open communication channels; (4) training support representatives on AI collaboration; and (5) conducting post-implementation reviews. Each step built on the previous one while creating natural incentives for continued participation.

FROM THEORY TO PRACTICE: BUILDING YOUR AI VALUE TOOLKIT

So how do you bridge the gap between understanding these principles and implementing them Monday morning? Research reveals that successful organizations don't choose between sophisticated and simple; they build layered toolkits that start simple and add sophistication as capabilities mature.

Based on analysis of successful implementations across industries, every effective AI value toolkit needs tools in five categories that address the core challenges identified in our opening: problem clarity, value measurement, cross-functional collaboration, implementation execution, and continuous evolution.

OVERVIEW OF KEY TOOLS AND WHEN TO USE THEM

This toolkit includes tools for every stage of the AI value journey:

Assessment and Planning Tools

- AI value opportunity assessment
- Data readiness evaluation
- AI value maturity diagnostic
- Stakeholder alignment analysis

Strategic Alignment Tools

- AI value-mapping canvas
- Cross functional value matrix
- Executive narrative builder
- Business case calculator

Implementation Tools

- Scaling blueprint template
- Adoption flywheel tracker
- Value realization dashboard
- Risk assessment framework

Sustaining Tools

- Continuous improvement checklist
- Performance evolution tracker
- Capability development planner
- Long-term roadmap builder

Rather than using every tool for every initiative, select those that address your specific challenges and context. A startup implementing first AI capabilities needs different tools than a global enterprise scaling across multiple business units.

Online resource integration: To support your implementation efforts, interactive versions of these tools are available at the online resource centre at www.theaivalues.org, including:

- AI value opportunity assessment (interactive diagnostic)
- Stakeholder alignment analysis (collaborative mapping tool)
- AI value mapping canvas (digital template)
- Cross-functional value matrix (dynamic framework)
- AI trust dashboard (digital tracking)
- Scaling blueprint template (customizable roadmap)
- Value realization dashboard (real-time tracking)

These digital tools allow you to collaborate with your team, save progress, and adapt frameworks to your specific organizational context.

Adapting Frameworks to Your Organization

Every organization brings unique context, constraints, and capabilities to their AI journey. Effective frameworks provide structure while remaining flexible enough to accommodate these differences.

When adapting these tools to your organization, consider five key dimensions:

- **Industry Context:** Regulatory requirements, competitive dynamics, and customer expectations vary dramatically across industries. A healthcare organization implementing clinical AI faces fundamentally different constraints than a retailer implementing personalization engines.
- **Organizational Scale:** A 50-person startup has different governance needs, resource constraints, and implementation capabilities than a 50,000-person multinational corporation. Scale affects not just resource availability but also communication patterns, decision-making processes, and change management requirements.

- **Technical Maturity:** Organizations with established data science teams and modern infrastructure can move faster than those building foundational capabilities. Your current technical maturity should guide both ambition and implementation approaches.
- **Cultural Characteristics:** Some organizations embrace experimentation and iteration while others prefer predictable, linear progress. Successful frameworks match organizational culture rather than fighting against it.
- **Strategic Context:** AI initiatives serving existential strategic needs require different approaches than those addressing operational improvements. Strategic urgency affects risk tolerance, resource allocation, and timeline expectations.

A pharmaceutical company exemplified effective adaptation when implementing a drug discovery AI. Taking the basic value mapping canvas, the organization modified it to include regulatory considerations, clinical trial implications, and research collaboration requirements – creating a tool that addressed the specific context while maintaining the core structure.

Implementation Considerations and Requirements

Before diving into specific tools, it's important to understand what successful implementation typically requires.

- **Leadership Engagement:** These tools work best when leadership actively participates rather than delegating to staff. Executive engagement signals importance and ensures frameworks address real priorities rather than perceived ones.
- **Cross-Functional Participation:** AI value creation crosses organizational boundaries. Effective use of these tools requires input from diverse perspectives: technical, business, operational, and strategic.
- **Iterative Application:** Unlike traditional planning tools that get used once, these frameworks benefit from repeated application as understanding and circumstances evolve. Build iteration into your implementation approach.
- **Action Orientation:** These tools are designed to drive decisions and actions, not just create documentation. Focus on using insights to make progress rather than achieving perfect analysis.
- **Context Adaptation:** While the frameworks provide structure, effective implementation requires adaptation to your specific situation. Don't force your reality to fit the tool – adapt the tool to fit your reality.

Building Your Customized Toolkit

Rather than attempting to use every framework, most organizations benefit from selecting and customizing a subset that addresses their specific needs and stage of development.
Consider this progression:

- **Getting Started:** If you're early in your AI journey, focus on assessment and planning tools that create clarity about opportunities and readiness.
- **Building Momentum:** Once you have clarity, emphasize strategic alignment tools that create shared understanding and support for moving forward.
- **Implementing Successfully:** During active implementation, prioritize execution tools that maintain focus on value creation and adoption.
- **Sustaining Value:** For mature initiatives, emphasize sustaining tools that ensure continued evolution and improvement.

A manufacturing company followed this progression effectively, starting with the opportunity assessment and maturity diagnostic, then moving to the value mapping canvas and business case calculator, followed by the scaling blueprint and adoption tracker during implementation, and finally the performance evolution tracker and capability development planner for sustained value creation.

This sequential approach prevented overwhelm while ensuring they had the right tools for each phase of their journey.

THE AI VALUE OPERATING SYSTEM (AIValueOS)

Beyond individual tools, successful AI value creation requires an integrated operating system – a coordinated set of practices that work together to create sustainable results.

The AI value operating system (AIValueOS) represents the culmination of lessons learned from successful implementations across industries. It's not a rigid methodology but a flexible framework that adapts to different organizational contexts while maintaining focus on value creation.

Vision Development and Communication

AIValueOS begins with creating and communicating a compelling vision that connects AI capabilities to organizational purpose and strategic objectives.

Effective AI visions have several characteristics:

- **Outcome-Focused:** describing business results rather than technical capabilities
- **Stakeholder-Relevant:** connecting to priorities that matter to different organizational constituencies
- **Inspirational Yet Realistic:** creating excitement while maintaining credibility
- **Action-Oriented:** setting clear directions for moving forward

A financial services organization exemplified effective vision development with the following statement: "We will use AI to become the most trusted financial partner for our customers by providing personalized insights, proactive support, and transparent decision-making that helps them achieve their financial goals while protecting their interests."

This vision connected AI to a core value proposition (trust), specified the customer benefit (personalized insights and support), and implied specific capabilities (decision-making transparency) without getting lost in technical details.

Vision Development Process

1. **Strategic Alignment Assessment:** Map AI potential to current strategic priorities
2. **Stakeholder Value Exploration:** Understand what different groups want from AI
3. **Capability-to-Outcome Translation:** Connect technical possibilities to business results
4. **Vision Statement Crafting:** Create concise, compelling language
5. **Communication Strategy Development:** Plan how to share and reinforce the vision

Values-to-Actions Translation

The next AIValueOS component translates high-level values and vision into specific actions and decisions.

This translation (see Table 16.1) addresses a common failure pattern: organizations that articulate inspiring visions but struggle to connect them to daily decisions and activities.

A retail organization demonstrated effective translation by connecting the vision of "personalized customer experiences" to specific actions: implementing recommendation engines, redesigning

TABLE 16.1
Translation Framework

Vision Level	Translation Questions	Output
Strategic Vision	What business outcomes do we want?	Strategic objectives and success metrics
Operational Principles	How should we approach achieving these outcomes?	Decision criteria and prioritization guidelines
Tactical Actions	What specific steps will we take?	Project plans, resource allocations, milestone definitions
Daily Decisions	How do these principles guide daily choices?	Decision frameworks, escalation criteria, review processes

customer journey touchpoints, training associates on AI-enhanced service approaches, and measuring success through customer satisfaction and engagement metrics.

Role-Specific Ownership Matrix

AIValueOS includes a clear ownership matrix that defines responsibilities across different roles and functions.

Traditional organizational charts often prove inadequate for AI initiatives that require new forms of collaboration and shared accountability. The ownership matrix (see Table 16.2) creates clarity about who does what while enabling necessary flexibility.

By creating clear accountability for results, this matrix prevents the responsibility ambiguity that often undermines AI initiatives.

Feedback and Evolution Infrastructure

AIValueOS includes systematic approaches for capturing feedback and evolving based on learning.

Unlike traditional projects with fixed requirements, AI initiatives benefit from continuous adaptation based on results, user feedback, and changing circumstances.

Feedback Collection Mechanisms

- **Quantitative Performance Tracking:** Automated metrics and dashboards

TABLE 16.2
Key Roles in AI Value Creation

Role	Primary Responsibilities	Key Relationships
Executive Sponsor	Strategic direction, resource allocation, organizational alignment	Board, C-suite, business unit leaders
Value Champion	Business case development, outcome tracking, stakeholder alignment	Executive sponsor, functional leaders, implementation teams
Technical Lead	Architecture, development, integration, performance	Value champion, data teams, IT infrastructure
Data Steward	Data quality, governance, accessibility, compliance	Technical lead, business users, privacy/compliance
Adoption Facilitator	Change management, training, support, feedback collection	Business users, HR, communications
Business Representative	Requirements definition, testing, feedback, outcome validation	Value champion, adoption facilitator, end users

- **Qualitative User Feedback:** Surveys, interviews, and observation
- **Stakeholder Pulse Checks:** Regular assessment of satisfaction and engagement
- **External Environment Monitoring:** Competitive intelligence and market changes
- **Technical Performance Analysis:** Model accuracy, system reliability, integration effectiveness

Evolution Decision Framework

1. **Signal Recognition:** What feedback indicates need for change?
2. **Impact Assessment:** How significant is the issue or opportunity?
3. **Option Evaluation:** What changes could address the situation?
4. **Decision Making:** Who decides what changes to implement?
5. **Implementation Planning:** How will changes be executed?
6. **Results Tracking:** How will we know if changes are effective?

A healthcare organization used this approach effectively when their clinical decision support system received mixed physician feedback. Rather than defending the original design, they systematically collected feedback, identified key pain points, developed solutions, and implemented improvements that dramatically increased adoption and satisfaction.

Transparent Narratives Framework

AIValueOS emphasizes ongoing communication through transparent narratives that build understanding and maintain support.

Effective AI initiatives require sustained stakeholder engagement throughout implementation. This means moving beyond periodic status reports to ongoing storytelling that connects progress to values and outcomes.

Narrative Components

- **Progress Against Vision:** How current activities advance stated objectives
- **Learning Integration:** What we've discovered and how it's informing our approach
- **Challenge Acknowledgement:** Difficulties we're facing and how we're addressing them
- **Value Demonstration:** Tangible benefits already realized and projected future impact
- **Future Direction:** Next steps and evolving priorities

Audience-Specific Adaptation

Different stakeholders need different narrative elements emphasized:

- **Executives:** Strategic progress, competitive implications, resource requirements
- **Technical Teams:** Implementation challenges, architectural decisions, performance metrics
- **Business Users:** Workflow impact, capability enhancement, support availability
- **End Customers:** Service improvements, experience changes, value delivery

A telecommunication company exemplified effective narrative management by creating quarterly "AI journey updates" that connected technical progress to customer experience improvements and competitive positioning, maintaining stakeholder engagement throughout a complex multi-year transformation.

Decision-Point Integration Model

The final AIValueOS component integrates AI considerations into standard organizational decision-making processes.

From Chaos to Clarity

Rather than treating AI as a separate initiative, mature organizations embed AI value considerations into their routine planning, budgeting, and strategic processes.

Integration Points

- **Strategic Planning:** AI capability assessment and opportunity identification
- **Budget Planning:** AI investment prioritization and resource allocation
- **Project Portfolio Review:** AI initiative evaluation and optimization
- **Performance Reviews:** AI outcome assessment and lesson capture
- **Risk Management:** AI-related risk identification and mitigation

DECISION INTEGRATION FRAMEWORK

For each decision point:

1. **AI Relevance Assessment:** Could AI create value in this context?
2. **Current Capability Evaluation:** What AI capabilities do we currently have?
3. **Opportunity Identification:** What new value could AI enable?
4. **Resource Requirement Analysis:** What would be required to capture this value?
5. **Priority Ranking:** How does this compare to other opportunities?
6. **Decision Documentation:** What did we decide and why?

This integration ensures that AI becomes part of routine organizational thinking rather than remaining a special initiative that competes for attention with "regular" business activities.

TOOL: AIVALUEOS MATURITY MAP

To help organizations assess their current state and plan development, AIValueOS includes a maturity assessment tool (see Table 16.3).

Organizations can use this maturity map to:

- Assess their current AIValueOS implementation
- Identify specific improvement opportunities
- Plan development priorities
- Track progress over time

Most organizations begin at Level 1 or 2 and gradually develop toward higher maturity as their AI initiatives mature and organizational capabilities grow.

TABLE 16.3
A Maturity Map

Dimension	Level 1: Reactive	Level 2: Managed	Level 3: Integrated	Level 4: Optimizing
Vision Clarity	Vague aspirations	Clear statements	Stakeholder alignment	Dynamic evolution
Values Translation	Ad hoc connection	Systematic process	Embedded practices	Continuous adaptation
Role Definition	Unclear accountability	Defined responsibilities	Shared ownership	Fluid collaboration
Feedback Systems	Anecdotal input	Structured collection	Integrated analysis	Predictive insights
Narrative Management	Sporadic communication	Regular updates	Stakeholder-specific stories	Adaptive storytelling
Decision Integration	Separate consideration	Periodic inclusion	Routine integration	Strategic embedding

STRATEGIC ALIGNMENT TOOLS

Successful AI initiatives require alignment across multiple dimensions – strategic priorities, stakeholder expectations, resource allocation, and implementation approaches. This section provides specific tools for creating and maintaining this essential alignment.

THE AI VALUE MAPPING CANVAS

The AI value mapping canvas serves as a comprehensive planning and communication tool that creates shared understanding of how AI initiatives connect to business value.

Canvas Structure

The canvas includes nine interconnected sections that collectively create a complete picture of the AI value opportunity:

1. **Strategic Context**
   - What strategic objectives are we pursuing?
   - How does this AI initiative advance these objectives?
   - What competitive or market factors create urgency?
2. **Stakeholder Ecosystem**
   - Who will be impacted by this initiative?
   - What does each stakeholder group value?
   - Where might interests conflict or align?
3. **Current State Assessment**
   - What challenges or limitations exist today?
   - What is the cost or impact of the status quo?
   - What constraints must we navigate?
4. **Future State Vision**
   - What outcomes are we targeting?
   - How will success be measured?
   - What will be different for each stakeholder?
5. **AI Capability Requirements**
   - What AI capabilities are needed?
   - What data and technical requirements exist?
   - What organizational capabilities must we develop?
6. **Value Creation Mechanisms**
   - What specific value will AI create?
   - What are the "cause and effect" relationships?
   - Where are the highest-leverage opportunities?
7. **Implementation Approach**
   - What sequence of activities will create value?
   - What resources and timeline are required?
   - How will we manage risks and dependencies?
8. **Success Metrics**
   - How will we measure progress and impact?
   - What leading and lagging indicators will we track?
   - How will we attribute value to AI capabilities?
9. **Communication Strategy**
   - How will we maintain stakeholder engagement?
   - What narratives will resonate with different audiences?
   - How will we adapt messaging as implementation progresses?

Using the Canvas Effectively

The canvas works best when completed collaboratively with diverse stakeholders. This process creates shared understanding while surfacing potential conflicts or misalignment early in the planning process.

A manufacturing company used the canvas effectively when planning a predictive maintenance initiative. The collaborative completion process revealed that operations leaders primarily valued uptime improvement while finance focused on cost reduction and maintenance teams wanted better work planning tools. This insight led to a multi-dimensional success framework that addressed all three priorities.

STAKEHOLDER VALUE ALIGNMENT MATRIX

The Stakeholder value alignment matrix, shown in Table 16.4, provides a structured approach for understanding and reconciling different stakeholder perspectives on AI value.

Conflict Resolution Process

When the matrix reveals conflicting stakeholder priorities, use this structured approach:

1. **Conflict Identification:** Explicitly name the competing priorities
2. **Impact Analysis:** Assess consequences of optimizing for each perspective
3. **Common Ground Exploration:** Identify shared interests and win-win opportunities
4. **Trade-off Evaluation:** Examine options for balancing competing priorities
5. **Decision Framework:** Establish criteria for resolving similar conflicts
6. **Communication Plan:** Explain decisions and reasoning to affected stakeholders

A financial services organization used this process effectively when a customer analytics initiative created tension between marketing (seeking maximum personalization) and privacy (requiring data minimization). The structured approach led to a solution that delivered meaningful personalization while exceeding privacy requirements – effectively turning potential conflict into competitive advantage.

STRATEGIC PRIORITIZATION FRAMEWORK

With multiple AI opportunities typically available, organizations need systematic approaches for prioritization that consider both potential impact and implementation feasibility.

TABLE 16.4
Matrix Structure

Stakeholder	Primary Value Drivers	Key Concerns	Success Metrics	Engagement Approach
Executive Leadership	Strategic advantage, competitive positioning	ROI uncertainty, resource requirements	Revenue impact, market share	Regular briefings, strategic framing
Business Unit Leaders	Operational efficiency, performance improvement	Implementation disruption, capability gaps	Department KPIs, process metrics	Collaborative planning, outcome focus
Technical Teams	Technical excellence, integration success	Resource constraints, complexity management	System performance, reliability	Technical involvement, recognition
End Users	Job enhancement, workflow improvement	Change burden, skill requirements	User satisfaction, adoption rates	Involvement, training, support
Customers	Service improvement, value delivery	Privacy concerns, experience disruption	Customer satisfaction, loyalty	Transparent communication, feedback

Prioritization Dimensions

Value Potential

- Financial impact magnitude
- Strategic importance
- Competitive advantage creation
- Risk-mitigation value

Implementation Feasibility

- Technical complexity
- Data availability and quality
- Organizational readiness
- Resource requirements

Strategic Alignment

- Connection to core objectives
- Stakeholder support level
- Timeline compatibility
- Risk profile acceptance

Portfolio Balance

- Quick wins vs. long-term capability building
- Innovation vs. optimization
- High certainty vs. high potential
- Internal vs. customer-facing impact

Prioritization Process

1. **Opportunity Inventory:** List all potential AI applications under consideration
2. **Dimensional Scoring:** Rate each opportunity across the four dimensions
3. **Portfolio Mapping:** Visualize opportunities in a balanced portfolio view
4. **Strategic Filtering:** Apply organizational constraints and preferences
5. **Resource Allocation:** Match priorities with available resources and capabilities
6. **Timeline Sequencing:** Create logical implementation sequence
7. **Review and Adjustment:** Plan regular reassessment as conditions change

A healthcare organization used this framework to prioritize among 23 potential AI applications across clinical, operational, and administrative domains. The systematic assessment revealed that while the most technically sophisticated applications scored highest on innovation, a combination of clinical decision support and administrative automation provided the best balance of impact and feasibility for their first implementation phase.

BUSINESS CASE TEMPLATES FOR DIFFERENT AUDIENCES

Different stakeholders need business case information presented in formats that match their priorities, decision-making processes, and information preferences.

Executive Summary Template (1–2 pages)

- **Strategic Context:** Brief description of business challenge and opportunity
- **Proposed Solution:** High-level description of AI approach and capabilities

- **Expected Impact:** Quantified benefits across multiple value dimensions
- **Investment Requirements:** Total cost and resource needs
- **Implementation Timeline:** Key milestones and decision points
- **Risk Assessment:** Primary risks and mitigation approaches
- **Recommendation:** Clear request for specific decisions or support

Financial Analysis Template (3–5 pages)

- **Investment Summary:** Detailed cost breakdown across implementation phases
- **Benefit Quantification:** Financial modelling of expected returns
- **Scenario Analysis:** Multiple projections with sensitivity analysis
- **ROI Calculations:** Various financial metrics and comparison benchmarks
- **Cash Flow Projections:** Timeline of investments and returns
- **Risk Analysis:** Financial implications of various risk scenarios
- **Funding Recommendations:** Optimal financing and resource allocation

Technical Architecture Template (5–10 pages)

- **Current State Analysis:** Existing systems and capabilities assessment
- **Solution Architecture:** Technical approach and system integration
- **Implementation Plan:** Detailed development and deployment timeline
- **Resource Requirements:** Technical skills, infrastructure, and tools needed
- **Performance Specifications:** Expected system capabilities and limitations
- **Risk Mitigation:** Technical risks and management approaches
- **Scaling Considerations:** Plans for enterprise expansion

Operational Impact Template (3–7 pages)

- **Process Analysis:** Current workflows and proposed changes
- **User Impact Assessment:** Effects on different user groups and roles
- **Change Management Plan:** Training, support, and adoption approaches
- **Performance Metrics:** Operational KPIs and measurement methods
- **Integration Requirements:** Connections with existing operational systems
- **Support Model:** Ongoing maintenance and user support plans
- **Success Criteria:** Specific operational outcomes and timeframes

A global retailer demonstrated effective audience-specific communication by creating tailored business cases for an inventory optimization AI. The executive version emphasized competitive positioning and customer experience impact, the financial version detailed ROI calculations and scenario modelling, the technical version addressed integration complexity and scalability, and the operational version focused on workflow changes and training requirements. This comprehensive approach secured support across all stakeholder groups.

IMPLEMENTATION AND EXECUTION FRAMEWORKS

Moving from planning to execution requires frameworks that maintain focus on value creation while navigating the inevitable complexities of implementation.

DATA READINESS ASSESSMENT

Since data quality and accessibility determine AI value potential, systematic data readiness assessment should precede significant AI investments.

Assessment Dimensions

Data Availability

- What data exists that could support AI applications?
- How comprehensive is coverage across relevant domains?
- What critical data gaps exist?
- How accessible is existing data for AI purposes?

Data Quality

- How accurate and consistent is available data?
- What quality issues exist and how significant are they?
- How reliable are data collection and maintenance processes?
- What validation and cleansing would be required?

Data Integration

- How fragmented is data across systems and sources?
- What integration capabilities currently exist?
- How complex would comprehensive integration be?
- What standards and governance exist for integration?

Data Governance

- What policies and procedures govern data use?
- How mature are privacy and compliance processes?
- What approval and oversight requirements exist?
- How adaptable is governance to AI applications?

Technical Infrastructure

- Can current systems support AI data requirements?
- What infrastructure investments would be needed?
- How scalable are current data management capabilities?
- What security and performance considerations exist?

Assessment Process

1. **Inventory Development:** Catalogue available data sources and characteristics
2. **Quality Evaluation:** Systematic assessment of data accuracy and consistency
3. **Gap Analysis:** Identification of missing data required for target AI applications
4. **Integration Assessment:** Evaluation of technical and process integration requirements
5. **Governance Review:** Analysis of policy and compliance implications
6. **Infrastructure Evaluation:** Assessment of technical capability and scalability
7. **Investment Planning:** Prioritization of data improvements based on AI value potential
8. **Roadmap Development:** Timeline for addressing critical data readiness gaps

A telecommunications company used this assessment to discover that while they had abundant customer interaction data, critical network performance data was fragmented across systems with inconsistent quality standards. This insight led them to prioritize data integration and quality improvements before launching their customer experience AI initiative, ultimately accelerating implementation and improving results.

TABLE 16.5
Maturity Dimensions

Dimension	Level 1: Exploring	Level 2: Implementing	Level 3: Integrating	Level 4: Optimizing
Strategy and Vision	Ad hoc experiments	Focused initiatives	Integrated strategy	Adaptive optimization
Data Capabilities	Fragmented sources	Basic integration	Enterprise platform	Strategic asset
Technical Infrastructure	Limited tools	Departmental systems	Enterprise architecture	Advanced platforms
Talent and Skills	Individual specialists	Dedicated teams	Enterprise capability	Innovation leadership
Governance and Ethics	Minimal oversight	Basic frameworks	Comprehensive governance	Dynamic optimization
Cultural Adoption	Sceptical resistance	Cautious acceptance	Broad engagement	Innovation mindset
Value Realization	Unclear impact	Departmental benefits	Enterprise value	Competitive advantage

AI Maturity Model

Organizations at different maturity levels require different approaches to AI implementation. This model (see Table 16.5) helps assess current state and plan appropriate development paths.

Maturity Assessment Process

1. **Current State Evaluation:** Honest assessment across all dimensions
2. **Gap Identification:** Comparison between current and desired state
3. **Development Planning:** Prioritization of maturity improvements
4. **Resource Allocation:** Investment in capability development
5. **Progress Tracking:** Regular reassessment and adjustment

Maturity-Appropriate Strategies

Level 1 Organizations should focus on

- Limited, low-risk proof-of-concepts
- Basic data and infrastructure improvements
- Foundational skill development
- Simple governance establishment

Level 2 Organizations can pursue

- Departmental implementations with clear ROI
- Selective technology platform investments
- Cross-functional team development
- Structured governance expansion

Level 3 Organizations should target

- Enterprise-wide implementations
- Advanced platform capabilities
- Sophisticated talent strategies
- Comprehensive governance frameworks

Level 4 Organizations can optimize through

- Continuous innovation programmes
- Advanced analytical capabilities
- Industry leadership positions
- Dynamic adaptation mechanisms

SCALING BLUEPRINT

Moving from successful pilots to enterprise value requires systematic scaling approaches that address both technical and organizational dimensions.

Scaling Phases

Phase 1: Foundation Validation

- Pilot results verification
- Stakeholder feedback integration
- Technical architecture validation
- Business case refinement

Phase 2: Controlled Expansion

- Limited deployment to additional contexts
- Process and system integration
- User training and support development
- Performance monitoring establishment

Phase 3: Organizational Integration

- Enterprise-wide deployment planning
- Cross-functional process integration
- Comprehensive training and change management
- Governance and compliance embedding

Phase 4: Optimization and Evolution

- Performance optimization based on scaled results
- Continuous improvement process establishment
- Advanced capability development
- Strategic value maximization

Scaling Decision Framework

At each phase transition, assess:

- **Value Demonstration:** Has current phase delivered expected value?
- **Technical Readiness:** Can technology support expanded deployment?
- **Organizational Capability:** Does organization have required skills and processes?
- **Stakeholder Support:** Do key stakeholders endorse continuation?
- **Resource Availability:** Are necessary resources available for the next phase?
- **Risk Assessment:** Are risks acceptable given potential rewards?

Scaling Success Factors

- **Clear Success Criteria:** Specific metrics for each scaling phase
- **Flexible Architecture:** Technical design supporting varied deployment contexts

- **Change Management:** Systematic approach to organizational adoption
- **Performance Monitoring:** Continuous visibility into technical and business performance
- **Stakeholder Engagement:** Ongoing communication and support building
- **Learning Integration:** Systematic capture and application of implementation insights

A manufacturing company exemplified effective scaling with a quality prediction AI. They moved through controlled expansion across three plants with different characteristics, learning from each deployment to refine both technology and processes before enterprise-wide rollout. This deliberate approach took longer than immediate enterprise deployment but achieved much higher success rates and ultimate value realization.

RISK ASSESSMENT FRAMEWORK

AI initiatives create unique risks that require systematic identification and management throughout implementation.

Risk Categories

Technical Risks

- Model performance degradation
- Integration complexity
- Scalability limitations
- Security vulnerabilities

Business Risks

- Value realization shortfalls
- Stakeholder resistance
- Competitive response
- Market changes

Operational Risks

- Process disruption
- User adoption failures
- Support capability gaps
- Performance monitoring failures

External Risks

- Regulatory changes
- Privacy concerns
- Ethical criticisms
- Public perception issues

Risk Assessment Process

1. **Risk Identification:** Systematic enumeration of potential risks across categories
2. **Probability Estimation:** Assessment of likelihood for each identified risk
3. **Impact Analysis:** Evaluation of consequences if risks materialize
4. **Risk Prioritization:** Focus on highest probability/impact combinations
5. **Mitigation Planning:** Development of specific responses for priority risks
6. **Monitoring Setup:** Establishment of early warning indicators
7. **Response Preparation:** Plans for rapid response if risks occur
8. **Regular Review**: Periodic reassessment as implementation progresses

Risk Mitigation Strategies
- **Prevention:** Eliminate or reduce risk probability through design and planning
- **Detection:** Early identification when risks begin materializing
- **Response:** Rapid corrective action when risks are detected
- **Recovery:** Plans for restoring value when risk impacts occur
- **Learning:** Integration of risk insights into future planning.

A financial services organization used this framework effectively during a credit-decision AI implementation. By systematically identifying regulatory, bias, and performance risks early, they developed mitigation strategies that prevented several potential issues and created rapid response capabilities for others. This proactive approach enabled aggressive implementation timelines while maintaining appropriate risk management.

COMMUNICATION AND ENGAGEMENT TOOLS

Sustainable AI value requires ongoing stakeholder engagement that maintains support and alignment throughout implementation.

EXECUTIVE BRIEFING TEMPLATES

Regular executive communication maintains support and enables course correction based on leadership feedback.

Monthly Progress Brief (2–3 pages)

Executive Summary

- Key accomplishments since last update
- Current status against major milestones
- Significant challenges or changes
- Decisions needed from executive leadership

Value Realization Update

- Quantified benefits delivered to date
- Progress against original projections
- Value attribution methodology and confidence
- Projected timeline for additional benefits

Implementation Progress

- Technical development and deployment status
- User adoption and engagement metrics
- Training and change management progress
- Integration with operational processes

Risk and Issue Management

- New risks identified or materialized
- Mitigation actions taken or planned
- Issues requiring executive intervention
- Changes to risk profile or management approach

Resource and Timeline Status

- Budget utilization and projected spending
- Resource allocation and capability needs
- Timeline adherence and any adjustments
- Upcoming decisions or milestone reviews

Next Steps and Support Needs

- Planned activities for next period
- Specific support required from leadership
- Upcoming decisions requiring executive input
- Strategic guidance needed for course correction

Quarterly Strategic Review (5–7 pages)

Quarterly reviews provide deeper strategic assessment and planning:

- **Strategic Alignment Assessment:** Connection between AI progress and organizational objectives
- **Competitive Positioning Update:** Market and competitor intelligence with strategic implications
- **Value Portfolio Analysis:** Performance across multiple AI initiatives and their interaction
- **Capability Development Progress:** Organizational skill and infrastructure advancement
- **Governance and Risk Evolution:** Changes to oversight and risk management approaches
- **Future Planning:** Emerging opportunities, threats, and strategic adjustments

CROSS-FUNCTIONAL COMMUNICATION PLAN

AI initiatives require coordinated communication across diverse stakeholder groups with different interests and information needs, as illustrated in Table 16.6.

TABLE 16.6
Stakeholder Communication Matrix

Stakeholder Group	Information Needs	Communication Frequency	Preferred Channels	Key Messages
Executive Leadership	Strategic progress, value realization, resource needs	Monthly	Briefings, dashboards	Business impact, competitive positioning
Business Unit Leaders	Operational impact, performance metrics, implementation support	Bi-weekly	Reports, meetings	Process improvement, capability enhancement
Technical Teams	System performance, integration status, development priorities	Weekly	Technical reviews, collaboration tools	Architecture progress, technical achievements
End Users	Training availability, workflow changes, support resources	As needed	Training sessions, help systems	Job enhancement, support availability
Customer-Facing Staff	Service changes, capability updates, customer communication	Monthly	Training, communication briefs	Customer benefit, service improvement

Communication Principles

- **Relevance:** Tailor information to stakeholder interests and needs
- **Transparency:** Share both successes and challenges honestly
- **Consistency:** Maintain regular communication rhythms and formats
- **Two-Way:** Create opportunities for feedback and questions
- **Action-Oriented:** Connect information to specific next steps or decisions

THE AI TRUST DASHBOARD

Trust represents perhaps the most critical factor in AI adoption success. This dashboard provides systematic visibility into trust indicators across the organization (see Table 16.7).

Trust Dimensions

System Reliability

- Technical performance consistency
- System availability and responsiveness
- Error rates and exception handling
- Integration stability with existing systems

Decision Transparency

- Explainability of AI recommendations
- Visibility into decision factors
- Ability to understand and validate outputs
- Clear boundaries of AI authority

User Experience

- Ease of interaction with AI systems
- Quality of user interface and support
- Responsiveness to user feedback
- Integration with existing workflows

Organizational Support

- Training and capability development
- Help desk and technical support quality

TABLE 16.7
Trust Measurement Framework

Trust Level	Indicators	Measurement Methods	Target Thresholds
High Trust	Enthusiastic adoption, advocacy behaviour, expansion requests	User surveys, usage analytics, referral tracking	>80% satisfaction, >90% usage rate
Moderate Trust	Cautious adoption, compliance usage, neutral feedback	Engagement metrics, feedback analysis	60–80% satisfaction, 70–90% usage
Low Trust	Minimal usage, work-around behaviour, negative feedback	Usage data, support tickets, informal feedback	40–60% satisfaction, 50–70% usage
Distrust	Avoidance behaviour, active resistance, undermining	Non-usage analytics, resistance indicators	<40% satisfaction, <50% usage

- Change management and communication
- Leadership commitment demonstration

Value Demonstration

- Tangible benefits to individual users
- Clear connection between AI and improved outcomes
- Recognition and reward for adoption
- Visible organizational commitment

Trust Building Interventions

Based on dashboard insights, implement targeted interventions:

- **For Low Reliability:** Focus on technical improvements and system stabilization
- **For Poor Transparency:** Enhance explainability and user education
- **For Bad User Experience:** Improve interface design and workflow integration
- **For Weak Organizational Support:** Strengthen training and change management
- **For Unclear Value:** Better communicate benefits and recognition programmes

A healthcare organization used this dashboard effectively to identify trust issues with their clinical decision support system. Low transparency scores led them to implement enhanced explainability features, while poor organizational support scores prompted expanded training programmes. These targeted interventions increased overall trust scores by 34% within six months.

VALUE STORYTELLING FRAMEWORKS

Beyond metrics and dashboards, sustained stakeholder engagement requires compelling narratives that make AI value tangible and relatable.

The Situation, Task, Action, Result STAR Framework

Situation: Describe the business context and challenge
- What problem or opportunity existed?
- Why was action needed?
- What were the stakes or consequences?

Task: Explain what needed to be accomplished
- What specific outcomes were targeted?
- What constraints or requirements existed?
- Who was involved and responsible?

Action: Detail the AI solution and implementation
- What AI capabilities were developed or deployed?
- How was implementation managed?
- What challenges were overcome?

Result: Quantify the impact and value created
- What specific outcomes were achieved?
- How do results compare to expectations?
- What broader implications exist?

Example STAR Story

Situation: Our customer service team was overwhelmed with 40% more support requests following a product launch, leading to three-day response times and declining satisfaction scores.

Task: We needed to reduce response times to under four hours while maintaining service quality, without significant staff increases.
Action: We implemented an AI-powered case classification and routing system that automatically prioritized urgent issues and provided agents with suggested responses based on similar historical cases.
Result: Response times dropped to 2.3 hours on average, customer satisfaction increased by 28%, and agents reported 35% improvement in job satisfaction due to more efficient workflows.

The Before/During/After Framework

This framework helps stakeholders understand the transformation journey:

Before: What was life like prior to AI implementation?
- Specific pain points and limitations
- Quantified baseline performance
- Individual and organizational frustrations

During: What was the implementation experience?
- Key milestones and changes
- Challenges encountered and overcome
- Learning and adaptation processes

After: What is the new reality with AI?
- Transformed workflows and capabilities
- Quantified improvements and benefits
- New possibilities that didn't exist before

Audience-Specific Story Adaptation

The same underlying success can be told differently for different audiences:

- **For Executives:** Emphasize strategic impact, competitive advantage, and organizational transformation
- **For Technical Teams:** Focus on solution elegance, technical achievements, and innovation aspects
- **For End Users:** Highlight workflow improvements, job enhancement, and daily experience benefits
- **For Customers:** Emphasize service improvements, value delivery, and experience enhancement

A manufacturing company developed multiple versions of their predictive maintenance success story: executives heard about 23% reduction in unplanned downtime and $4.7M annual savings; maintenance technicians heard about 40% more effective work prioritization and 60% reduction in emergency calls; customers heard about 99.7% reliability improvement and faster service delivery.

MEASUREMENT AND EVALUATION RESOURCES

Sustainable AI value requires comprehensive measurement approaches that track both technical performance and business impact while providing insights for continuous improvement.

VALUE MEASUREMENT FRAMEWORK

Effective AI value measurement (see Table 16.8) addresses multiple dimensions and timeframes while creating clear attribution between AI capabilities and business outcomes.

TABLE 16.8
Measurement Architecture

Measurement Level	Time Horizon	Primary Purpose	Key Metrics
Real-Time Operational	Immediate	System monitoring and optimization	Performance, usage, errors
Short-Term Impact	Weekly/Monthly	Tactical adjustment and improvement	User adoption, process efficiency
Medium-Term Value	Quarterly	Business performance and ROI	Financial returns, operational KPIs
Long-Term Strategic	Annual	Strategic positioning and transformation	Market position, capability advancement

Measurement Dimensions

Financial Impact

- Direct cost savings from efficiency improvements
- Revenue increases from enhanced capabilities
- Cost avoidance from risk reduction
- Investment returns and payback periods

Operational Performance

- Process efficiency and speed improvements
- Quality and accuracy enhancements
- Resource utilization optimization
- Error reduction and consistency gains

Strategic Advancement

- Competitive positioning improvement
- Market share and customer acquisition
- Innovation capability development
- Organizational learning and adaptation

Stakeholder Experience

- Customer satisfaction and engagement
- Employee experience and productivity
- Partner and supplier relationship improvement
- Investor and board confidence

Capability Development

- Technical skill advancement
- Process maturity evolution
- Data asset quality improvement
- Organizational AI readiness

ROI CALCULATION MODELS

Different AI applications require tailored ROI approaches that reflect their specific value creation mechanisms and timeframes.

Direct ROI model (for automation and efficiency applications):

$$ROI = (\text{Annual Benefits} - \text{Annual Costs})/\text{Total Investment} \times 100$$

Where:

- Annual Benefits = Labour savings + Error reduction + Speed improvements
- Annual Costs = System maintenance + Support + Training
- Total Investment = Development + Infrastructure + Implementation

Value-based ROI model (for customer-facing applications):

$$ROI = (\text{Customer Lifetime Value Increase} \times \text{Customer Base})/\text{Total Investment} \times 100$$

Factors:

- Increased retention rates from better service
- Higher transaction values from personalization
- Acquisition improvements from reputation enhancement
- Cross-sell/up-sell effectiveness from recommendations

Risk-adjusted ROI model (for risk management applications):

$$ROI = (\text{Risk Losses Avoided} - \text{Implementation Costs})/\text{Total Investment} \times 100$$

Considerations:

- Historical loss patterns and frequencies
- Probability reduction from AI implementation
- Secondary benefits from improved risk visibility
- Compliance and reputation value protection

Portfolio ROI model (for enterprise AI platforms):

$$\text{Portfolio ROI} = \sum (\text{Individual Application ROIs} \times \text{Strategic Weight})/\text{Total Platform Investment}$$

Elements:

- Multiple applications sharing common infrastructure
- Strategic value weights reflecting organizational priorities
- Option value for future applications
- Learning and capability development benefits

ROI Calculation Template

Investment Components

- Initial development costs (internal + external)
- Infrastructure and technology investments
- Training and change management expenses
- Ongoing operational costs

Benefit Components

- Direct financial returns (quantified)
- Operational efficiency gains (valued)
- Risk reduction benefits (calculated)
- Strategic value creation (estimated)

Risk Adjustments

- Implementation risk probability
- Performance shortfall potential
- Market change implications
- Technology evolution impacts

Sensitivity Analysis

- Best case scenario projections
- Expected case realistic estimates
- Worst case downside protection
- Break-even threshold identification

KPI Development Guide

Key performance indicators for AI initiatives must balance technical performance, business outcomes, and leading indicators of sustainable success.

KPI Categories and Examples

Technical Performance KPIs

- Model accuracy and precision rates
- System response time and availability
- Data quality and completeness scores
- Integration stability and error rates

Adoption and Usage KPIs

- User engagement and activity levels
- Feature utilization rates
- Training completion and proficiency
- Support ticket volume and resolution

Business Impact KPIs

- Process efficiency improvements
- Quality and accuracy enhancements
- Customer satisfaction changes
- Financial performance impacts

Strategic Progress KPIs

- Capability maturity advancement
- Competitive positioning improvement
- Innovation pipeline development
- Organizational learning indicators

KPI Selection Criteria

- **Actionable:** Metrics that can influence decisions and behaviours
- **Attributable:** Clear connection between AI and measured outcomes
- **Accessible:** Data that can be collected reliably and efficiently
- **Aligned:** Connection to strategic objectives and stakeholder priorities
- **Appropriate:** Suitable for the specific AI application and context

KPI Development Process
1. **Objective Mapping:** Connect each KPI to specific business objectives
2. **Baseline Establishment:** Measure current state before AI implementation
3. **Target Setting:** Define specific improvement goals and timeframes
4. **Data Source Identification:** Determine where and how metrics will be collected
5. **Calculation Methodology:** Specify exact formulas and measurement approaches
6. **Reporting Frequency:** Establish appropriate measurement and review cycles
7. **Responsibility Assignment:** Define who owns each metric and its improvement
8. **Review and Evolution:** Plan regular assessment and refinement of KPIs

Balanced KPI Portfolio Example

A retail organization implementing customer analytics developed this balanced portfolio:

Technical KPIs

- Recommendation engine accuracy: >85%
- System response time: <200 ms
- Data completeness score: >95%

Adoption KPIs

- Store manager engagement: >80% weekly usage
- Customer service utilization: >70% of interactions
- Training completion: 100% within 30 days

Business KPIs

- Average transaction value: +12% improvement
- Customer satisfaction: +15% increase
- Cross-sell conversion: +25% improvement

Strategic KPIs

- Personalization capability maturity: Level 3 within 12 months
- Competitive differentiation score: Top quartile in market research
- Employee AI confidence: >75% comfort level

Continuous Improvement Templates

Sustainable AI value requires systematic approaches to learning and evolution based on performance data and stakeholder feedback.

Monthly Improvement Review Template

Performance Analysis

- KPI performance against targets
- Trend analysis and pattern identification
- Variance investigation and root cause analysis
- User feedback and satisfaction assessment

Issue Identification

- Technical problems and their frequency
- User experience pain points

From Chaos to Clarity

- Process integration challenges
- Stakeholder concern emergence

Improvement Opportunities

- Performance optimization possibilities
- Feature enhancement requests
- Process streamlining options
- Capability expansion potential

Action Planning

- Priority improvement selection
- Resource allocation for improvements
- Implementation timeline development
- Success measurement approach

Learning Capture

- Key insights from current period
- Best practices for replication
- Lessons learned for future initiatives
- Knowledge sharing opportunities

Quarterly Strategic Review Template

Value Realization Assessment

- Financial impact against projections
- Strategic objective advancement
- Stakeholder satisfaction evolution
- Competitive positioning changes

Capability Evolution

- Technical capability advancement
- Organizational skill development
- Process maturity improvement
- Cultural adaptation progress

External Environment Analysis

- Market condition changes
- Competitive response assessment
- Regulatory development implications
- Technology evolution impact

Strategic Alignment Review

- Continued relevance to strategic priorities
- Resource allocation optimization
- Portfolio balance assessment
- Future opportunity identification

Roadmap Adjustment

- Initiative priority rebalancing

- Timeline and milestone revision
- Resource reallocation needs
- New opportunity integration

PUTTING IT ALL TOGETHER: THE INTEGRATED APPROACH

Rather than using these tools in isolation, maximum value comes from integrated application that connects assessment, planning, implementation, and evolution into a coherent approach.

Creating Your Customized Roadmap

Every organization brings unique context, constraints, and capabilities to their AI journey. The key to success lies in selecting and sequencing tools that match your specific situation and objectives.

Roadmap Development Process
  Step 1: Current State Assessment: Use assessment tools to understand your starting position:
  - AI Value Opportunity Assessment
  - Data Readiness Evaluation
  - AI value maturity diagnostic
  - Stakeholder Alignment Analysis

  Step 2: Future State Visioning: Apply strategic alignment tools to create shared direction:
  - AI Value Mapping Canvas
  - Cross-Functional Value Matrix
  - Strategic Prioritization Framework

  Step 3: Implementation Planning: Leverage execution frameworks to guide development:
  - Scaling Blueprint Template
  - Risk Assessment Framework
  - Business Case Calculator

  Step 4: Communication and Engagement: Implement ongoing stakeholder management:
  - Executive Briefing Templates
  - Cross-Functional Communication Plan
  - Value Storytelling Frameworks

  Step 5: Measurement and Evolution: Establish systematic improvement approaches:
  - Value Measurement Framework
  - KPI Development Guide
  - Continuous Improvement Templates

Customization Considerations
  Industry Context: Regulatory requirements, competitive dynamics, and customer expectations vary dramatically across industries. A healthcare organization implementing clinical AI faces fundamentally different constraints than a retailer implementing personalization engines.
  Organizational Scale: Scale introduces complexity, but the real tension lies in **decision latency** – the time it takes to coordinate, approve, and act. Startups move fast not just because they're small, but because they **decide** faster. Large enterprises must find new ways to shrink that gap.
  Technical Maturity: Early maturity solves foundational problems – but it also **raises the bar**. Once you've conquered integration and governance, the next question then becomes: *Are you solving the right problems with AI?* Technical maturity is only an asset if it's paired with strategic clarity.

Cultural Characteristics: Some organizations embrace experimentation while others prefer predictable progress. Match tools to cultural preferences.

Strategic Context: AI initiatives serving existential strategic needs require different approaches than those addressing operational improvements.

Sequencing and Prioritization

Effective tool application follows logical sequences that build capability and momentum over time.

Phase 1: Foundation Building (Months 1–3)

- Complete maturity and readiness assessments
- Develop initial value mapping and stakeholder alignment
- Create basic communication and governance frameworks
- Establish baseline measurements

Phase 2: Pilot Implementation (Months 4–9)

- Apply implementation frameworks to initial use cases
- Test and refine measurement approaches
- Build stakeholder engagement and trust
- Capture and document early lessons

Phase 3: Scaling and Integration (Months 10–18)

- Use scaling frameworks for broader deployment
- Implement comprehensive measurement systems
- Establish continuous improvement processes
- Develop advanced capabilities and governance

Phase 4: Optimization and Evolution (Months 19+)

- Focus on strategic value maximization
- Implement advanced measurement and optimization
- Build innovation and adaptation capabilities
- Establish industry leadership positions

BUILDING ORGANIZATIONAL CAPABILITIES

Beyond specific tool application, sustainable AI value requires developing organizational capabilities that support ongoing success.

Essential Capabilities

- **Strategic Thinking:** Ability to connect AI possibilities to business strategy and priorities
- **Cross-Functional Collaboration:** Skills for working effectively across organizational boundaries
- **Technical Translation:** Capability to connect technical possibilities with business value
- **Change Management:** Expertise in guiding organizational adaptation and adoption
- **Value Measurement:** Competency in tracking and attributing multidimensional impact
- **Continuous Learning:** Mindset and processes for ongoing improvement and adaptation

Capability Development Approaches

- **Formal Training:** Structured programmes building specific skills and knowledge
- **Experiential Learning:** Hands-on experience through actual AI implementations

- **Mentoring and Coaching:** Guidance from experienced practitioners and advisors
- **Community Building:** Networks for sharing experiences and best practices
- **External Partnerships:** Relationships with vendors, consultants, and other organizations

Creating Sustainable Value through AI

The ultimate goal extends beyond successful tool application to sustainable value creation that continues evolving and improving over time.

Sustainability Factors

- **Embedded Practices:** AI value creation becomes part of routine organizational practices rather than special initiatives
- **Cultural Integration:** AI thinking and approaches become natural parts of organizational culture
- **Capability Evolution:** Organizations continuously develop more sophisticated AI capabilities and applications
- **Ecosystem Development:** Relationships with partners, vendors, and community create ongoing value
- **Innovation Mindset:** Continuous exploration of new AI possibilities and applications
- **Value Acceleration:** Returns from AI investment compound over time rather than plateauing

The Long-Term Vision

Organizations that successfully integrate these tools and frameworks don't just implement AI – they become AI-native organizations that naturally leverage artificial intelligence to create ongoing value for all stakeholders.

This transformation requires patience, persistence, and continuous learning. But organizations that commit to the journey position themselves for sustained competitive advantage in an increasingly AI-driven world.

The tools in this chapter provide the roadmap. Your organization's unique context, creativity, and commitment will determine how far that roadmap takes you toward realizing AI's transformational potential.

17 Conclusion
The Courage to Create Value That Matters

In writing these final words, I'm thinking about that boardroom moment from our opening chapter: the one where silence filled the room after a simple question. That awkward pause wasn't just about one failed AI project. It was a mirror reflecting something far deeper: the gap between what we build and what we value.

If you've made it this far, you've travelled with me through the messy, complex, often frustrating reality of making AI work for real people in real organizations. We've dissected the value paradox, mapped the three languages that unlock AI's potential, and built frameworks for turning technical possibility into genuine impact.

But here's what I've learned after working with over 35 organizations and watching countless AI initiatives rise and fall: the tools in this book aren't the transformation – you are.

When I started writing this book, I thought I was documenting proven frameworks for AI value creation. What I discovered was something more fundamental: a guide for navigating one of the most significant leadership challenges of our time.

You now understand that value isn't singular but multidimensional, that strategy must connect to operations through the bridge of influence, and that successful AI requires fluency in languages that most leaders have never learned to speak. You've seen how data foundations either enable or constrain every AI dream, and why implementations fail not from technical limitations but from human misalignments.

Most importantly, you've witnessed transformation in action – where cutting-edge AI capabilities create deeply human outcomes.

That financial services firm didn't just turn a $150K project into $9.5M in value. Their agentic AI systems now work around the clock, spotting market opportunities that human analysts would miss, while generative AI (GenAI) helps advisors explain complex investment strategies in plain English. But here's what really matters: the young couple who got pre-approved for their first home in 20 minutes instead of two weeks. The small business owner whose loan application was processed overnight, letting her secure inventory before her biggest sales season. The retiree whose advisor could instantly model dozens of scenarios, turning a stressful financial conversation into a confident plan for the future.

The insurance company that reduced claims processing from three weeks to three hours? They're piloting agentic AI that investigates straightforward claims autonomously – but the real story is the mother whose car was totalled on Friday who had a replacement vehicle arranged before Monday morning. The family whose house fire claim was approved before they finished filling out the paperwork. The small contractor who went from dreading insurance headaches to recommending his carrier to other businesses.

The healthcare system using GenAI to free doctors from documentation? Emergency room physicians who used to spend more time typing than talking now look patients in the eye while AI captures every detail. Parents don't wait three hours wondering what's wrong with their child – they get immediate attention from doctors who can focus on healing instead of paperwork.

These weren't just efficiency gains. They were moments when the latest AI capabilities – agentic systems, generative models, autonomous decision-making – finally delivered on the promise that technology should make life more human, not less.

And here's what really struck me most: these transformations didn't happen because of superior algorithms or bigger budgets. They happened because leaders ultimately made a choice – to stop settling for AI that impressed technologists and start demanding AI that served the people they were actually trying to help.

THE LEARNING NEVER STOPS

I'll be honest with you – I'm still learning too. Every engagement teaches me something new about the intricate dance between human intentions and artificial intelligence. Just last month, working with a healthcare organization, I discovered how cultural context can completely reshape what "efficiency" means for patient care. Only a couple of weeks before that, a conversation with a manufacturing CEO revealed a blind spot in how I was thinking about workforce transformation.

The humbling truth is that AI value creation is an evolving discipline. The frameworks in this book represent the best of what I've learned so far, but they're not the final word. They're tools for thinking and starting points for your own discoveries: foundations for building something better than what exists today.

What gives me hope is watching leaders like you take these frameworks and make them your own, adapt them to contexts I've never considered, and create value in ways that surprise even experienced practitioners.

THE LEADERSHIP MOMENT

We're living through a remarkable inflection point in human history. From ChatGPT sparking global conversations about AI capabilities to agentic systems that can plan, execute, and learn autonomously, artificial intelligence isn't just another technology wave – it's a fundamental shift in how intelligence itself can be applied to human challenges.

The emergence of GenAI has democratized access to sophisticated AI capabilities, putting powerful tools in the hands of every employee. Meanwhile, agentic AI systems are beginning to operate in decision loops that were once exclusively human domains – from autonomous customer service agents that can solve complex problems to AI systems that independently manage supply chain disruptions.

These aren't distant futures. They're reshaping organizations today. Yet in boardrooms around the world, I still see that same awkward silence when someone asks: "But where's the value from our GenAI pilots?" or "How do we govern AI agents that make decisions faster than humans can review them?"

This is your moment to break that silence. You now have something most leaders lack: a clear understanding of how to connect AI capabilities – whether generative, agentic, or predictive – to genuine value creation. You understand the languages that bridge technical possibility and strategic reality. You've seen the frameworks in action and witnessed their power to transform not just organizations, but the lives of the people they serve.

But here's what no framework can give you: your decision to actually use them.

THE COURAGE TO CHOOSE WHAT MATTERS

The organizations thriving in our AI-driven world aren't necessarily those with the most sophisticated algorithms or the largest datasets. They're led by people who had the courage to ask hard questions: What outcomes do we actually want? Who benefits and who ultimately bears the cost? How do we ensure AI serves our deepest values rather than undermining them?

These questions matter whether you're pursuing profit or purpose, serving shareholders or citizens, optimizing efficiency or advancing equity. The context changes, but the fundamental challenge remains: ensuring that artificial intelligence amplifies what you most want to preserve and advances human potential.

I've seen this courage in action. The CEO who chose transparency over technical mystique when explaining AI decisions to her board. The government leader who prioritized citizen benefit over bureaucratic efficiency in designing AI services. The healthcare administrator who refused to let cost optimization override quality of care in implementing AI diagnostics.

Each of these leaders faced the same choice you face now: letting AI happen to their organization, or choosing to shape how AI serves their deepest purposes.

FROM POTENTIAL TO PROOF

Every day, while you're still figuring out how to get value from your GenAI pilots or wondering how to govern agentic AI systems, others are shaping the AI-driven world you'll have to operate in. This isn't meant to create panic – it's meant to spark action.

The financial services firm I mentioned? They didn't just implement AI tools. They fundamentally redefined what's possible in customer service, risk assessment, and regulatory compliance – setting standards that force competitors to play catch-up. The insurance company with radically enhanced claims processing? They've moved beyond efficiency gains to reimagining what insurance means when AI can predict, prevent, and resolve issues faster than traditional processes ever could.

The healthcare system freeing doctors from paperwork? They're not just using GenAI – they're pioneering a new model of medical practice where technology amplifies human connection rather than replacing it.

This is what happens when leaders move from GenAI experiments to generative transformation, from agentic AI pilots to autonomous value creation, from AI projects scattered across departments to AI capabilities woven into the fabric of how value gets created.

You now have everything you need to make this leap: the understanding that value is multidimensional, the languages for translating between stakeholder perspectives, the frameworks for aligning the latest AI capabilities with strategic priorities, and the tools for measuring and communicating impact in ways that most matter to real people.

THE WORLD WE'RE BUILDING TOGETHER

As I reflect on the organizations I have worked with and the leaders I have learned from, I'm genuinely optimistic about what we are building together. Yes, AI presents risks and challenges we're still learning to navigate. Yes, the gap between hype and reality remains frustratingly large. Yes, the technical complexity can feel overwhelming.

But I've also seen GenAI help doctors explain complex diagnoses in ways patients can truly understand, agentic AI systems give teachers real-time insights into student's learning patterns, and AI agents accelerating scientific discovery by running thousands of hypothesis tests simultaneously. I've watched organizations become more human, not less, as they learned to use artificial intelligence in service of their deepest values.

The future belongs to leaders who understand that AI's greatest power isn't in its algorithms – it is in the collective voices of human wisdom that guides them. Whether you're optimizing supply chains or serving citizens, analysing markets or advancing medical research, the same truth applies: AI amplifies not just our capabilities, but our choices about what matters.

YOUR NEXT CHAPTER STARTS NOW

This book ends, but your transformation story is just beginning. The frameworks you've learned will evolve as you apply them. The tools will improve as you use them in your unique context. The insights will deepen as you face challenges I haven't yet encountered.

That's not just okay, it is essential. The goal was never to give you a perfect blueprint, but to help you become the kind of leader who can navigate an uncertain future with clarity about what truly matters.

So, here is my final challenge: Take one framework from this book and apply it this week. Have one conversation that translates your GenAI pilots into measurable business value. Ask one hard question about whether your agentic AI systems are optimizing what matters to your customers. Start small – but move forward.

The courage to create value that matters begins with the courage to start.

The tools are in your hands. The frameworks are ready to be adapted to your context. The moment to act isn't coming someday – it is here right now.

Your organization, your customers, your community, and perhaps the world itself will be shaped by the choices you make about AI value in the coming days, months, and years.

The question isn't whether AI will shape our future—it's whether we will have the courage to shape it wisely.

So, what kind of future will you now choose to build?

The conversation doesn't end here. I would love to learn about your transformation journey and the insights you discover along the way. You can find me — and many others continuing this work — at www.theaivalues.org, where new frameworks, stories, and lessons are shared from leaders creating AI value in ways that continue to surprise and inspire.

Because the truth is, we're all still learning – and that's exactly as it should be.

Index

Note: Page numbers in **bold** refer to tables.

3B-Fibreglass, 116
30-day scaling action plan, 117–118

A

Access constraints, 48
Adoption, employee perspectives, 175
    advocacy, 178–179
    awareness, 176
    barriers, addressing, 179–180
    belief, 178
    executive adoption, 180–181
    frontline adoption, 181–182
    manager adoption, 181
    role-specific adoption techniques, 180–182
    supportive environment, 192–195
    trust in AI systems and decisions, 174
    understanding, 176–177
    use, 177
    value, 177–178
Adoption-specific recognition, 194
Advanced AI applications, 305
Africa, building foundational capacity, 296
Agentic AI, 294
    and autonomous systems, 91
    in supply chain management, 295
AI, *see* Artificial intelligence
AI adoption paradox, 55
    false starts, 55–56
AI advisory scale, 111
    personal leverage, 111
    practice scale, 111
    team amplification, 111
AI asset management framework, 285–286
AI assets, protecting and leveraging, 285
AI collaboration, 149–150
AI content layer, 298
AI credit scoring system, 11
AI governance for value creation, 88
    for agentic AI and autonomous systems, 91
    Apple's business model and AI, 94
    balanced, 88–89
    contributes to value creation, 91
    cross-agency AI implementation, 94–95
    decision rights and accountability frameworks, 89–90
    effectiveness, 90
    ethical compliance with business value realization, 92–93
    in financial services, 96–97
    gaps in real-time decision-making systems, 92
    global regulatory trends, 93
    in healthcare, 95–96
    innovation with risk management, balancing, 89
    strategic integration, 93–94
AI implementation, 53
    challenges, 45
    pilot purgatory, 46
    reality gap, 45–46
    scaling challenges, 46
    traditional approaches, 46–47
    financial services, 53–54
    healthcare, 54
    manufacturing, 54–55
    retail, 53
AI initiatives alignment with corporate strategy, 73, **73**
AI insights and learnings, 283
AI leadership
    assessment framework, **378**
    development gap, 376–377
    literacy, 358–359
        curriculum, 358–359
AI maturity model, 393
    maturity-appropriate strategies, 393–394
    scaling blueprint, 394
AI measurement, problems of, 244–245
    metric hierarchy framework, 245
AI narrative framework, 125
    boardroom storytelling stack, 131
    fill-in-the-blank Narrative Builder Template, 128
        context-sensitive framework questions, 129–130
        dynamic narrative examples, 130
        impact projection, 128–129
        iteration and refinement process, 130
        narrative adaptation guidelines, 129
        shift explanation, 128
        structured ask, 129
        tension statement, 128
    framework, 125–126
        impact, 127
        shift, 126–127
        steps for decision-makers, 128
        tension, 126
AI ownership framework, 242–243
AI pilots and scaled implementations, 99
    3B-Fibreglass and real-world value, 116
    30-day scaling action plan, 117–118
    AI advisory scale, 111
        personal leverage, 111
        practice scale, 111
        team amplification, 111
    framework, 112
        evolution planning, 114
        feedback integration mechanisms, 113
        organizational learning loops, 113
        value measurement systems, 113
    Glasgow, game-changing public sector impact, 116–117
    invisible middle, 101–102
    scale, technical foundations for, 110
        e-commerce scale-or-fail example, 110
        gaming guru's intentional silos, 110
    scale-ready assessment framework, 107–108

scaling, horizons of, 108
    capability building, 109
    transformation, 109–110
    value realization, 108–109
scaling challenge, 102
    CNH industrial approach, 103–104
    European steel manufacturer's people-first scale, 103
    hypothesis-driven value creation, 103–104
    infrastructure investment paradox, 104
    "More Is Better" trap, 102–103
    "scale-or-fail" reality check, 104
    Tesla's mission-driven approach, 102
    "value multiplication" approach, 103
    value multiplication principles, 104–105
scaling obstacles, 114
    data quality gap, 114
    governance gap, 115
    process disconnect, 115
    skill mismatch, 114–115
    value dilution challenge, 116
scaling success, 105
    organizational readiness, 106
    people factor, 107
    strategic alignment, 105
    sustainable implementation, 106
    value translation, 105–106
strategic tension, 107
success patterns from leaders, 112
    capability building in parallel, 112
    ecosystem thinking, 112
    value-first implementation, 112
success trap, 100
transformations, 100–101
value alignment loop, 118
value mapping canvas, 118
AI portfolio management framework, 280–281
AI-powered customer service, 11
AI-powered marketing system, 13
AI-powered personalization, 50
AI-powered trade promotion optimization, 72
AI readiness, 29–30
    accuracy, 29
    completeness, 29
    consistency, 29
    readiness, governance, 30–31
    relevance, 29
    timeliness, 29
AI trust-building playbook, 182
    creating psychological safety, 185–186
    involvement and co-creation approaches, 183–184
    skills development and career path clarity, 184–185
    transparent communication strategies, 182–183
AI trust dashboard, 398–399
AI value, 27
    advisor, 346–347
    creation, **385**
    dashboards, 248, 326–327
        decision-driven dashboard design process, 249–250
        layered dashboard architecture, 249
    failure, preventing, 56–57
    glossary, 14
    mapping canvas, 314–315, 388
    capability mapping, 63
    current state assessment, 63
    implementation requirements, 63
    in practice, 63
    predictive quality initiative in manufacturing, 64
    stakeholder value expectations, 63
    strategic objectives, 63
    structure, 388
    structure and components, 63
    using effectively, 389
    value realization plan, 63
    team, 13
AI value fog, 5
    breaking through, 7
    continuous narrative connection, 7
    cross-functional alignment, 7
    explicit value definition, 7
    narrative gaps, 5–6
        prevent effective collaboration, 6
    operational drift, 5
        diverts resources from value creation, 6
    perception mismatch, 6
        destroy adoption and trust, 7
    strategic vagueness, 5
        creates misdirected effort, 6
    value translation tools, 7
AI value operating system (AIValueOS), 384
    decision-point integration model, 386–387
        decision integration framework, 387
    feedback and evolution infrastructure, 385–386
    maturity map, 387
    role-specific ownership matrix, 385
    transparent narratives framework, 386
        audience-specific adaptation, 386
    values-to-actions translation, 384–385
    vision development and communication, 384
Ambidextrous portfolio model, 281, **281**
Anticipation-alarm balance, 209
    concern response matrix, 209
Apple's business model and AI, 94
Application track, 85
Artificial intelligence (AI); *see also specific entries*
    capabilities, 3
    effective integration for, 30
    strategic value
        competitive advantage, 2
        innovation acceleration, 3
        market differentiation, 3
    tools, 26
Asana, building cross-functional AI literacy, 151
Augmentation initiatives, 132–133
Australian government's copilot trial, 324

B

Balanced capability development, 146
Balancing innovation and responsibility, 207
    false dichotomy problem, 207–208
    innovation-trust cycle, 208
Balancing speed and responsible implementation, 375–376
Barriers to AI value, **356**
Baseline alternative theory (BAT), 242
Bipartisan value framing, 144
Blockchain, 307

Index

BMW's cross-functional AI integration, 101
Board-level AI trust dashboard, 139
  building executive confidence, 139
    real-world, 139–140
    risk radar, 140–142
    value journey visualization, 140
    visual representation of progress and impact, 139–140
Boardroom storytelling stack, 131
Bridging data and business strategies, 35
  aligning data investments with AI value targets, 35–36
  data priorities based on business impact, 36
  return on data investments, 36
  value-driven data roadmap, 37
Buffer planning, 85
Business case for AI value, 64
  financial frameworks for AI investments, 65
  Global Bank's hybrid business case, 65
  global manufacturer's pivot, 67
  healthcare provider's scenario-based approach, 66
  retailer's dual-track business case, 66
  short-term and long-term value considerations, 66
  telecommunications company's stage-gated approach, 65–66
  traditional ROI *vs.* capability investment, 64–65
Business language, 125
Business process integration, 32

C

Capability
  building, 91
    in parallel, 112
  compounding effect, 257–258
  metrics, 36
  teams, 84
  track, 85
  transformation framework, 262
  valuation approaches, 65
Capture mechanism implementation, 279
Career-path integration, 194
Causal ambiguity, 239
Centralized AI, 76
Centralized foundation, distributed application, 78
CEO and Board's role in AI value creation, 78, 354
  accountability for value, 355–356
  behaviour impact matrix, **357**
  conflicting value propositions between departments, 81–82
  cross-functional collaboration, 79–80
  environment for AI success, 79
  geographic and cultural contexts, 81
  leadership strategies, 80–81
  modelling data-driven leadership, 80
  modelling leadership behaviours, 357–358
  organizational barriers, 356–357
  securing executive buy-in for long-term AI investments, 82
  strategic context, 354–355
  "value trough" period, 82–83
  value vision and priorities, 78–79
Change leadership, 345–346, 349–350
Change resistance, 46
Chief Finance Officer (CFO), 1
Chief Marketing Officer (CMO), 1

Chief Operating Officer (COO), 1
Chief Technology Officer (CTO), 1, 26
China, state-led strategy, 296
Citizen-centred framing, 143
Clinical leadership model, 369
CNH Industrial, hypothesis-driven value creation, 103–104
Coalition for AI Success, 87–88
Collaborative recognition systems, 20
Common vocabulary, 6
Commonwealth Bank of Australia (CBA) document AI, 61–62
  H2O.ai, 72
Communication, 364
  gaps, 51
  managing expectations, 366–367
  planning, 64
  technical and business languages, 364–365
  value narratives, 365–366
Communication and engagement tools, 396
  AI trust dashboard, 398–399
  cross-functional communication plan, 397–398
  executive briefing templates, 396
    monthly progress brief, 396–397
    quarterly strategic review, 397
  value storytelling frameworks, 399
    audience-specific story adaptation, 400
    before/during/after framework, 400
    situation, task, action, result star framework, 399–400
Communication framework, 206
  addressing concerns proactively, 208
    concern horizon framework, 209
  anticipation-alarm balance, 209
    concern response matrix, 209
  balancing innovation and responsibility, 207
    false dichotomy problem, 207–208
    innovation-trust cycle, 208
  credibility, building, 210
    transparency dimensions framework, 210
  ethical AI messaging, principles for, 206
  FACTS framework for responsible AI communication, 206–207
  strategic transparency advantage, 211
Community and public outreach, 214
  community engagement spectrum, 214–215
Community engagement spectrum, 213–215, 237
  collaborate, 214
  consult, 214
  inform, 214
  involve, 214
Comparative benchmarking, 82
Competencies, leadership, 347
  change leadership, 349–350
  cross-functional leadership, 348–349
  learning agility, 350–351
  strategic thinking with technical awareness, 347–348
  value translation, 348
Competitive implications, 52
Comprehensive value measurement, 242
  Microsoft copilot SMB value case, 244
  secondary benefit paradox, 243
  total value of AI ownership framework, 242–243
  triple BAT framework for lifecycle measurement, 242

Concern response matrix, 209, **210**, 237
Conflicting incentives and competing priorities, 50
Conflict resolution process, 389
Conflict to collaboration, 162–163
    mutual benefit and impact, 164
    shared ownership of outcomes, 163
    territorial concerns and resource competition, 163
Consent management, 35
Consent triangle, 229–230
Context loss, 48
Continuous monitoring, 92
Control contradiction, 340–341
Convergence tensions, 307
Coordination complexity, 103
Core business processes, 264–265
    process integration framework, 265–266
Core coalition, 87
Cost of AI value failure, 51
    competitive implications, 52
    direct financial impacts, 52
    opportunity costs, 52
    organizational momentum and confidence erosion, 52
    regulatory and reputational risks, 52
Costs of poor data foundations, 27
    opportunity cost, 27
    resource drain, 27
    scale limitation, 27
    trust erosion, 27
CP AXTRA
    prompt engineering challenge, 327
    strategic early adopters, 324
Creative professional's dilemma, 172–173
Credibility, building, 210
    transparency dimensions framework, 210
Critical data gaps and quality issues, 321
Critical gaps for AI success, 38
Critical path analysis, 85
Cross-agency AI implementation, 94–95
Cross-agency coalition building, 143
Cross-cutting regulatory themes, 204
Cross-function(al)
    AI literacy at Asana, 151
    AI roadmap, 83
        aligning technical and business timelines, 85
        cross-functional ownership of outcomes, 84–85
        integrated enterprise AI strategy, 83
        managing dependencies across departments, 85–86
    alignment, 7, 316
        financial services risk vs. experience conflict, 318
        manufacturing "value pods," 319
        Norwegian Government's five-step alignment process, 317
        retail "value pairs," 319
        shared accountability for outcomes, 319
        stakeholder value alignment matrix, 317–318
        WEX's integration success, 316
    challenges, 19
    collaboration, 79–80, 351–352
    collaborative spaces, 156–157
    communication plan, 397–398
    coordination, 91
    governance, 23–24, 51
    imperative, 75–76
    leadership, 348–349
    measurement opportunities, 258
    measurements, 248
    reviews, 85
    steering committees, 155–156
    success, collaboration, 165–166
        financial services, 168
        healthcare, 167
        manufacturing, 166–167
        retail, 166
        values, 168–169
    support, 148
    value clarity, 23
        establish cross-functional governance, 23–24
        executive learning journeys, 24–25
        executive value storytelling, 24
        multi-dimensional success reviews, 24
        value expectation audit, 23
        value-inclusive culture, 24
        value translation mechanisms, 23
        value translators, 24
    values canvas, collaboration, 161
        components, 161
        step-by-step guide, 162
Cross-industry value transformation, 306
Cross-organizational AI strategy, 75
    AI governance for value creation, 88
        for agentic AI and autonomous systems, 91
        Apple's business model and AI, 94
        balanced, 88–89
        contributes to value creation, 91
        cross-agency AI implementation, 94–95
        decision rights and accountability frameworks, 89–90
        effectiveness, 90
        ethical compliance with business value realization, 92–93
        in financial services, 96–97
        gaps in real-time decision-making systems, 92
        global regulatory trends, 93
        in healthcare, 95–96
        innovation with risk management, balancing, 89
        strategic integration, 93–94
    centralized and decentralized approaches, 76
    balancing, 77–78
    CEO's role in value creation, 78
        conflicting value propositions between departments, 81–82
        cross-functional collaboration, 79–80
        environment for AI success, 79
        geographic and cultural contexts, 81
        leadership strategies, 80–81
        modelling data-driven leadership, 80
        securing executive buy-in for long-term AI investments, 82
        "value trough" period, 82–83
        value vision and priorities, 78–79
    cross-functional AI roadmap, 83
        aligning technical and business timelines, 85
        cross-functional ownership of outcomes, 84–85
        integrated enterprise AI strategy, 83
        managing dependencies across departments, 85–86
    cross-functional imperative, 75–76
    governance, 98
    manages inevitable tensions, 98

Index

scaled success, 98
shared ownership, 77
stakeholder value alignment matrix, 86
    coalition for AI success, 87–88
    conflicts and contradictions, 86–87
    mapping stakeholder objectives and concerns, 86
    win-win value propositions, 87
strategic integration, 97
structural changes and integration, 97
traditional boundaries, 76–77
C-suite leaders, 84
C-suite lens, 120–125
C-suite value alignment, 11–12
    leaders, 12, 14
        efficiency advocate, 12
        funding models, 14
        revenue champion, 12
        technology catalyst, 12
    modeling, 15
Cultural boundaries, 76
Cultural leadership, 371–372
    framework, 372–373
Cultural leadership framework, 372
Cultural resistance to data sharing, 34
Cultural signals, 79
Customer and partner communication, 212–213
    disclosure-usability balance, 213
    progressive disclosure framework, 213–214
Customer churn prediction system
    implementation metrics, 62
    leading indicators, 62
    primary business metrics, 62
Customer intelligence platform, 96
Customer service, 69
Customer value and AI, 3–4
    experience enhancement, 3
    personalization, 4
    trust building, 4

D

Data
    abstraction, 34
    accessibility, 37
    availability, 37
    black hole phenomenon, 32
    boundaries, 76
    challenges, 31
        cultural resistance to data sharing, 34–35
        data black hole phenomenon, 32
        data quality and consistency issues, 32–33
        data regulation and privacy restrictions, 35
        governance and compliance complexities, 33
        legacy systems and technical debt, 33–34
        siloed data environments, 31–32
    heterogeneity, 54
    investments and AI value targets, 35–36
    minimization, 35
    network effects, 308
    priorities based on business impact, 36
    regulation, 35
    transparency card system, 226–227
Data-AI value connection, 26
    AI value, 27
    costs of poor data foundations, 27
    executive-practitioner disconnect, 26–27
    value-driven data, 28
Data as value foundation, 40
    financial services, 40
    healthcare, 40–41
    public sector, 41–42
    retail, 41
DataCrowd, 117
Data-driven culture, 352–353
"Data excellence" centre, 322
Data foundation
    framework, 42–43
    imperative, 43–44
Data foundations for AI value, 29
    data accessibility and democratization, 29
    governance and ethical use, 30–31
    integration across disparate sources, 30
    quality and usability considerations, 29–30
    scale and performance foundations, 31
Data governance, 38, 322
Data maturity, 27, **27**
    vs. AI success, **28**
    assessment for AI readiness, 37
        critical gaps for AI success, 38–39
        current data capabilities, 37–38
        expectations and timelines, 39–40
        investments based on value potential, 39
Data privacy and innovation, 225
    building trust through transparent practices, 227
    clear communication, 225–226
    data transparency card system, 226–227
    innovation-protection balance, 227
        graduated value exchange framework, 228
    managing consent and control expectations, 228–229
        consent triangle, 229–230
        turning privacy into competitive advantage, 230
    privacy abundance mindset, 230–231
Data quality, 37
    and consistency, 32–33
    gap, 114
Data readiness, 26, 320
    assessment, 391
        assessment dimensions, 392–393
        diagnostic tool, **38**
Data reality gap, 320
Data-to-value integration, 313
    cross-functional alignment, 316
        financial services risk *vs.* experience conflict, 318
        manufacturing "value pods," 319
        Norwegian Government's five-step alignment process, 317
        retail "value pairs," 319
        shared accountability for outcomes, 319
        stakeholder value alignment matrix, 317–318
        WEX's integration success, 316
    data foundations, 320
        critical data gaps and quality issues, 321
        "data excellence" centre, 322
        data governance, 322
        data readiness, 320
        data reality gap, 320
        financial services progressive implementation, 321
        healthcare risk prediction gap, 320–321

retail "AI data product teams," 322
Siemens' predictive maintenance success, 321
sustainable data capabilities, 322
vision-obstacle-value framework, 322
implementation guide and toolkits, 329–330
financial services, 333–334
healthcare, 335–336
healthcare visual roadmap, 333
knowledge multiplication at Barclays Group, 331
manufacturing, 336–337
pharmaceutical capability development roadmap, 331
retail, 334–335
retail implementation risk register, 332–333
measure, learn, and evolve, 326
AI value dashboard, 326–327
CP AXTRA's prompt engineering challenge, 327
financial services risk prediction pivot, 328
healthcare "AI value rounds," 327
Makro Wholesale's transformation approach, 328–329
manufacturing multilevel dashboard, 327
retail "AI value council," 328
self-sustaining capabilities, 328
scaling gap, 323
Australian government's copilot trial, 324
CP AXTRA'S strategic early adopters, 324
financial services "AI factory," 325
gaming company's proof of value, 326
healthcare "impact clinics," 324
pipeline *vs.* platform tension, 323
quick-win opportunities, 323
retail "value capture fund," 325
Siemens' industrial copilot ecosystem, 325
success to fund future phases, 325
technical and organizational foundations, 325
telecommunications triple-win approach, 324
shared value goals, 314
AI value mapping canvas, 314–315
healthcare patient experience transformation, 316
value realization roadmap, 315–316
Telstra's strategic approach, 314
Decentralized AI, 76
Decision-driven dashboard design process, 249–250
Decision rights and accountability frameworks, 89–90
Delegation trap, 340
Departmental perspectives on collaboration, 152, 158
adoption and engagement, 160–161
finance, 152
game theory to navigate conflicting interests, 159
HR and people, 153–154
incentive structures, 159–160
IT and technology, 153
legal, risk, and compliance, 154
marketing and sales, 153
operations, 152–153
Departmental thinking, 75
Dependency mapping, 36, 39
Different risk perspectives, 51
Direct return models, 65
Direct value metrics, 36
Disclosure-usability balance, 213
Distrust in AI, 6
Documentation-utilization balance, 284
Dual operating model, 263–264

Dual-track leadership model, 368–369
Dual-track roadmap, 85
Duplicated effort, 49
Dynamic narrative examples, 130
iteration and refinement process, 130
visual, 131

E

Economic value and AI, 2
Economic value from AI
cost reduction, 2
margin expansion, 2
revenue growth, 2
Ecosystem orchestration, 308
Ecosystem thinking, 112
Effort estimation, 39
Employee perspectives on AI, 172
adoption, 175
advocacy, 178–179
awareness, 176
barriers, addressing, 179–180
belief, 178
executive adoption, 180–181
frontline adoption, 181–182
manager adoption, 181
role-specific adoption techniques, 180–182
supportive environment, 192–195
understanding, 176–177
use, 177
value, 177–178
AI trust-building playbook, 182
creating psychological safety, 185–186
involvement and co-creation approaches, 183–184
skills development and career path clarity, 184–185
transparent communication strategies, 182–183
creative professional's dilemma, 172–173
human foundation of AI value, 198–199
human side of AI transformation, 174–175
job displacement and skills relevance, 173
managing resistance, building support, 189
champions at all levels, 191–192
"First Five" programme in healthcare, 191
informal influencers and networks, 189–190
positive AI experiences, 190–191
resistance patterns, 189
organizational adoption, 195
citizen service agent support, 196–197
clinical AI integration success, 198
digital transformation, 195–196
shop floor AI adoption in manufacturing, 197–198
organization-wide AI literacy
common AI language, 188
core concept understanding, 186
experiential learning approaches, 187–188
role-specific AI educational needs, 186–187
real concerns, 172
trust in AI systems and decisions, 174
Employee value and AI, 4
augmentation, 4
satisfaction, 4
skill development, 4
Enablement metrics, 36
End-to-end accountability, 50
Energy-value connection, 362

Index

Engagement priorities, 87
Enhancement prioritization, 279
Enterprise resource planning (ERP), 29
Enterprise-wide AI transformation, 96
Escalation protocols, 85
Ethical AI, 231
    attracting and retaining talent, 233–234
    and competitive advantage, 222
        financial services, 223
        healthcare, 223–224
        retail, 224–225
    ethical talent lifecycle, 234–235
    messaging, principles for, 206
    regulatory requirements, getting ahead of, 235–236
    short-term *versus* long-term value balance, 233
        sustainable value hierarchy, 233
    trust and loyalty, 231
        trust ROI framework, 231–232
Ethical communication, 236; *see also* Communication framework
Ethical compliance with business value realization, 92–93
Ethical responsibility, 373–375
Ethics leadership imperative, 374
Europe, regulatory-first framework, 295–296
European Steel Manufacturer's people-first scale, 103
Evolution planning, 114
Evolving AI value landscape, 293
    agentic AI, 294
        in supply chain management, 295
    cross-industry value transformation, 306
    current capabilities, 294
    data network effects, 308
    ecosystem orchestration, 308
    ethical and regulatory challenges, 300
        adaptable governance frameworks, 301–302
        ethical stress testing, 303
        ethics evolution framework, 303
        governance failures, 302–303
        regulatory developments, 300–301
    future-proof value matrix, 310–311
    future value opportunities, 297
        AI value sprints, 299
        creating organizational flexibility, 298
        decision-making processes, 298
        human-AI talent spectrum, 299
        innovative infrastructure approaches, 298
        technical architecture trap, 297–298
    generative AI, 294
    human-AI collaboration models, 309
    industry-specific value evolution, 296–297
        AI in humanitarian aid, 297
    innovation and stability, 304
        emerging capabilities into core systems, 305–306
        governance spectrum debate, 304
        managing risk, 305
        space for experimentation, 304–305
    outcome-based models, 308
    regional approaches to AI development, 295
        Africa: building foundational capacity, 296
        China: state-led strategy, 296
        Europe: regulatory-first framework, 295–296
        India: public-private partnership, 296
        Middle East: leadership-driven adoption, 296
        US: private-sector innovation, 295
    scenario planning for AI evolution, 300
    space for innovation, 299
    value creation through technology convergence, 306
        Blockchain, 307
        convergence tensions, 307
        Internet of Things (IoT), 306–307
        spatial computing, 307
Exception proliferation, 53
Executive adoption of AI, 180–181
Executive and board priorities, 120
    board-level concerns and expect, 122–123
    boardroom, language of, 125
    C-suite lens, 120
        CEO perspective, 120–121
        CFO perspective, 121
        CHRO perspective, 122
        CIO/CTO perspective, 121
        CMO perspective, 122
        COO perspective, 121
        CRO/CLO perspective, 122
    executive incentives and AI incentives, 124
    geopolitical tension in AI governance, 123–124
Executive briefing templates, 396
    monthly progress brief, 396–397
    quarterly strategic review, 397
Executive business case, 131
    financial framing, 131–132
    flexibility into value projections, 134–135
    option value recognition, 135
    ROI models, 132–133
    short-and long-term value considerations, 133–134
Executive buy-in, 120
Executive education, 39
Executive engagement success, 142
    board-level AI governance, 146
        balanced capability development, 146
        balanced reporting, 147
        dedicated AI committee structure, 146
        forward-looking governance evolution, 147
        risk-calibrated governance framework, 146
    evidence-based credibility, 142
    executive-specific value narratives, 143
    funding, 142
    healthcare-specific strategies, 144
        autonomy-sensitive implementation, 145
        clinical leadership co-development, 145
        evidence-based validation, 145
        patient outcome orientation, 145
        physician experience enhancement, 145–146
    multi-dimensional value demonstration, 142
    phased implementation, 142–143
    political support, public sector initiative, 143
        bipartisan value framing, 144
        citizen-centred value definition, 143
        cross-agency coalition building, 143
        political cycle alignment, 144
        risk mitigation through phased transparency, 144
    strategic framing, 142
Executive learning journeys, 24–25
Executive-specific value translations, 21
Executive sponsors, 32, 51, 84, 343–344
Executive value storytelling, 24
Expectation management framework, 367
Explicit mapping, 68
Explicit value, 7
Exploitation-exploration balance, 280–281

External perceptions, 211
    community and public outreach, 214
        community engagement spectrum, 214–215
    customer and partner communication, 212–213
        disclosure-usability balance, 213
        progressive disclosure framework, 213–214
    media engagement strategies, 211–212
        media preparedness toolkit, 212
    regulatory and policy engagement, 215
    regulatory engagement advantage, 215–216
External stakeholder, 201
    concerns and priorities, 203
    expectations gap, 203–204
    public perception challenges and opportunities, 205
    regulatory environment, 204–205
    specificity shield, 205–206
    stakeholder ecosystem, 201–202
    stakeholder paradox, 202
External trust, 200–201

F

FACTS framework for responsible AI communication, 206–207, 237
Failed collaboration, costs of, 165
False dichotomy problem, 207
Feasibility analysis, 36
Feedback
    integration mechanisms, 113
    limitations, 48
Finance, 69
Financial frameworks for AI investments, 65
    capability valuation approaches, 65
    direct return models, 65
    risk-adjusted frameworks, 65
Financial framing, 131–132
Financial impacts, 52
Financial services, 40, 96–97
    "AI factory," 325
    progressive implementation, 321
    risk prediction pivot, 328
    risk vs. experience conflict, 318
    value integration scorecard, 246
Forrester's AI validation process, 255
Fragmented data, 49
Fraud detection AI, 48
Frontline adoption of AI, 181–182
Function representatives, 84
Funding
    model evolution, 276
    transformation
        framework, 278–279
        mechanisms, 278
Future-proof value matrix, 310–311
Future scenario planning, 82

G

Game theory
    AI value alignment, 18–19
    conflicting interests navigation, 159
Gamification, 19
    collaborative recognition systems, 20
    cross-functional challenges, 19
    value translation competitions, 20
    visualization dashboards, 19
Gaming company
    proof of value, 326
    small-scale experiment, 60–61
Gaming Guru's intentional silos, 110
Generative AI, 294
    leadership challenge, 373
Glasgow, game-changing public sector impact, 116–117
Global regulatory divergence, 204
Global regulatory trends, 93
Go/no-go decision, 64
Governance, 98
    accessibility, 29
    effectiveness, 90
    gap, 115
    hurdles, 46
    leadership model, 370
Graduated value exchange framework, 228
Great silo dilemma, 50

H

Healthcare, 18, 40–41
    AI governance for value creation in, 95–96
    "AI value rounds," 327
    "impact clinics," 324
    patient experience transformation, 316
    risk prediction gap, 320–321
    system, measurement frameworks, 255
Healthcare providers
    scenario-based approach, 66
    structured prioritization, 68
Hidden process dependencies, 53
Honesty-Anxiety Balance, 218–219
Honesty-confidence paradox, 273–274
Human-AI collaboration models, 309
Human-AI talent spectrum, 299
Human foundation of AI value, 198–199
Human-in-the-loop (HITL) methodology, 92
Human side of AI transformation, 174–175
Hybrid architectures, 34
Hypothesis-driven value creation, 103–104

I

Impact analysis, 68
Impact assessment, 36
Implementation and execution frameworks, 391
    AI maturity model, 393
        maturity-appropriate strategies, 393–394
        scaling blueprint, 394
    customized roadmap, 406
        customization considerations, 406–407
        organizational capabilities, 407–408
        roadmap development process, 406
        sequencing and prioritization, 407
    data readiness assessment, 391
        assessment dimensions, 392–393
    risk assessment framework, 395–396
    scaling blueprint, 394–395
Implementation planning, 64
Improvement velocity framework, 267, **267**
Incentive alignment, 32, 79

Index

Incentive conflicts, 50
Inconsistent experiences, 49
India, public-private partnership, 296
Industry-specific measurement approaches, 245
    financial services, 245–246
    healthcare, 246–247
    retail, 247–248
Industry-specific regulatory focus, 204
Industry-specific value evolution, 296–297
    AI in humanitarian aid, 297
Inevitable tensions, 98
Infrastructure investment paradox, 104
Initial-to-sustained value gap, 45
Initiative owners, 84
Innovation and governance, 353–354
Innovation-protection balance, 227
    graduated value exchange framework, 228
Innovation-stability balance, 262–263
Innovation-trust cycle, 208
Innovation with risk management, balancing, 89
Innovation zone, 89
Insight-generating initiatives, 133
Institutional knowledge, 284–285
Integration
    complexity, 48
    leadership model, 371
    points, 85
Intelligent operations system, 96
Internet of Things (IoT), 306–307
Inventory management, 50
Investments
    based on value potential, 39
    proof problem in, 238–239
    staging, 35
Invisible middle, 101–102

J

Job displacement
    concerns, 218
        honesty-anxiety balance, 218–219
        job impact communication framework, 218
        staged disclosure approach, 219
    and skills relevance, 173
Jurisdictional awareness, 35

K

Knowledge accessibility, 29
Knowledge management and intellectual property, 283
    AI asset management framework, 285–286
    AI insights and learnings, 283
    documentation-utilization balance, 284
    institutional knowledge, 284–285
    knowledge management system, 283–284
    protecting and leveraging AI assets, 285
Knowledge management system, 283–284
Knowledge sharing, 91
KPMG model, 361

L

Labour market transformation, 309–310
Lab-to-field performance gap, 45
Lasting collaborative mechanisms, 164–165
Layered dashboard architecture, 249
Layered transparency system, 221
Leadership, 339–340
    action plan, 377–379
    balance, 368
    behaviours, 193
        authentic engagement, 193
        consistent messaging, 193
        supportive accountability, 193
        transparent decision-making, 193
    case studies, 367
        clinical leadership model, 369
        dual-track leadership model, 368–369
        governance leadership model, 370
        integration leadership model, 371
        WellSpan Health's approach, 369
    CEO and Board's role in AI value creation, 354
        accountability for value, 355–356
        modelling leadership behaviours, 357–358
        organizational barriers, 356–357
        strategic context, 354–355
    certainty illusion, 340
    communication, 364
        managing expectations, 366–367
        technical and business languages, 364–365
        value narratives, 365–366
    competencies, 347
        change leadership, 349–350
        cross-functional leadership, 348–349
        learning agility, 350–351
        strategic thinking with technical awareness, 347–348
        value translation, 348
    conditions for AI success, 351
        cross-functional collaboration, 351–352
        data-driven culture, 352–353
        innovation and governance, 353–354
        vision, 351
    control contradiction, 340–341
    cultural leadership, 371–372
        framework, 372–373
    delegation trap, 340
    development continuum, 360–361
    development imperative, 358
        AI leadership literacy, 358–359
        leadership community, 361–362
        leadership development pathways, 360–361
        value-focused mindset, 359–360
    future of AI leadership, 373
        balancing speed and responsible implementation, 375–376
        ethical responsibility, 373–375
        in generative AI era, 373
    imperative, 376
        AI leadership development gap, 376–377
    misalignment, 47
    model for AI value, 341
        capability-building, 343
        change acceleration, 342
        cross-functional integration, 341–342
        strategic vision, 341
        value translation, 342
    modelling, 34, 185

roles in AI transformation, 343
 AI value advisor, 346–347
 change leader, 345–346
 executive sponsor, 343–344
 technical leader, 345
 value orchestrator, 344
strategies, 80
 commercial organizations, 80
 government organizations, 80–81
 non-governmental organizations (NGOs), 81
sustainable energy, 362
 daily success pattern, 362–363
 weekly success structure, 363–364
Learning agility, 350–351
Learning-focused metrics, 185
Legacy constraints, 34
Lightful's AI squad approach, 169
Limited scale economics, 49
Long-term AI value roadmap, 286
 adaptable architecture framework, 287–288
 building adaptability, 287
 continuous evolution, 286
 culture of ongoing innovation, 290
 evolution roadmap framework, 286–287
 innovation culture framework, 290–291
 lifecycle management framework, 289
 lifecycle of AI capabilities, 288–289
 stability-evolution balance, 287

M

Maintenance
 challenges, 48
 practice inconsistency, 54
Maintenance-enhancement balance, 270–271
Makro Wholesale's transformation approach, 328–329
Manager adoption of AI, 181
Managing resistance, building support, 189
 champions at all levels, 191–192
 "First Five" programme in healthcare, 191
 informal influencers and networks, 189–190
 positive AI experiences, 190–191
 resistance patterns, 189
  experience-based resistance, 189
  identity-based resistance, 189
  knowledge-based resistance, 189
  personal impact resistance, 189
Manufacturing, 18
 multilevel dashboard, 327
 "value pods," 319
Manufacturing firms
 measurement frameworks, 253–254
 risk transparency, 71
Market-driven innovation, 8
Marketing, 69
Maturity-appropriate strategies, 393–394
Maturity dimensions, **393**
Measurement and evaluation resources, 400
 continuous improvement templates, 404
  monthly improvement review template, 404–405
  quarterly strategic review template, 405–406
 ROI calculation models, 401–402
  KPI categories and examples, 403
  KPI development guide, 403
  KPI development process, 404
  KPI portfolio example, 404
  KPI selection criteria, 403
  template, 402–403
 value measurement framework, 400–401
  measurement dimensions, 401
Measurement architecture, **401**
Measurement frameworks, 239
 AI measurement, problems of, 244–245
  metric hierarchy framework, 245
 AI value dashboards, 248
  decision-driven dashboard design process, 249–250
  layered dashboard architecture, 249
 challenges with AI, 239
 comprehensive value measurement, 242
  Microsoft copilot SMB value case, 244
  secondary benefit paradox, 243
  total value of AI ownership framework, 242–243
  triple BAT framework for lifecycle measurement, 242
 cross-functional measurements, 248
 early capturing and communicating, 250
  quick win discovery process, 250
  self-funding AI framework, 251–252
  value narrative framework, 251
 industry-specific measurement approaches, 245
  financial services, 245–246
  healthcare, 246–247
  retail, 247–248
 measurement-value paradox, 239–240
 in practice, 257
  capability compounding effect, 257–258
  cross-functional measurement opportunities, 258
  validation-decision balance, 257
 precision-relevance trade-off, 240
 pre-implementation measurement design, 241
  checklist, 241
  "measuring-the-wrong-thing" trap, 241–242
  value-pathway-mapping approach, 241
 quantitative-qualitative balance, 240
 success stories, 253
  Forrester's AI validation process, 255
  healthcare system, 255
  manufacturing company, 253–254
  retail bank, 254
 sustainable value measurement, 252–253
  portfolio value premium, 253
 value-measurement dashboard toolkit, 255
  metric-to-decision mapping tool, 256
  staged value expectations framework, 256–257
  technical-to-business translation matrix, 256
Measurement-value paradox, 239–240
"Measuring-the-wrong-thing" trap, 241–242
Media engagement strategies, 211–212
 media preparedness toolkit, 212
Metadata management, 33
Metric boundaries, 76
Metric-to-decision mapping tool, 256
Microsoft copilot SMB value case, 244
Microsoft's approach to AI governance, 140
Middle East, leadership-driven adoption, 296
Milestone-based funding, 82
Misaligned stakeholder expectations, 47

Index

Mistake normalization, 185
Monday morning problem, 380–381
    AI value toolkit, 381–382
        assessment and planning tools, 381
        implementation tools, 382
        strategic alignment tools, 382
        sustaining tools, 382
    customized toolkit, 383–384
    frameworks to organization, 382–383
    game theory solution, 381
    implementation considerations and requirements, 383–384
Monitoring inadequacies, 48
"More Is Better" trap, 102–103
Multi-dimensional success reviews, 24
Multidimensional value in practice
    customer impact, 17
    financial impact, 17
    healthcare example, 18
    manufacturing, 18
    marketing team values, 17
    operational impact, 17
    primary cross-functional metrics, 17
    retail transformation, 16
    store operations team values, 17
    supply chain team values, 16
Multi-faceted value story, **68**
Multi-level governance model, 370
Multi-modal explanation approach, 222
Mutual understanding and respect, 157

N

Narrative adaptation guidelines, 129–160
Narrative connection, 7
Narrative gaps, 5–6
    prevent effective collaboration, 6
Narrative sustainability framework, 274
Nash equilibrium, 18–19
Net promoter score (NPS), 240
Network
    engineering, 69
    optimization system, 69
Norwegian Government's five-step alignment process, 317
Norwegian government's implementation of chatbots, 317
NPS, *see* Net promoter score

O

Object recognition program, Google, 55
Operational drift, 5
    diverts resources from value creation, 6
Operational value and AI, 3
    decision improvement, 3
    efficiency, 3
    quality, 3
Opportunity costs, 27, 52
Optimization conflicts, 81
Organizational adoption, 195
    citizen service agent support, 196–197
    clinical AI integration success, 198
    digital transformation, 195–196
    shop floor AI adoption in manufacturing, 197–198
Organizational belief, 272–273
    belief ritual framework, 273–274
    narrative sustainability framework, 274–275
    storytelling, 274
    traction measurement, 275
    usage-value balance, 276
Organizational capabilities, 267–268
    capability development system, 268–269
Organizational learning loops, 113
Organizational misalignments, 49
    conflicting incentives and competing priorities, 50
    great silo dilemma, 50
    siloed approaches to AI implementation, 49
    technical-business communication gap, 51
    unclear ownership of value creation, 50–51
Organizational momentum and confidence erosion, 52
Organizational readiness, 106
    assessment, 108
Organizational silos
    build shared language, 14
    competing priorities, 13
    cross-functional value teams, 13
    cross-pollination, 13–14
    c-suite modelling, 15
    data fragmentation, 13
    executive leadership approaches, 14
    implementation resistance, 13
    incentives and metrics, 13
    strategic budget allocation, 14–15
    strategic hiring and partnerships, 15
    value blindness, 13
    value translation tools, 13
Organizational structure, 79
Organization-wide AI literacy
    common AI language, 188
    core concept understanding, 186
    experiential learning approaches, 187–188
    role-specific AI educational needs, 186–187
Outcome measurement, 35

P

Parallel processes, 53
Pattern recognition, 91
People transparency, 210
People trust, 174
Perception mismatch, 6
    destroy adoption and trust, 7
Performance
    constraints, 48
    -metric alignment, 194
    transparency, 210
    trust, 174
Physical variations, 54
Pilot purgatory, 46
Pilot-to-scale capability gap, 45
Pipeline *vs.* platform tension, 323
Political cycle alignment, 144
Poor data quality, 48
Portfolio balancing, 36, 82
Portfolio categorization, 280
Portfolio management for ongoing investment, 280
Portfolio model, 281
Portfolio value premium, 253
Precision-relevance trade-off, 240

Predictable failures, 46
Predictive quality initiative in manufacturing, 64–65
Pre-implementation measurement design, 241
    checklist, 241
        "measuring-the-wrong-thing" trap, 241–242
        value-pathway-mapping approach, 241
Principled trade-offs, 68, 92
Priority conflicts, 82
Privacy by design, 33
Privacy-enhancing technologies, 35
Privacy restrictions, 35
Privacy-utility trade-off, 308
Process boundaries, 76
Process disconnect, 115
Process inconsistencies, 46
Process transparency, 210
Process trust, 174
Production zone, 89
Progressive disclosure framework, 213, 237
Project-based funding, 277–278
Project-to-capability transformation, 261–262
Proof of concept (PoC), 28
Protection transparency, 210
Public sector, 41–42
Public trust for AI, 219
    function transformation trust framework, 219–220
    societal value exchange, 220
Purpose transparency, 210
Purpose trust, 174

Q

Quality degradation, 103
Quality for AI, 33
    automated monitoring, 33
    clear standards, 33
    proactive assessment, 33
    root cause remediation, 33
Quality inconsistencies, 48
Quick win
    discovery process, 250
    identification, 39
    opportunities, 323

R

RACI+, 89–90
RAND research, 6
Range-based business case, **66**
Realistic assessment, 34
Realistic value expectations, 69
    communicating constraints, 70
        building credibility through honest assessment, 71
        healthcare provider's transparent limitations, 70–71
    Hype Cycle, 69
    retailer's expectations management failure, 69–70
    timeframe communication, 70
    timeframes for value delivery, 70
Real-time decision-making systems, 92
Recognition systems, 34
Regional approaches to AI development, 295
    Africa: building foundational capacity, 296
    China: state-led strategy, 296
    Europe: regulatory-first framework, 295–296
    India: public-private partnership, 296
    Middle East: leadership-driven adoption, 296
    US: private-sector innovation, 295
Regional regulatory differences, 216
    regional regulatory patterns, 216–217
    regulatory alignment matrix, 217–218
Regional regulatory patterns, 216
    China, 217
    European Union, 216
    United States, 216–217
Regulatory alignment matrix, 217, 237
Regulatory and policy engagement, 215
Regulatory and reputational risks, 52
Regulatory engagement advantage, 215–216
Regulatory environment, 216
    job displacement concerns, 218
        honesty-anxiety balance, 218–219
        job impact communication framework, 218
        staged disclosure approach, 219
    public trust for AI, 219
        function transformation trust framework, 219–220
        societal value exchange, 220
    regional regulatory differences, 216
        regional regulatory patterns, 216–217
        regulatory alignment matrix, 217–218
    specificity shield, 220–221
    transparency around data usage and model decision-making, 221
        layered transparency system, 221
        multi-modal explanation approach, 222
        transparency-comprehensibility trade-off, 221–222
Regulatory tracking, 33
Reinvestment system development, 279
Representativeness issues, 48
Resource allocation, 64, 79
Resource allocation optimization, 280
Resource clarity, 39
Resource conflicts, 82
Resource drain, 27
Resource exhaustion, 102
Resource prioritization, 91
Retail, 16
    "AI data product teams," 322
    "AI value council," 328
    company's cautionary tale, 62
    customer data platforms, 41
    transformation, 16
        customer impact, 17
        financial impact, 17
        marketing team values, 17
        operational impact across functions, 17
        primary cross-functional metrics, 17
        store operations team values, 17
        supply chain team values, 16
    "value capture fund," 325
    "value pairs," 319
Retail bank
    crisis in, 175
    measurement frameworks, 254
Retailer's dual-track business case, 66
Retailer's expectations management failure, 69–70
Return on data investments (ROI), 36
Risk-adjusted frameworks, 65
Risk mitigation, 34, 64

Index

Risk reduction metrics, 36
Risks
    assessment framework, 395–396
    -based governance, 33
    -calibrated governance framework, 146
    governance matrix, **88**
    management ecosystem, 96
    and sustainability management, 280
Risk tolerance, 79

S

Scale, technical foundations for, 110
    e-commerce scale-or-fail example, 110
    gaming guru's intentional silos, 110
Scale and performance foundations, 31
Scaled success, 98
Scale limitation, 27
"Scale-or-fail" reality check, 104
Scale-ready assessment framework, 107–108
Scaling, horizons of, 108
    capability building, 109
    transformation, 109–110
    value realization, 108–109
Scaling blueprint, 394–395
Scaling challenges, 46, 102
    CNH industrial approach, 103–104
    European steel manufacturer's people-first scale, 103
    hypothesis-driven value creation, 103–104
    infrastructure investment paradox, 104
    "More Is Better" trap, 102–103
    "scale-or-fail" reality check, 104
    Tesla's mission-driven approach, 102
    "value multiplication" approach, 103
    value multiplication principles, 104–105
Scaling obstacles, 114
    data quality gap, 114
    governance gap, 115
    process disconnect, 115
    skill mismatch, 114–115
    value dilution challenge, 116
Scaling success, 105
    organizational readiness, 106
    people factor, 107
    strategic alignment, 105
    sustainable implementation, 106
    value translation, 105–106
Self-funding AI framework, 251–252
Self-sustaining capabilities, 328
Self-sustaining value cycles, 279
Shared accountability for outcomes, 319
Shared metrics and success definitions, 156
Shared outcome ownership model, 77
Shared ownership, 77
Shared success metrics, 51
Shifting priorities, 47
Siemens' industrial copilot ecosystem, 325
Siemens' predictive maintenance success, 321
Signals of longevity *vs.* signs of slow failure, 271–272
Siloed approaches to AI implementation, 49
Skill mismatch, 114–115
Skills accessibility, 29
Skills distribution, 54
Skills gaps, 46
Skill transition failure, 53
Societal value and AI, 4
    community impact, 4
    ethical considerations, 4
    sustainability, 4
Spatial computing, 307
Specificity shield, 220–221
Staged value expectations framework, 256–257
Stage-gated investment approach, 65–66
Stakeholders
    alignment techniques, 67
        conflicting priorities, 67
        cross-functional team, 67
        healthcare provider's structured prioritization, 68
        manufacturing firm's multi-faceted value story, 68
        telecommunications company's shared metrics, 69
        value, 68
    identification matrix, **23**
    map, **86**
    participation, 92
    value alignment matrix, 86, 317–318, 389
        coalition for AI success, 87–88
        conflict resolution process, 389
        conflicts and contradictions, 86–87
        mapping stakeholder objectives and concerns, 86
        prioritization dimensions, 390
        strategic prioritization framework, 389
        win-win value propositions, 87
Strategic AI communication, principles for, 236
Strategic alignment, 105
    assessment, 107
Strategic alignment tools, 408
    AI value mapping canvas, 388
        structure, 388
        using effectively, 389
    business case templates, 390–391
    creating sustainable value through AI, 408
        long-term vision, 408
    stakeholder value alignment matrix, 389
        conflict resolution process, 389
        prioritization dimensions, 390
        strategic prioritization framework, 389
Strategically scaling stage, 100
Strategic hiring and partnerships, 15
Strategic modernization, 34
Strategic positions, 355, **355**
Strategic priorities mapping, **61**
Strategic tension, 107
Strategic thinking with technical awareness, 347–348
Strategic transparency advantage, 211
Strategic vagueness, 5
    creates misdirected effort, 6
Strategic value and AI, 2–3
    competitive advantage, 2
    innovation acceleration, 3
    market differentiation, 3
Strategy misalignment, 47
Structural boundaries, 76
Structural changes and integration, 97
Structured collaboration frameworks, 157–158
Success measurement differences, 54
Success patterns from leaders, 112
    capability building in parallel, 112

ecosystem thinking, 112
  value-first implementation, 112
Success trap, 100
Support accessibility, 185
Supportive environment of AI, 192–195
Support network, 87
Sustainable AI value, 260–261, 291
  adaptability by design, 291
  AI portfolio management framework, 280–281
  ambidextrous portfolio model, 281
  capability transformation framework, 262
  continuous improvement mechanisms, 266–267
    improvement velocity framework, 267
    system, 266–267
  core business processes, 264–265
    process integration framework, 265–266
  dual operating model, 263–264
  early warning system, 272
  exploitation-exploration balance, 281
  funding model evolution, 276
  funding transformation framework, 278–279
  funding transformation mechanisms, 278
  innovation-stability balance, 262–263
  institutional knowledge, 291–292
  knowledge management and intellectual property, 283
    AI asset management framework, 285–286
    AI insights and learnings, 283
    documentation-utilization balance, 284
    institutional knowledge, 284–285
    knowledge management system, 283–284
    protecting and leveraging AI assets, 285
  long-term AI value roadmap, 286
    adaptable architecture framework, 287–288
    building adaptability, 287
    continuous evolution, 286
    culture of ongoing innovation, 290
    evolution roadmap framework, 286–287
    innovation culture framework, 290–291
    lifecycle management framework, 289
    lifecycle of AI capabilities, 288–289
    stability-evolution balance, 287
  maintenance-enhancement balance, 270–271
  organizational belief, 272–273
    belief ritual framework, 273–274
    narrative sustainability framework, 274–275
    storytelling, 274
    traction measurement, 275
    usage-value balance, 276
  organizational capabilities, 267–268
    capability development system, 268–269
  portfolio management for ongoing investment, 280
  project-based funding, 277–278
  project-to-capability transformation, 261–262, 291
  self-sustaining value cycles, 279, 291
  signals of longevity *vs.* signs of slow failure, 271–272
  sustainable funding framework, 277
  value-based deals and partnerships, 282
  value cycle framework, 279–280
  value erosion pattern, 269–270
  value partnership framework, 282–283
  value sustainability index, 271
Sustainable capability model, 276
Sustainable data capabilities, 322
Sustainable energy
  daily success pattern, 362–363
  weekly success structure, 363–364
Sustainable executive support, 147
  credible results, 148
  governance, 147
  implementation reality, 148
  language of leadership, 147
  multi-dimensional value cases, 147
  ongoing relationships, 148
  stakeholder-specific priorities, 148
Sustainable funding framework, 277
Sustainable implementation, 106
Sustainable implementation assessment, 108
Sustainable values
  framework, 112
    evolution planning, 114
    feedback integration mechanisms, 113
    organizational learning loops, 113
    value measurement systems, 113
  hierarchy, 233, 237
  measurement, 252–253
    portfolio value premium, 253

T

TAG's data pipeline transformation, 60
Team-based incentives, 194
Technical accessibility, 29
Technical and business languages, 364–365
Technical-business communication gap, 51
Technical connectivity, 32
Technical debt accumulation, 48
Technical infrastructure, 38
Technical language, 125
Technical leader, 345
Technical leaks, 48
Technical limitations, 46
Technical objectives masquerading, 47
Technical-to-business translation
  gap, 45
  matrix, 256
  patterns, 364
Technical *vs.* business language, 51
Technology-first approach, 58–59, **59**
Telecommunications triple-win approach, 324
Telstra's partnership approach, 164
Telstra's strategic approach, 314
Tension triangle (business, finance and IT), 155
Tesla's mission-driven approach, 102
Three-tiered governance model for escalating ethical issues, 354, **354**
Tiered implementation, 68
Time-lag complexity, 239
Timeline
  conflicts, 82
  disconnects, 51
  honesty, 39
Total value of AI ownership (TVAO), 242
Traditional boundaries, 76–77
Traditional project model, 276
Transformations, 100–101
Transitional service model, 276
Transition zone, 89
Translation framework, **385**

Index

Transparency around data usage and model decision-making, 221
   layered transparency system, 221
   multi-modal explanation approach, 222
   transparency-comprehensibility trade-off, 221–222
Transparency-comprehensibility trade-off, 221–222
Transparency dimensions framework, 210, 237
Transparent communication, 92
Triple BAT framework for lifecycle measurement, 242
True cost of AI value failure, 51
   direct financial impacts, 52
   opportunity costs and competitive implications, 52
   organizational momentum and confidence erosion, 52
   regulatory and reputational risks, 52
Trust erosion, 27
Trust imperative, 310
Trust ROI framework, 231–232, 237

U

Unclear objectives, 47
Unclear ownership of value creation, 50–51
Unprecedented collaboration, 150–151
   Asana, building cross-functional AI literacy, 151
   conflict to collaboration, 162–163
      mutual benefit and impact, 164
      shared ownership of outcomes, 163
      territorial concerns and resource competition, 163
   costs of failed collaboration, 165
   cross-functional collaborative spaces, 156–157
   cross-functional steering committees, 155–156
   cross-functional success, 165–166
      financial services, 168
      healthcare, 167
      manufacturing, 166–167
      retail, 166
      values, 168–169
   cross-functional values canvas, 161
      components, 161
      step-by-step guide, 162
   departmental perspectives, 152, 158
      adoption and engagement, 160–161
      finance, 152
      game theory to navigate conflicting interests, 159
      HR and people, 153–154
      incentive structures, 159–160
      IT and technology, 153
      legal, risk, and compliance, 154
      marketing and sales, 153
      operations, 152–153
   lasting collaborative mechanisms, 164–165
   Lightful's AI squad approach, 169
   mutual understanding and respect, 157
   shared metrics and success definitions, 156
   structured collaboration frameworks, 157–158
   Telstra's partnership approach, 164
US, private-sector innovation, 295
Use case mapping, 35

V

Vague value propositions, 47
Validation-decision balance, 257
Value alignment challenge, 22
   executive-driven AI success factors, 22–23
Value alignment loop, 118
Value and performance assessment, 280
Value-based deals and partnerships, 282
Value-based funding, 277
Value-chain complexity, 239
Value creation through technology convergence, 306
   Blockchain, 307
   convergence tensions, 307
   Internet of Things (IoT), 306–307
   spatial computing, 307
Value cycle framework, 279–280
Value demonstration, 34
Value dilution, 103
   challenge, 116
Value-driven AI on limited budget, 72
Value-driven data, 25, 28, 33–34, 110
   roadmap, 37
Value-driven prioritization, 35
Value erosion pattern, 269–270
Value-ethics mapping, 92
Value expectation audit, 23
Value-first framework, 56, 59, **59**, 71–72, 359–360
   action plans, 74
   advantages, 60
   build flexibility into business cases, 74
   Commonwealth Bank of Australia's, 72
   comprehensive metrics, 74
   create explicit connections to strategy, 73
   elements, 58–61
   gaming company's small-scale experiment, 60–61
   global financial institution, 73
   implementation, 73–74
      success metrics, 61–62
   limited budget, 72
   multidimensional value, 73
   outcomes, 60
   principles of, 56
   public sector, 72
   realistic expectations, 74
   retail company's cautionary tale, 62–63
   reverse the planning sequence, 73
   start strategically small, 74
   structured planning tools, 74
   TAG'S data pipeline transformation, 60
      high-impact use cases, 60
Value-first implementation, 112
Value-focused mindset, 359–360
Value-inclusive culture, 24
Value integration imperative, 25
Value leakage, 48
Value leakage throughout the AI lifecycle, 47
   data quality and accessibility limitations, 48
   strategy misalignment and unclear objectives, 47
   technical implementation challenges, 48–49
Value mapping canvas, 118
Value measurement
   dashboard toolkit, 255
      metric-to-decision mapping tool, 256
      staged value expectations framework, 256–257
      technical-to-business translation matrix, 256
   differences, 51
   principles of, 258
   systems, 113

Value multiplication
  approach, 103
  principles, 104–105
Value multiplier effect, 15
  cross-functional value amplification, 16
  enabling new capabilities, 16
  transcending traditional trade-offs, 15
Value narrative framework, 251
Value narratives, 365–366
Value orchestrator, 344
Value partnership framework, 282–283
Value-pathway-mapping approach, 241
Value pods, 319
Value quantification, 39
Value realization roadmap, 315–316
Values
  Board and CEO perspectives, 9
  business unit leader perspectives, 10
  customer perspectives, 11
  dimensions of, 1
    customer value, 3–4
    economic value, 2
    employee value, 4
    operational value, 3
    societal value, 4
    strategic value, 2–3
  employee perspectives, 10
  external stakeholder perspectives, 11
  organizational silos
    build shared language, 14
    cross-functional value teams, 13
    cross-pollination, 13–14
    c-suite modelling, 15
    executive leadership approaches, 14
    incentives and metrics, 13
    strategic budget allocation, 14–15
    strategic hiring and partnerships, 15
    value translation tools, 13
  technical team perspectives, 10
  vs. values, 7–8
    regional and cultural dimensions, 8–9
Values-based initiatives, 8
Value staging, 39, 82
Value storytelling frameworks, 399
  audience-specific story adaptation, 400
  before/during/after framework, 400
  situation, task, action, result star framework, 399–400
Value stream mapping, 279
Value sustainability index, 271, **271**
Value to build credibility, 135
  meaningful metrics for early-stage reporting, 136–138
  quick-win strategies, 135–136
  "Value Trough" Period, 138–139
Value translation, 105–106, 348
  assessment, 108
  competitions, 20
  framework, 20
  mechanisms, 23
  tools, 7
Value translators, 24
Value trough period, 82
Vendor-partner inversion, 282–283
Vision-obstacle-value framework, 322
Visual dependency mapping, 85
Visualization dashboards, 19

W

WellSpan health leadership development model, 360
WellSpan Health's approach, 369
WEX
  alignment approach, 158
  integration success, 316
Win-win value propositions, **87**, 87–88
World Vision International's tiered education approach, 187–188

For Product Safety Concerns and Information please contact our EU representative GPSR@taylorandfrancis.com Taylor & Francis Verlag GmbH, Kaufingerstraße 24, 80331 München, Germany

Batch number: 09524505

Printed by Printforce, the Netherlands